Springer Series in Synergetics   Editor: Hermann Haken

Synergetics, an interdisciplinary field of research, is concerned with the cooperation of individual parts of a system that produces macroscopic spatial, temporal or functional structures. It deals with deterministic as well as stochastic processes.

Temporal Order

Proceedings of a Symposium on Oscillations
in Heterogeneous Chemical and Biological Systems,
University of Bremen, September 17–22, 1984

Editors: L. Rensing and N. I. Jaeger

With 201 Figures

Springer-Verlag
Berlin Heidelberg New York Tokyo

Professor Dr. Ludger Rensing

Fachbereich 2 (Zellbiologie) der Universität Bremen,
Leobener Straße, NW 2,
D-2800 Bremen, Fed. Rep. of Germany

Professor Dr. Nils I. Jaeger

Fachbereich 2 (Forschungsgruppe Angewandte Katalyse) der Universität Bremen,
Bibliothekstraße, NW 2,
D-2800 Bremen, Fed. Rep. of Germany

Series Editor:
Professor Dr. Dr. h. c. Hermann Haken
Institut für Theoretische Physik der Universität Stuttgart, Pfaffenwaldring 57/IV,
D-7000 Stuttgart 80, Fed. Rep. of Germany

ISBN-13: 978-3-642-70334-8   e-ISBN-13: 978-3-642-70332-4
DOI: 10/1007/978-3-642-70332-4

Softcover reprint of the hardcover 1st edition 1985

Offsetprinting: Beltz Offsetdruck, Hemsbach. Bookbinding: J. Schäffer OHG, Grünstadt
2153/3130-543210

Preface

This volume contains the lectures and contributions presented at the International Symposium on Temporal Order held in Bremen, September 17-22, 1984.

Temporal order, such as a more or less regularly repeated temporal sequence of events, can evolve in open systems far removed from equilibrium. Progress during the last decade in the analysis and the modelling of this complex phenomenon in both biological and chemical systems gave rise to the idea of a joint conference. The purpose of the symposium was to stimulate future work by enhancing the exchange of experimental and theoretical results between neighbouring disciplines.

Theoretical work in general, and mathematical models in particular, provided the basis for a mutual discussion and, thus, helped to overcome difficulties in understanding the results of different experimental fields. Chemical systems, for example, are more rigorously controllable through their experimental conditions in comparison to biological systems, which maintain highly effective autonomous control against environmental influences. Therefore, different states such as bistability, oscillations and chaos can be defined and, hence, described better in chemical systems. Chemical systems may thus provide some insights into functional structures that also exist in more complex biological systems.

Oscillations in heterogeneous chemical and biological systems, i.e. in chemical and biochemical reactions at interfaces, such as the surface of a solid catalyst or a cellular membrane, were chosen as the main topic of the symposium in an attempt to investigate similarities in properties and mechanisms. Heterogeneous catalytic reactions, enzyme catalyzed glycolysis and protein synthesis-dependent circadian (= endogeneous diurnal) rhythms and cell cycles - all driven by fluxes of matter and energy - were therefore given special attention.

Even though we emphasized heterogeneous systems, recent progress in the experimental and theoretical treatment of the dynamics of homogeneous systems was necessarily included.

The analysis of oscillations, multiple stable steady states or chaos may lead to insights into the basic mechanisms or modes determining the macroscopic behaviour of complex systems. Mathematical models based on empirical data and assumptions about reaction mechanisms and rates may serve as guidelines for detecting common functional structures in different experimental systems. The treatment of chemomechanical instabilities and the periodic movement of amoebae might serve as such an example.

Macroscopic temporal order is based on the co-operative interaction of individual parts of the system (HAKEN's concept of synergetics). In catalyzed chemical reactions the co-operative interactions may be mediated by temperature or concentration changes and/or by phase transitions. In biological systems, concentrations of metabolites or membrane potentials may play this role. This indicates the particular importance of interfaces and membranes in heterogeneous systems.

If the specific nature of elements of an oscillatory system is not known, they might be analyzed by means of external perturbations of the oscillation. A variety of different physical or chemical modes of perturbation may be considered, for example pulse-wise changes of temperature or concentration changes. Those changes that alter properties of the oscillation, such as phase, may be considered to affect oscillatory elements. A similar line of reasoning applies to the effect of periodic inputs on the oscillator (forced oscillations, resonance, entrainment of chemical and circadian rhythms): they can be effective only if they interact with the oscillator.

Questions regarding the significance of oscillatory temporal order were only briefly discussed during the symposium. The role of circadian rhythms is obvious: they function as clocks that co-ordinate many events in cells and multicellular organisms with the geophysical 24h periodicity. Other examples for the significance of oscillations are the directed movement of cells or organisms in space by means of rhythmic contraction and relaxation of contractile locomotory elements, or the transport of blood by means of the heart beat. The question, however, whether or not an oscillatory chemical or biochemical reaction system under a given set of conditions is more flexible or effective compared to a non-oscillatory state is still open to speculation.

The interdisciplinary nature of the symposium became possible through the help and support of our colleagues H.O. PEITGEN from the Mathematics Department, P.J. PLATH from the Chemistry Department and P. RICHTER from the Physics Department.

We are indebted to Stiftung Volkswagenwerk, the University of Bremen, Die Gesellschaft der Freunde der Universität and Siemens AG for financial support and gratefully acknowledge the effective help of Mrs. TIN and Mrs. THOMSEN in preparing the proceedings.

Bremen
December 1984   *Ludger Rensing · Nils Jaeger*

Contents

Part II   Dynamics of Biological Systems

Movement in Space

Cellular Metabolism and Transport

Circadian Rhythms

Part I

Dynamics of Chemical Systems

Spontaneous Temporal and Spatial Phenomena in Physicochemical Systems

U.F. Franck

Institut für Physikalische Chemie, R.W.-Technische Hochschule Aachen, Postfach D-5100 Aachen, Fed. Rep. of Germany

1. Structural Aspects

From the very beginning chemists dealing with oscillations have been concerned with the question whether or not heterogeneity is a necessary prerequisite for the occurrence of oscillatory behaviour. Looking more closely into such systems, we can distinguish clearly two kinds of spatial structures:

Permanent structures forming interfaces between different kinds of phases . The interfaces, as a rule, are the sites of the oscillatory processes, as   is the case at electrodes, heterogeneous catalysts, membranes, biological tissues etc..

Dynamic structures arising from unstable situations which are present necessarily in all chemical and physicochemical oscillators. They manifest themselves in the form of propagating conversions of state or as dissipative structures. They may occur in all kinds of systems containing unstable states, either in the form of spatial or interfacial dynamic structures, the varying circumstances of which are illustrated in more detail in Fig. 1 and 2.

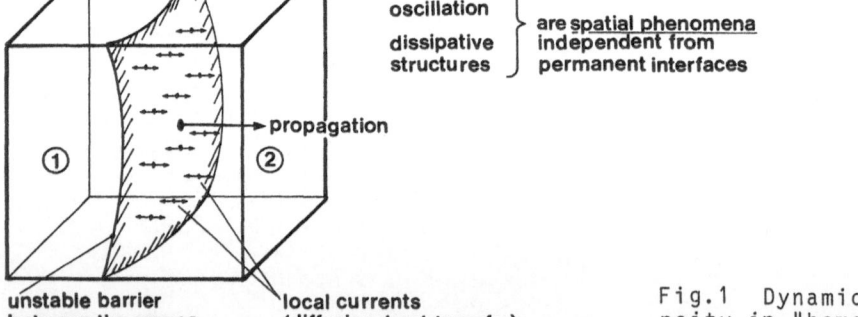

propagation ⎫
oscillation ⎬ are spatial phenomena independent from
dissipative ⎭ permanent interfaces
structures

→ propagation

① ②

unstable barrier between the ranges of different states (1 and 2)

local currents (diffusion, heat-transfer) releasing the change of state (2→1)

Fig.1 Dynamic inhomogeneity in "homogeneous" oscillatory systems

Oscillatory chemical systems without permanent interfacial structures (Fig. 1) may actually have the appearance of being homogeneous. However, because of their intrinsic instability, they can exist in two different kinetic states simultaneously. In a situation such as this, the system consists of distinct ranges of different kinetic and thermodynamical properties. Hence, in actual fact, the system in question is really heterogeneous. This spontaneous heterogeneity happened to exist, as already mentioned, mostly in transient form of propagating dynamic structures.

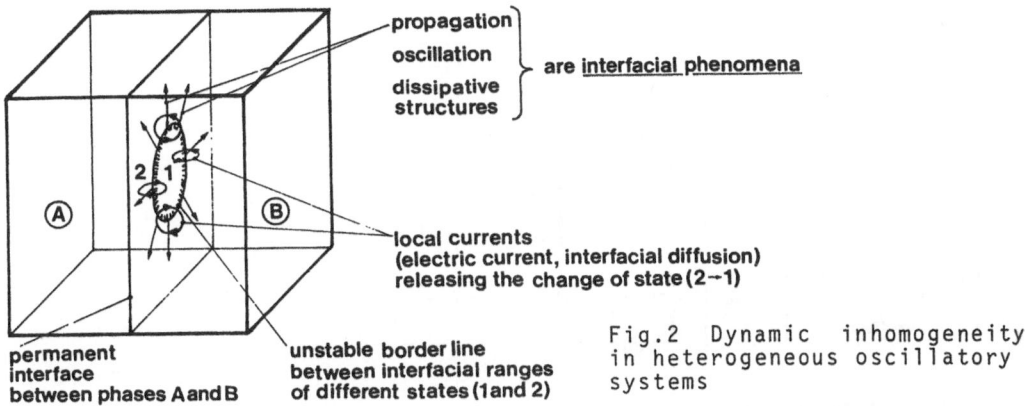

Fig.2 Dynamic inhomogeneity in heterogeneous oscillatory systems

The divergency of the thermodynamic situation on both sides of the barrier between the different ranges acts as driving force for local diffusion or heat fluxes across the barrier. It is easy to show that these eddy currents have the tendency to trigger the conversion of state on both sides of the barrier. It depends upon on which side the respective conversion of state is triggered first. In the direction of that side, the barrier is moving successively through the space of the system. This is the well-known and well-established kinetic mechanism of the propagation phenomena in physicochemical systems containing unstable states [1, 2, 5] .

At this point it is important to state that, strictly speaking, oscillations are absolutely impossible in perpetually homogeneous chemical systems. The so-called "homogeneous" oscillatory systems produce their (transient) inhomogeneity themselves as a direct result of their intrinsic instability.

The kinetic/spatical situation in systems with permanent heterogeneity is somewhat different (Fig. 2). Here, the instability concerns different kinetic states of the interface between two permanent phases (A and B). The border line between differing areas of the interface (1 and 2) is the site of local currents, which may be electric currents, interfacial diffusion or, possibly, heat fluxes.

Comparing both classes of chemical oscillatory systems, it is worth noting that electric local currents can exist only in interfacial systems. This is because electric currents require closed circuits which are, in principle, not realizable in systems without permanent interfaces [6] . For this reason, electrochemical and electrobiological oscillations are strictly bound to electrode interfaces or membranes resp.. In this way, the nature of the effective local currents is decisive for the structural requirements of oscillatory systems.

2.   Propagation of Dynamic Structures in Chemical Systems

The occurring propagating dynamic heterogeneity can be observed in most of the oscillating chemical systems by means of suitable technics applying synchronous measurements at distant sites of the system in question. Figure 3 gives two typical examples of such measurements. Figure 3a concerns the BELOUSOV-ZHABOTINSKY-reaction having no permanent interfacial structure. The oscillation is usually traced here by recording the bromide concentration by means of a bromide-sensitive membrane electrode. This electrode is extremely

BZ - reaction

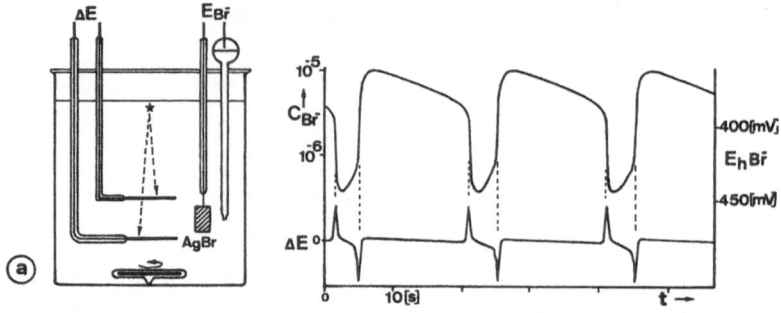

oscillating cobalt in CrO₃/HCl

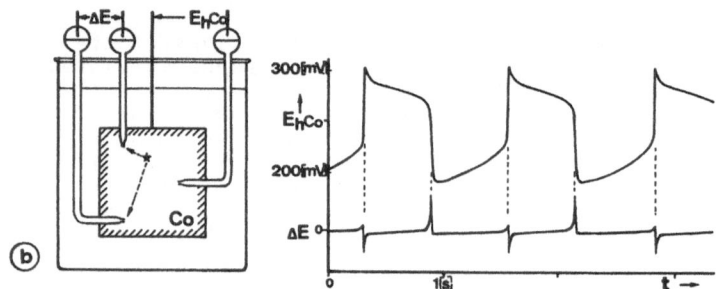

Fig.3 Transient inhomogeneity in oscillatory chemical systems
a) BZ-system: $KBrO_3$ (0.55 mole/l), malonic acid (0.8 mole/l),
         $Ce(SO_4)_2$ ($2 \cdot 10^{-3}$ mole/l) H_2SO_4 (0.5 mole/l)
b) cobalt electrode: CrO_3 (0.4 mole/l), HCl (1 mole/l)   [5]

slow in its voltage response (about half a minute), much too slow
for tracing the propagation wave, which normally has a velocity of
about 10 cm/s. The silverbromide electrode, being sufficiently fast,
is generally unsuited for bromide measurements in the BZ-system because of the presence of bromate. In the particular case given here,
the recordings concern only the qualitative temporal changes of the
voltage difference between two electrodes of the same kind. Freshly
prepared silverbromide electrodes could be used here without serious
problems. The lifetime of such electrodes is about 20 minutes, being
quite sufficient for the purpose in question.

The propagation waves of conversion of state originate, as a rule,
from the upper surface of the solution. Travelling downwards, they
reach the horizontal electrodes at different time according to their
propagation velocity, giving rise to transient voltage differences
lasting as long as they proceed from the upper to the lower electrode.
There occur alternatively two kinds of waves, triggering sudden conversions of the bromide concentration downwards and upwards respectively.

In electrochemical oscillatory systems (Fig. 3b) the situation
is quite similar, but with the distinction that the propagation
waves are bound to the interface of the electrode, and that they are
driven by local electric currents and not by local diffusion fluxes
as is the case in the BZ-system.

4

Because oscillations in chemical systems positively concern a phenomenon of temporal and spatial nature, they possess like all propagating physical oscillations not only a definite frequency but also a definite propagation velocity, and hence a definite wavelength.

3. Self-influence of Processes in Oscillatory Systems

At this point the fundamental question arises, how oscillatory behaviour is brought forth in detail and how can it be understood on the basis of elementary chemical kinetics. There are, of course, many possibilities for attacking this problem. The initial idea of the approach being pointed out here is the fundamental statement that: in chemical systems oscillatory behaviour and instability are primarily the result of self-influence of the reactions involved.

Self-influence is nothing other than what is called "feedback". The self-influence may lead by coupling mechanisms to self-enhancement or self-inhibition of the reaction or process in question. Accordingly, these kinetic effects are designated as positive or negative feedback respectively.

As is well-known, positive feedback labilizes the system and may bring about instability. Negative feedback acts antagonistic to the positive feedback. It stabilizes the system and causes recovery and adaptation behaviour.

4. The Principle of Antagonistic Feedback

It is an essential fact that all chemical and physicochemical oscillators known so far obviously possess positive and negative feedback properties simultaneously. That can be demonstrated easily by studying the temporal behaviour after perturbations by means of defined stimuli [5]. Obviously, oscillations such as these can be regarded as a result of an antagonistic interaction or interplay between a labilizing positive feedback releasing a conversion of state and a coupled negative feedback, which recovers that conversion and restores the initial state, so that the closed chain of reactions can start again. This is what is called "the principle of antagonistic feedback of physicochemical oscillations" [6].

Actually, it concerns the same kinetic situation as is well-known from feedback-induced oscillations in electric circuits, in technical control equipments or in other realistic feedback systems.

5. Fundamentals of Feedback Kinetics

In the following, it shall be shown that the principle of antagonistic feedback is a really useful tool for analyzing the kinetic mechanisms of oscillatory systems and, moreover, that it is possible to predict the occurrence of oscillations in systems of known kinetics and to "synthesize", so to speak, kinetic models of chemical oscillators.

Before doing this, it is necessary to look more closely into the details of feedback kinetics.

Feedback in chemical and physicochemical systems can arise in two ways (Fig. 4). If, for instance, the product of a reaction acts upon the very same reaction as reactant, then a feedback situation arises in which the "effect" of the reaction becomes its own "cause". That kind of feedback has no effect upon the property of the "transmitting" reaction. For this reason it is called "non-systemic" feed-

back. A typical example is the stoichiometric autocatalytic reaction
$A + X \rightarrow 2X$. There is no influence upon the rate constant k of the
transmitting reaction.

In most cases of physicochemical oscillators the existing feedback concerns the properties of the transmitting reaction itself.
This situation is called accordingly "systemic" feedback. It is
realized mostly in systems with permanent interfaces and it is, as
a rule, non-stoichiometric.

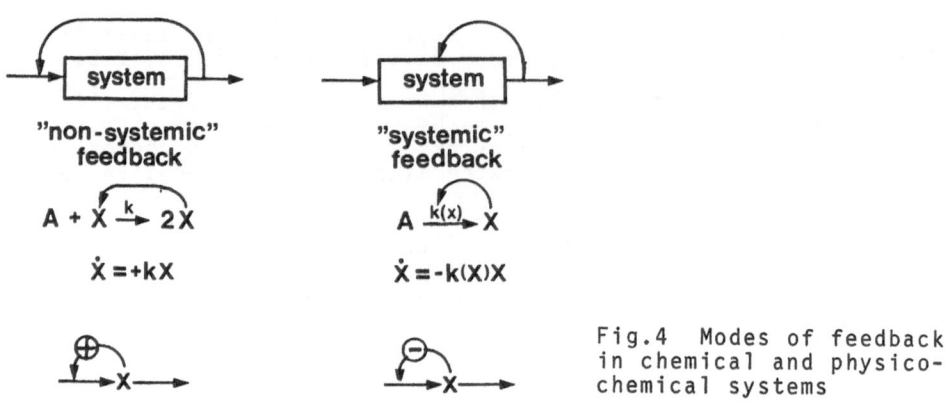

"non-systemic"
feedback

$A + X \xrightarrow{k} 2X$

$\dot{X} = +kX$

"systemic"
feedback

$A \xrightarrow{k(x)} X$

$\dot{X} = -k(X)X$

Fig.4 Modes of feedback
in chemical and physico-
chemical systems

In open systems,feedback may concern the formation or the consumption of the species in question (symbolized by horizontal arrows in Fig. 5). Hence there arise four different feedback situations,which were designated by J. HIGGINS (1967) [3] as "backward
activation", "forward inhibition", "backward inhibition" and "forward activation", exhibiting positive feedback in the first two cases
and negative feedback in the last two cases.

Feedback may occur in processes of only one degree of kinetic
freedom. This is the case of "self-feedback" (Fig.5 above). Generally, however, feedback arises as a result of coupled reactions in
systems of more than one degree of freedom. This is the case of
"cross-feedback" which was introduced into oscillation kinetics by
K.F. BONHOEFFER as early as 1946 [4] . There exist formally 16 different possibilities of cross-feedback in systems of two degrees of
freedom (Fig. 5), 8 for positive and 8 for negative feedback [5]. However, in stoichiometric chemical systems with non-systemic feedback
only the positive coupling arrows are to be taken into consideration
(indicated by exclamation marks in Fig. 5). Hence,only four cases of
cross-feedback remain to be discussed here. For a later discussion,
it is important to keep in mind that crossed positive arrows always
mean positive feedback,and parallel arrows always mean negative feedback. Figure 6 gives examples of simple cases of stoichiometric non-
systemic feedback.

There is one case only of self-feedback. It is the backward activation of the autocatalytic reaction already mentioned. There exist
four simple cases of cross-feedback including only second-order reactions. Among these cases,those of negative feedback are of particular interest, because they indicate that also negative feedback
may be connected to autocatalytic reactions.

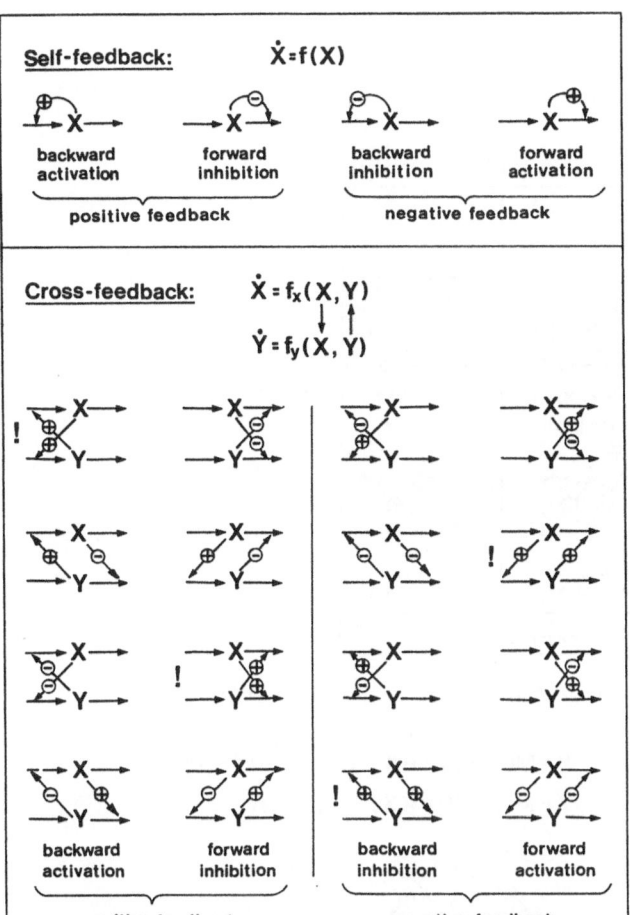

Fig.5 Classification of
chemical feedback

6. Models of Chemical Oscillators "Synthesized" on the Basis of Antagonistic Feedback

In order to exemplify the applicability of the principle of antagonistic feedback,a simple model of a chemical oscillator shall now
be synthesized by assembling suitable reactions of positive and
negative feedback,as given in the synoptic table of Fig. 6.

1977 W. GEISELER and H.H. FÖLLNER [7] showed that open chemical
systems containing two degrees of freedom are capable of exhibiting
bistability under appropriate circumstances. The GEISELER/FÖLLNER-
system (Fig. 7) consists of four simple reactions of second or first
order resp. concerning two concentration variables (X and Y). The
system has,accordingly,two degrees of freedom, and its temporal behaviour can be described by two simple simultaneous differential
equations. M, B, K, E, P and N are the kinetic coefficients. M in
particular is the flow-rate coefficient,and Xi and Yi are the influx
concentrations of X and Y. The feedback diagram (in the middle of
Fig.7) indicates that this system contains three kinds of positive
feedback, a strong autocatalytic self-feedback, a weak linear backward activation and a hyperbolic forward inhibition.

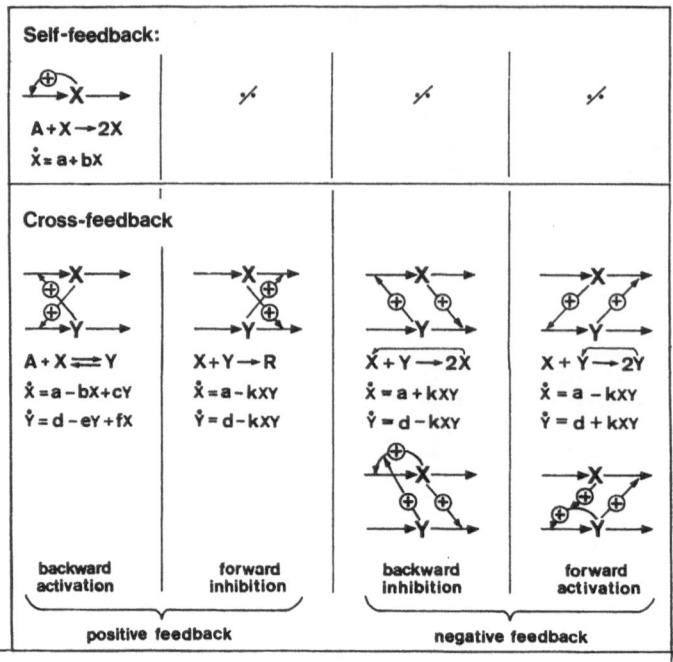

Fig.6 Simple cases of stoichiometric non-systemic feedback

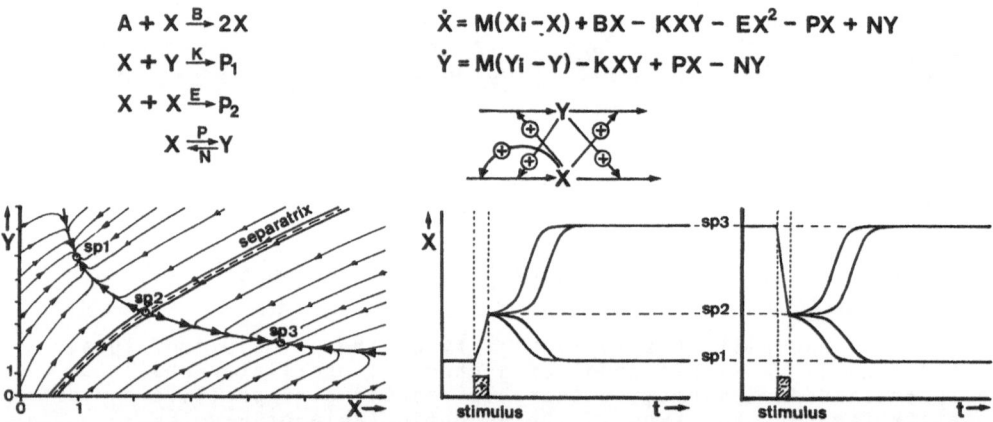

$$A + X \xrightarrow{B} 2X$$
$$X + Y \xrightarrow{K} P_1$$
$$X + X \xrightarrow{E} P_2$$
$$X \underset{N}{\overset{P}{\rightleftharpoons}} Y$$

$$\dot{X} = M(X_i - X) + BX - KXY - EX^2 - PX + NY$$
$$\dot{Y} = M(Y_i - Y) - KXY + PX - NY$$

Fig.7 Bistability of a chemical system with two degrees of freedom

The portrait of trajectories (Fig.7 l.h.s) shows three singular points (sp1, sp2, sp3). The middle one is unstable. It divides the plane of state, with its separatrix, into two separate ranges, each of which contains a stable steady state (sp1, sp3 resp.). These singular points attract all the trajectories of the respective range. In this way the separatrix with the unstable state sp2 is the factor which decides into which of both stable states a given non-steady state finally becomes steady. The mutual conversion of the stable states can be triggered by means of adequate stimuli (Fig.7 r.h.s) shifting the state of the system beyond the unstable state or its separatrix. This is the well-known mechanism of threshold behaviour

Table 1 Special values of the coefficients used in the computations of Fig.7 to Fig.11

Fig.	M	Xi	Yi	Zi	B	K	E	L	N	P	Xi'
7	0.1	8.5	120		6.4	1.56	0.5		0.6	1.1	
8	0.1	10	120	0.1	6.4	1.56	0.5	0.0312	0.6	1.0	8
9	0.1	1.75	3.0	0.01	0.404	1.539	0.115	0.192			1.64
10	0.1	1.8	3.5	0.001	0.153	0.18		0.055			1.76
11	0.1	1.6	2.35	3.04		11.8		0.55	0.396		1.59

xi': influx concentration of X in case of pulse stimulation

of bistable systems containing two degrees of freedom. (The diagrams shown in the Figures 7 to 11 are copies of computer graphics computed with coefficient values listed in Table 1).

If the principle of antagonistic feedback actually applies to the achievement of oscillatory behaviour, the bistable system discussed here should be transformable into an oscillatory one, merely by introduction of an additional reaction with negative feedback property with respect to X. A suitable reaction, for instance, is available for this purpose from Fig. 6 (at the bottom, right). It is an autocatalytic reaction of a third species Z coupled with X: $Z + X \rightarrow 2Z$. Although this reaction exhibits a positive feedback with respect to Z it produces an effective negative feedback (the coupling arrows being parallel!) with respect to X. Figure 8 shows the result of the introduction of this negative feedback reaction. The original bistable system has now become an oscillatory one. There are three degrees of freedom now, and the oscillation takes place accordingly in the 3-dimensional space of X, Y and Z. The resulting limit cycle is considerably stable. Inside of it there exists a 3-dimensional singular point having the property of an unstable focus.

By alteration of the influx concentration Xi (from 10 to 8, for instance, in the given example of Fig.8) the oscillatory activity ceases and the system remains in a stable state. It is now triggerable for pulses by ΔX-stimuli analoguously to the excitation of the action-potential of the living nerve. From the feedback point of view, the conversion of state triggered by a superthreshold stimulus is recovered now by the coupled, and therefore delayed, action of the intro-

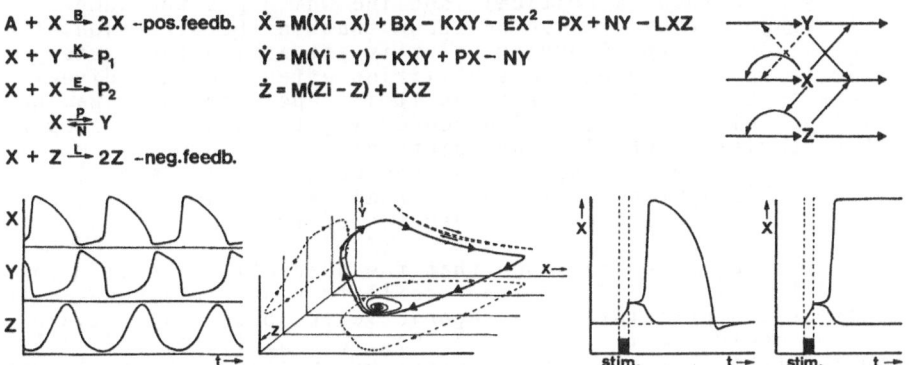

$A + X \xrightarrow{B} 2X$ –pos.feedb. $\dot{X} = M(Xi - X) + BX - KXY - EX^2 - PX + NY - LXZ$

$X + Y \xrightarrow{K} P_1$ $\dot{Y} = M(Yi - Y) - KXY + PX - NY$

$X + X \xrightarrow{E} P_2$ $\dot{Z} = M(Zi - Z) + LXZ$

$X \underset{N}{\overset{P}{\rightleftharpoons}} Y$

$X + Z \xrightarrow{L} 2Z$ –neg.feedb.

Fig.8 Chemical oscillator with 3 degrees of freedom (oscillator 1)

duced negative feedback. By diminution of the negative feedback effect the system restores its bistability behaviour according to Fig.7.

On the basis of the principle of antagonistic feedback it is possible to find out which reactions are essential for the occurrence of oscillations and which reactions may be omitted without losing the oscillatory property. In that way the oscillator discussed here can be simplified something more.

As a first step, the linear positive feedback may be eliminated from the dynamic equations of Fig.8. With that, the system becomes somewhat simpler, without changing its general kinetic behaviour. The resulting oscillation and its limit cycle are shown in Fig.9.

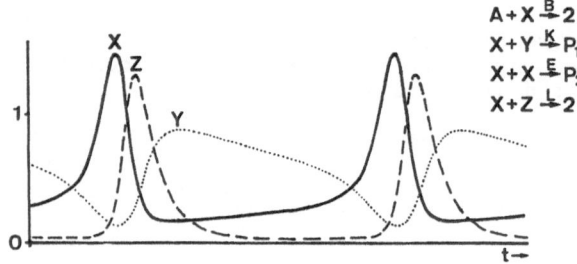

$$A+X \xrightarrow{B} 2X \qquad \dot{X}=M(X_i-X)+BX - KXY - EX^2 - LXZ$$
$$X+Y \xrightarrow{K} P_1 \qquad \dot{Y}=M(Y_i-Y) - KXY$$
$$X+X \xrightarrow{E} P_2 \qquad \dot{Z}=M(Z_i-Z) + LXZ$$
$$X+Z \xrightarrow{L} 2Z$$

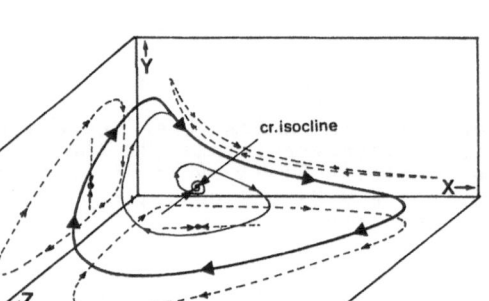

Fig.9 Oscillator 2
The critical isocline, having the same slope as its trajectories, is drawn by the computer. The arrows indicate that the singular point is attracting the trajectories in that direction, but it is repelling them in the plane of the focus

A short remark concerning the singular point in the 3-dimensional space should be made here. In the examples of chemical oscillators discussed here, the singular points inside the limit cycle possess, besides an unstable focus, a critical isocline which has the same slope as its trajectories. It can be shown that in the direction of this critical isocline, the singular point is attracting the states on the trajectories. That means a stabilizing effect in that direction, whereas in the plane of the focus it is repelling them, leading to instability. This mixed situation concerning the stability properties is characteristic for singular points of oscillatory systems with more than two degrees of freedom. As a consequence, it is impossible to stabilize such singular points by reversal of time, as   is the case in systems with less than three degrees of freedom.

In the next step, it can be shown that the quadratic term EX^2 can be eliminated too without losing the oscillatory property, because the requirements of the principle of antagonistic feedback still remain fulfilled. The newly simplified system (Fig.10) oscillates perfectly along a quite stable limit cycle. However, by omitting the non-linear term EX^2 the system no longer possesses bistability.

The system still exhibits triggerable pulse response in non-oscil-
lating states (Fig.10 r.h.s). Pulses of conversions of state are the
result of a kinetic competition between the action of the enhancing
positive feedback and the recovering negative feedback. As soon as
the latter outweighs the effect of the positive feedback, the recovery sets in. If, however, the negative feedback is too weak for this
compensation, no upper stable state will be attained, because the
system has turned now into a triggerable exploding one. Actually, bistability is no necessary prerequisite for the occurrence of oscillations or pulse response.

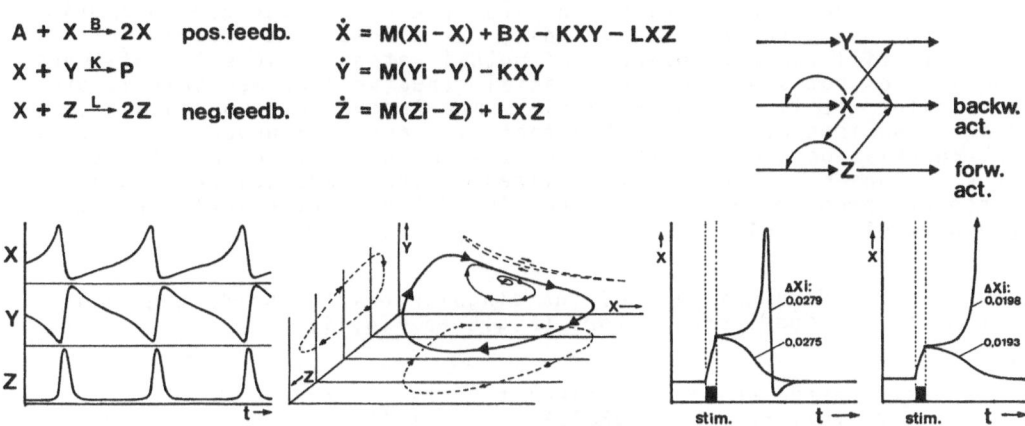

Fig.10 Chemical oscillator without bistability (oscillator 3)

The last example illustrates how the principle of antagonistic
feedback is capable of predicting unexpected possibilities of oscillatory behaviour in extremely simple chemical systems. For instance,
the autocatalytic reaction $X + Z \rightarrow 2X$ combines, as already mentioned,
a positive feedback with respect to X and a negative feedback with
respect to Z. There exists in fact a situation of antagonistic feedback in one single reaction only. If X is, in addition, cross-coupled
with Y independently from Z, then a sufficient phase-shift between X
and Z can be achieved in order to get sustained oscillations. This is
actually the case as shown in Fig. 11.

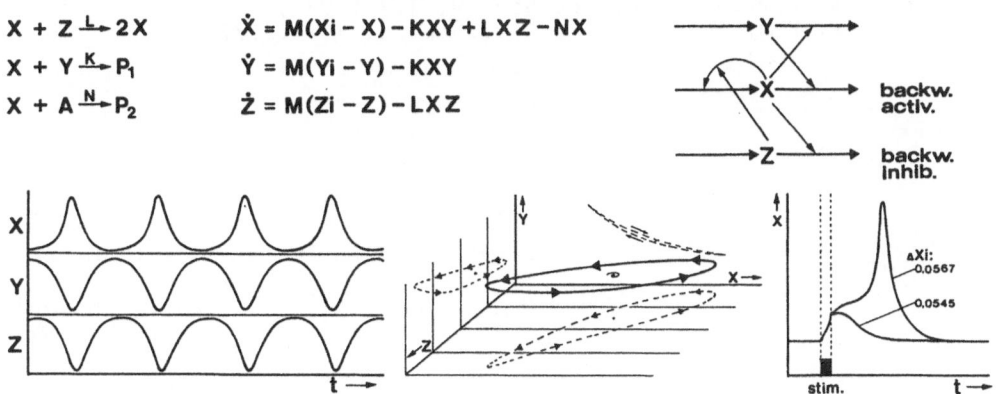

Fig.11 Chemical oscillator containing only one autocatalytic
reaction (oscillator 4)

Using the coefficient values given in Table 1 that extremely simple system oscillates along a stable limit cycle. In non-oscillatory states it is triggerable for pulses in the same way as all other oscillators discussed here. The third reaction $X + A \rightarrow P_2$ has no feedback function. It is just a necessary consumption reaction for X to avoid autocatalytic explosion.

7. The Generality of the Principle of Antagonistic Feedback

The applicability of the principle of antagonistic feedback demonstrated in the previous chapter could have also been realized by using any other kind of chemical or physical processes, because this principle obviously has general validity for all coupling mechanisms which fulfill the requirements of mutual antagonistic self-influence, and it does not depend on any special structural or mechanistic prerequisites. It is only of essential importance that two counteracting mechanisms of sufficient effectivity and an appropriate temporal correlation are present in the system in question. How these feedback mechanisms are actually realized in their special details is a secondary problem which has to be elucidated, particularly in every given practical case.

Table 2 Coordinate values of the singular points inside the limit cycles of the model oscillators (Fig.8 to Fig.11)

Fig.	X"	Y"	Z"
8	3.1763166	2.6836732	11.12445
9	0.50573038	0.34156156	0.34485418
10	1.7953519	0.82710378	0.079640312
11	0.15733509	0.12010913	1.6297271

References

1 L. Hermann: Handbuch der Physiologie, Bd.2/1 (1879)
2 U.F. Franck: Z.Elektrochem. 55, 154 (1951)
3 J. Higgins: Ind.Eng.Chem. 59, 19 (1967)
4 K.F. Bonhoeffer: Z.Elektrochem. 51, 24 (1948)
5 U.F. Franck: Ber.Bunsenges.Phys.Chem. 84, 334 (1980)
6 U.F. Franck: Faraday Symp.Chem.Soc. 9, 137 (1974)
7 W. Geiseler, H.H. Föllner: Biophys.Chem. 6, 107 (1977)

Some Factors Generating Temporally Ordered Behavior

Richard M. Noyes

Department of Chemistry, University of Oregon, Eugene, OR 97403, USA

1. Introduction

For the purposes of this paper, Temporal Order is taken to mean regular periodic variation of some property of a system. Authenticated (but generally ignored) examples of such behavior have existed for over sixty years. However, thorough understanding of their mechanisms began only a little over a decade ago. That understanding has now progressed sufficiently that it has been possible not only to identify the mechanistic features responsible for temporal order in several specific systems, but also to say with varying degrees of confidence that no alternative features will need to be invoked to explain systems yet to be discovered.

Much of the initial theoretical work was concerned with stabilities of steady states, and that work provides a framework for the discussion in Section 2. However, interesting features of many systems arise only when they are considerably removed from any steady state; some such situations are discussed in Section 3. Finally, Section 4 briefly mentions some examples in which a kind of temporal order is associated with boundary conditions which cause compositional nonuniformity to develop spontaneously.

2. Factors Affecting Behavior Near a Steady State

2.1 Definition of a Steady State

A closed constrained [1] chemical system will decay ultimately to a unique stationary state called chemical equilibrium in which all composition variables have become invariant. Even if the system is pumped [1] with a continuous flux of matter and/or energy, as in a continuously stirred tank reactor (CSTR), there will be one or more experimentally or at least computationally attainable stationary states, such that all composition variables are invariant.

The constraints on a truly stationary state are severe and involve the system as a whole. Often, even though concentrations of the major reactants and products change slowly with time, concentrations of some intermediate species relax to values whose absolute rates of change are very much smaller than those of the reactants and products. These intermediates are then said to be in a steady state, such that their concentrations can be treated as invariant over a considerable time. For the purposes of this Section, composition at a steady state is also uniform throughout the space occupied by the system.

A steady or stationary state is locally stable or unstable depending upon its response to the small composition fluctuations which are inevitable in any real system. If the system is stable, any fluctuation will be followed by a decay to the original state. The decay may involve damped oscillations, but it will not constitute temporal order as defined for this paper. If the system is unstable, at least some kinds of infinitesimal perturbations will cause the system to leave the immediate neighborhood of the steady state and move to another region of composition space. That development may be accompanied by temporal order. The important conclusion is that if a system is initially in a steady state, a necessary (but not necessarily sufficient) condition for it to develop temporal order is that the steady state is unstable.

13

2.2 The Community Matrix as a Qualitative Indicator of Steady State Stability

The argument developed here is based on an excellent presentation by TYSON [2].

Let $[X_i]_{ss}$ be the concentration of species X_i which is invariant in a steady or stationary state. Let x_i be the deviation in any other state of the concentration $[X_i]$ from the steady state value. Then

$$x_i = [X_i] - [X_i]_{ss}. \tag{1}$$

Although rates of chemical processes may be very complex functions of composition, in a uniform homogeneous solution those functions are continuous and have continuous derivatives. Then, for compositions sufficiently close to the steady state, all rates can be approximated as linear functions of deviations like x_i.

$$d[X_i]/dt = dx_i/dt = \sum_j \alpha_{ij} x_j \tag{2}$$

The summation is taken over all X_j species (including X_i) whose concentrations are invariant in the steady state. The matrix of the α_{ij} coefficients is called the community matrix for that steady state. If the rates of change of all concentrations are known as functions of composition, the α_{ij} can be written by inspection as functions of the rate constants and the composition of the steady state. Often at least the signs of the α_{ij} can be evaluated by inspection without even calculating that composition.

TYSON [2] has discussed this formulation, and has shown that those signs may be excellent qualitative indicators of steady state stability. For any set of conditions corresponding to a valid chemical mechanism, a steady state will be unequivocally stable to perturbations unless at least one of the following three situations applies:

(1) The mechanism involves direct autocatalysis. Then some α_{ii} is sufficiently positive.

(2) The mechanism involves indirect autocatalysis. Then some $\alpha_{ij}\alpha_{ji}$ is sufficiently positive. If α_{ij} and α_{ji} are both positive, the effect is called synergism; if they are both negative, it is called competition.

(3) The mechanism involves a feedback loop. Then some sequential product of the form $\alpha_{ij}\alpha_{jk}\alpha_{km}\cdots\alpha_{pq}\alpha_{qi}$ is sufficiently different from zero for a sequence involving three or more coefficients.

The above statements do not define sufficiency. In practice, most chemical steady states are stable. As is discussed in Section 2.3 below, direct autocatalysis is very unlikely to occur in any chemical mechanism described in terms of elementary processes. An expanded Oregonator model can illustrate both kinds of indirect autocatalysis as discussed in Section 2.4. A Bubbelator model serves to illustrate an instability created by a feedback loop as discussed in Section 2.5. Therefore, two well-studied models based on established mechanisms of real reactions are sufficient to illustrate all the types of steady state behavior which could lead to temporal order in any chemical system initially in a steady state.

2.3 Direct Autocatalysis in Biological and Chemical Systems

It is of the nature of living things to reproduce their own kind. Therefore, any model of population dynamics of interacting biological organisms will include terms based on direct autocatalysis. The steady state in a particular example may be stable, unstable, or even conservative as in the LOTKA [3] mechanism.

14

In contrast to biological systems, chemical changes are the consequences of elementary processes each of which takes place in a single step without the intervention of metastable intermediates.

I have not been able to devise an elementary process exhibiting direct autocatalysis which should contribute significantly to a real chemical system. A trivial exception might involve process (A1).

$$H + H + H_2 \rightarrow 2H_2 \tag{A1}$$

If hydrogen atoms were ineffective as third bodies for the recombination of such atoms, and if there were no walls or other molecules, the presence of hydrogen molecules would generate autocatalytic recombination of atoms in a single step process.

However, at least for the present, it appears appropriate to ignore the possibility of direct autocatalysis in chemical systems if they are described to the detail of invoking elementary processes.

2.4 The Expanded Oregonator Model as an Illustration of Indirect Autocatalysis

The use of the community matrix to assess the stability of a steady state can be illustrated with a well known model which can exhibit temporal order. The expanded Oregonator model [4] is represented by six irreversible steps.

$$A + Y \rightarrow X + P \tag{O1}$$

$$X + Y \rightarrow 2P \tag{O2}$$

$$A + X \rightarrow 2W \tag{O3a}$$

$$W + C \rightarrow X + Z' \tag{O3b}$$

$$2X \rightarrow A + P \tag{O4}$$

$$Z' \rightarrow gY + C \tag{O5}$$

Let A and P be regarded as major reactants and products, respectively, and let C_o be the initial concentration of C. If it is recognized that Z' and C are connected by the relation $C = C_o - Z'$, dynamic behavior can be described in terms of the four variables W, X, Y, and Z'. A subscript ss can be used to designate a steady state concentration of one of these variables. The α coefficients can be generated from inspection, and the algebraic matrix is presented in Table 1. All of the signs except for α_{xy} are immediately obvious, and this sign is easily evaluated. The qualitative community matrix is presented in Table 2.

All diagonal terms in Table 2 are negative, just as asserted in the discussion in Section 2.2. The terms α_{xw} and α_{wx} have been circled to emphasize that steps (O3a) and (O3b) exhibit a strong synergetic autocatalysis, in which X reacts to produce sufficient W to regenerate more X than that initially consumed.

The α_{wz} and α_{zw} terms also imply a mild synergetic autocatalysis which arises because increasing Z' decreases C and thereby slows step (O3b) to generate a larger steady-state concentration of W. The effect is much less important than the XW autocatalysis.

The other destabilizing features are also exhibited by the more conventional [5] 3-variable Oregonator generated if steps (O3a) and (O3b) are replaced by (O3).

$$A + X (+ 2C) \rightarrow 2X + 2Z' \tag{O3}$$

15

Table 1. Algebraic Community Matrix of Expanded Oregonator Model

w	$-k_{3b}(C_0-Z'_{ss})$	$2k_{3a}A$	0	$k_{3b}W_{ss}$
x	$k_{3b}(C_0-Z'_{ss})$	$-k_2Y_{ss} - k_{3a}A - 4k_4X_{ss}$	$k_1A - k_2X_{ss}$	$-k_{3b}W_{ss}$
y	0	$-k_2Y_{ss}$	$-k_1A - k_2X_{ss}$	gk_5
z	$k_{3b}(C_0-Z'_{ss})$	0	0	$-k_{3b}W_{ss} - k_5$

Table 2. Qualitative Community Matrix of Expanded Oregonator Model

w	$-$	$\boxed{+}$	0	$+$.
x	$\boxed{+}$	$-$	$-$	$-$
y	0	$-$	$-$	$+$
z	$+$	0	0	$-$

──

The 2C term has been placed in parentheses, because it must be included for stoichiometry but is irrelevant to the dynamics if the steps are irreversible and if $W_{ss} \ll X_{ss}$.

The algebraic community matrix for the 3-variable model has none of the W terms from Table 1, and the details of the coefficients which do not contain Y are modified. The qualitative community matrix for this model is presented in Table 3.

Table 3. Qualitative Community Matrix of Conventional Oregonator Model

x	$-$	$\boxed{-}$	0
y	$\boxed{-}$	$-$	$+$
z	$+$	0	$-$

The diagonal terms in Table 3 all remain negative even though step (O3) is no longer an elementary process. The α_{xy} and α_{yx} terms have been circled to emphasize the strong competitive indirect autocatalysis between X and Y which is a major factor in the remarkable behavior of this model.

The matrix in Table 3 also exhibits a feedback loop of $\alpha_{xy}\alpha_{yz}\alpha_{yx}$. However, the ability of feedback to destabilize a steady state is better illustrated by another model, which exhibits no autocatalysis and whose community matrix is developed in the next section.

2.5 The Bubbelator Model as an Illustration of a Destabilizing Feedback Loop

If molecules of a volatile substance are being produced continuously in solution, the final gaseous product may be evolved in pulses which exhibit temporal order.

Let M denote product molecules, let N denote nuclei capable of growing to become bubbles, and let B_j denote a bubble differing in size from a nucleus by j increments of radius. The size of an individual increment may be made as small as is desired. This nomenclature is equivalent to $N \equiv B_0$.

Then the Bubbelator model is generated by equations (B1) to (B4j).

$$A \rightarrow M \tag{B1}$$

$$M \rightleftarrows N \equiv B_0 \tag{B2}$$

$$M + B_j \rightleftarrows B_{j+1} \qquad j = 0,1,2,\ldots \tag{B3j}$$

$$B_j \rightarrow G \tag{B4j}$$

In this model, A denotes the chemical reactant and G the product gas. Step (B2) is not the simple first-order reaction implied but is an almost discontinuous step function, such that nuclei are formed at a significant rate only after the concentration of M has grown to a critical value.

If a steady-state distribution of bubble sizes has been attained, the concentration of B_j decreases monotonically with increasing j and all steps undergo net change from left to right. The qualitative community matrix can be written easily by inspection and is presented in Table 4.

Table 4. Qualitative Community Matrix of Bubbelator Model [6]

m		-	-	-	-	-	-	.	.
n		+	-	0	0	0	0	.	.
b_1		+	+	-	0	0	0	.	.
b_2		+	0	+	-	0	0	.	.
b_3		+	0	0	+	-	0	.	.
b_4		+	0	0	0	+	-	.	.
.		.	.	.	.	.	.	.	
.		.	.	.	.	.	.	.	

The diagonal terms in Table 4 are all negative as required by Section 2.2; there is no direct autocatalysis. All nonzero terms to the left of and below the diagonal are positive, while all nonzero terms above and to the right are negative; there is no indirect autocatalysis. However, there are many conceivable nonzero feedback loops based on one negative term and the remaining terms positive. The observed temporal order is clearly associated with destabilization of the steady state by negative feedback loops. A sufficient concentration of molecules must be attained before it is possible to form nuclei which grow to bubbles, but that growth depletes the molecules which are necessary to form more nuclei. The steady state is unstable because growth of bubbles depletes the reserve of molecules necessary to create new bubbles.

2.6 Summary of Possible Types of Steady State Instabilities

If a chemical mechanism is described entirely in terms of elementary processes, it is unlikely that any of them will involve direct autocatalysis. However, steps exhibiting synergetic indirect autocatalysis can behave equivalently to direct autocatalysis as in step (O3) of the conventional Oregonator mechanism. That mechanism generates an unstable steady state because of the strong competitive indirect autocatalysis of the X and Y species. Steady states can also be destabilized by feedback loops, and the Bubbelator mechanism illustrates a strong destabilization by means of negative feedback loops. Therefore, previously established mechanisms illustrate all of the possible ways in which a steady state can be made locally unstable to perturbation.

3. Temporal Order in Uniform Systems Far Removed from a Steady State

The treatment in Section 2 used the methods of linear algebra to show that a system in a steady state could not spontaneously change to exhibit temporal order unless a few specific types of behavior were involved. Those types could all be illustrated by the Oregonator and Bubbelator mechanisms previously developed to explain particular chemical systems.

However, even if a steady state is stable to infinitesimal perturbation, that steady state is not necessarily approached from every accessible region of composition space. Let the composition be defined in terms of n dimensions. Then local stability may be restricted by a surface of n-1 dimensions enclosing a finite region which includes the steady state, and within which every point would decay to the steady state. Outside that region, there may be different behavior including temporal order. The methods of linear algebra are not applicable to behavior far removed from a steady state. General criteria are not yet developed to indicate whether temporal order is or is not possible in such systems. In this Section, we shall examine the limit cycle trajectories traversed by the Oregonator and Bubbelator models for some parameter values, which always keep the systems far from their steady states. Those trajectories exhibit features which seem to be beyond those inherent in the linear stability analysis of Section 2.

3.1 Relaxation Oscillations and Delayed Feedback in the Oregonator Model

Temporal order implies that evolution of the system in composition space follows a closed continuous trajectory. If that trajectory never gets far from a locally unstable steady state, it is likely to approximate a sine function. However, at greater distances it may resemble what is called a relaxation oscillator. The system will exist for an appreciable time in one region of composition space and will then switch almost discontinuously to a second region. After some time in this region, it will discontinuously return to the first region.

For certain combinations of parameters, the Oregonator provides a fine example of this kind of behavior. The behavior of interest is obtained when $k_2 \gg k_1$ and $k_4 \gg k_3$ so that the concentration of X is always much less than those of the species which are forming and consuming it. Then we may employ a steady state approximation and write

$$dX/dt = k_1AY - k_2XY + k_3AX - 2k_4X^2 \simeq 0 \tag{3}$$

$$X = \frac{k_3A - k_2Y + \sqrt{(k_3A - k_2Y)^2 + 8k_1k_4AY}}{4k_4} \tag{4}$$

Adopt the conventions (5) to (7).

$$Q = \frac{8k_1k_4AY}{(k_3A - k_2Y)^2} \tag{5}$$

$$Y_{crit} = k_3A/k_2 \tag{6}$$

18

$$X_{crit} = \sqrt{\frac{k_1 k_3}{2k_2 k_4}} \, A \qquad (7)$$

If $Y > Y_{crit}$, $X = X_{small} < X_{crit}$.

$$X_{small} = \frac{k_2 Y - k_3 A}{4k_4}(-1 + \sqrt{1 + Q}) = \frac{k_1 AY}{k_2 Y - k_3 A}(1 - \frac{1}{4}Q + \frac{1}{8}Q^2 - \ldots)$$

$$\approx \frac{k_1 AY}{k_2 Y - k_3 A} \approx k_1 A/k_2 \qquad (8)$$

If $Y < Y_{crit}$, $X = X_{large} > X_{crit}$.

$$X_{large} = \frac{k_3 A - k_2 Y}{4k_4}(1 + \sqrt{1 + Q}) = \frac{k_3 A - k_2 Y}{2k_4}(1 + \frac{1}{4}Q - \frac{1}{16}Q^2 + \ldots)$$

$$\approx \frac{k_3 A - k_2 Y}{2k_4} \approx k_3 A/2k_4 \qquad (9)$$

The expressions for X_{large} and X_{small} are exact if enough terms are included in the series expansion of $\sqrt{1 + Q}$. If Y deviates from Y_{crit} enough that $Q \ll 1$, the approximate equations become satisfactory. Given the parameter values selected for most modeling computations, the limiting X_{large} is about 10^5 times the limiting X_{small} and Q will be less than 0.01 if Y deviates from Y_{crit} by more than about 3%.

When $Y \gg Y_{crit}$, $X_{small} = k_1 A/k_2$ and Z' will also be small. Only steps (01) and (02) are important then, and most chemical change is described by (0α).

$$A + 2Y \rightarrow 3P \qquad (0\alpha)$$
$$(0\alpha) \equiv (01) + (02)$$

The values of X and of Z' remain small while (0α) depletes Y and drives it toward Y_{crit}. When the relaxation occurs, Y is driven to a small fraction of Y_{crit} before Z' can change significantly.

When $Y \ll Y_{crit}$, $X_{large} = k_3 A/2k_4$ and steps (03) and (04) will become most important. Most chemical change is now described by (0β) with increasing contributions from (05) and (02).

$$A + 4C \rightarrow P + 4Z' \qquad (0\beta)$$
$$(0\beta) \equiv 2(03) + (04)$$

The values Y and of Z' will now be approximately proportional and will increase together until Y approaches Y_{crit}. Again, there will be a rapid relaxation as Y passes through Y_{crit} and overshoots to much greater values while Z' remains moderately large.

Therefore, whether Y is large or small, chemical behavior causes Y to approach Y_{crit} until the sudden relaxation causes a major overshoot. At the instant when $Y = Y_{crit}$, the value of Z' will be very different depending upon whether Y_{crit} was approached from above or from below. The rapid relaxation of Y is completed before Z' can respond enough to influence behavior significantly. After the relaxation is completed, the tendency for Y to approach Y_{crit} results from a <u>delayed</u> <u>feedback</u>.

TYSON [7] has carried out a more sophisticated analysis of relaxation oscillations in this model. He has shown that relaxation oscillations like these can even occur for parameter values such that the steady state itself is locally stable to infinitesimal perturbations. Therefore, the stability analysis of a steady state as in Section 2 does not provide a sure criterion as to the possibility of temporal order traversing a trajectory far removed from that steady state.

3.2 Self-Destruction of a Population Distribution

A gas evolution oscillator as modeled by the Bubbelator can also exhibit almost discontinuous changes in dynamic behavior superficially resembling the relaxation oscillations discussed in Section 3.1. However, the Bubbelator generates unidirectional pulses of reaction, while the Oregonator creates conditions so that Y undergoes almost discontinuous transitions in <u>alternating</u> directions.

A major factor in the Bubbelator behavior is the potential reversibility of steps (B2) and (B3j). If the system is maintained in a steady state by the pumping of step (B1), M is large and all steps are driven from left to right. However, if M decreases, step (3j) is driven toward the left for small j and toward the right for large j. For any M smaller than that needed to drive step (B2) to the right, there will be a critical j such that step (B3j) goes to the left for smaller j and to the right for larger j.

The physical basis of this behavior arises because for any concentration, C, of dissolved molecules there is an equilibrium bubble radius, r_{eq}, calculable by

$$r_{eq} = \frac{2\kappa\sigma}{C - C_{\infty}}. \tag{10}$$

In this equation, σ is the surface tension, κ is the Henry's law constant, and C_{∞} is the concentration that would be in equilibrium with gas at the hydrostatic pressure exerted on the solution. If any bubble is smaller than r_{eq}, it will shrink and ultimately disappear. If any bubble is larger than r_{eq}, it will grow by evaporation into it of molecules from solution.

The distribution of bubble sizes therefore undergoes a bifurcation such that small bubbles shrink and disappear while large bubbles grow until they leave the solution. If bubbles are physically unable to escape, any population of bubbles, no matter what it is, is unstable with respect to one single large bubble. This instability of every population distribution is a separate feature not inherent in the steady state population distribution where the system is steadily pumped by step (B1), but it greatly influences dynamic behavior if the system is removed from the steady state. Populations of bubbles of different sizes will change in pulses that never come close to approximating the steady state distribution populations.

A somewhat similar behavior would be exhibited by a population of vibrationally excited diatomic molecules in which the total number of quanta of vibrational excitation was constant but in which such quanta could be exchanged among molecules. Because of the usual anharmonicity, the larger the quantum number the smaller the spacing between energy levels. Then (B1) describes the usual direction of reaction.

$$A_m + A_n \rightarrow A_{m+1} + A_{n-1} \qquad m \geq n \tag{B1}$$

If several vibrationally excited molecules were suddenly created as by a laser pulse, the initial distribution would self-destruct. Most of the molecules would drop to the ground state while a few would become highly excited. This apparent violation of the Boltzmann H theorem is dependent upon the constancy of quanta of vibrational excitation and will vanish if vibrational excitation can be converted

to translational and rotational kinetic energy. However, sudden vibrational excitations can be followed by unanticipated effects.

Whenever an unperturbed population distribution is inherently unstable, a steady
state will only be attainable if the system is pumped by a continuous flux of matter or energy. Even if the resulting steady state is locally stable to fluctuations, different pumping rates will probably generate pulsed instabilities in which
population distributions differ greatly from the one in the steady state. This
kind of temporal order will be associated with a trajectory in multi-dimensional
composition space,which is very unlike the sort of trajectory exhibited by the limit cycle in an Oregonator model.

3.3 Delayed Feedback in the Bubbelator Model

The repeated pulses in a gas evolution oscillator involve more than the inherent
tendency for the bubble size distribution to self-destruct. Other important contributions to the behavior involve the almost discontinuous onset of nucleation
and the slowness at which growing small bubbles can deplete the concentration of
dissolved molecules.

VOLMER [8] first pointed out that the rate of creation, J, of bubbles in equilibrium with concentration C is given by (11) where α and β are independent of
bubble size.

$$J = \alpha \exp[-\beta/(C-C_\infty)^2] \tag{11}$$

Values of parameters are such that as C increases J behaves almost as a discontinuous step function. Formation of bubble nuclei will be initiated suddenly when
molecules are being created by chemical reaction.

Equations developed elsewhere [9] indicate that over a considerable range of
sizes these bubbles will grow in such a way that dr/dt is almost constant. However,
if n is the number of moles of gas in those bubbles, dn/dt will be approximately
proportional to r^2 and hence to τ^2 where τ is the time since the onset of nucleation.

There is therefore a delay before the growing bubbles seriously deplete the concentration of dissolved molecules, and this delay is essential to the tendency of
the system to emit bubbles in pulses rather than continuously.

3.4 Phenomenological Classification of Types of Behavior

The preceding discussion has been concerned with mechanistic features such as autocatalysis, overshoot, and delayed feedback, at least some of which are associated
with all presently known examples of,or models for,temporal order in uniform systems. An alternative classification can be based on the purely phenomenological
types of observed oscillations. The following classification has been found to be
useful even though the distinction between types is not always sharp:

Sinusoidal oscillations approximate the familiar "simple harmonic motion" of
mechanical systems. They are observed when the trajectory of a chemical oscillation traverses a region of composition space,which is never far removed from a conservative or unstable steady state.

Relaxation oscillations have been illustrated with the Oregonator model in Section 3.1. They involve sudden gross overshoots in opposite directions of the values of some composition variable. Because of delayed feedback, the values of another composition variable will differ for the two directions of overshoot.

Pulses have been illustrated with the Bubbelator model in Sections 3.2 and 3.3.
A sudden burst of reaction is shut off by a delayed feedback. In the example con-

sidered, this feedback results from the self-destruction of a population distribution, but there should be other feedback mechanisms by which an initial fast reaction could be shut off quickly. There does not seem to be any sharp distinction between a pulse and a relaxation oscillation in which the time in one of the regions of composition space is much longer than the time in the other region.

Sawtooth oscillations would occur with a relaxation oscillator if the transition in one direction was much faster than that in the other. Although I am not aware of a good example, there does not seem to be any reason that such oscillations might not be generated.

Complex oscillations occur when a single period involves repetitive relative minima and maxima of different amplitudes or separations. GANAPATHISUBRAMANIAN and NOYES [10] have shown that such behavior can be generated by an expanded Oregonator model.

Chaotic oscillations occur when no reproducible period is generated. RÖSSLER [11] has obtained such behavior by means of a model having two autocatalytic processes with different rate constants.

Although there may be other types of temporal order, the above phenomenological classification seems to cover the major types of behavior with which we are presently familiar.

4. Impacts of Spatial Gradients on Temporal Behavior

All of the systems considered in Sections 2 and 3 were maintained virtually uniform in composition and in temperature, whether they were closed or open, and no matter how far their compositions might be removed from those of any steady states. The possibilities for temporal order may be greatly enhanced if boundary conditions are such that gradients in temperature or composition can develop spontaneously in the system during chemical reaction.

Although diffusion of heat and of matter always operate in directions to restore uniformity, diffusive processes are relatively slow and may not be able to compete with gradients established by local differences in rates of reaction. Especially if the rate of autocatalytic growth of a species is nearly balanced to the rate of destruction of the same species, many dramatic spatial and temporal effects may manifest themselves.

One example of such a situation is the observation of BUSSE [12] that an ever-increasing number of bands is generated if a composition gradient exists along a tube containing an oscillatory Belousov-Zhabotinsky solution.

Perhaps another example may be provided by the repetitive luminous glows sometimes observed in a closed flask containing a mixture of carbon monoxide and oxygen with about 10^{-5} mole fraction of hydrogen or water. This system has been studied carefully by GRAY and associates [13]. Although they believe the observed effects will occur even in a system truly uniform in composition and in temperature, it is very difficult to be sure that such conditions have been attained in a closed system undergoing a very exothermic reaction.

We can anticipate that many other types of temporal order will be encountered in systems which are not constrained to uniformity.

5. Concluding Remarks

This manuscript started with uniform systems and showed that autocatalysis and feedback were the only conceivable mechanisms by which a chemical system in a steady state could spontaneously develop temporal order. These types of behavior are well illustrated by the already established Oregonator and Bubbelator models.

The manuscript then considered uniform systems far removed from a steady state and showed that occurrence of temporal order might be irrelevant to local steady state stability and that such order might exhibit features that could not be anticipated from linearized dynamic equations. Such features include relaxation oscillations and pulses which can also be modeled effectively by the Oregonator and Bubbelator models, respectively. Other types of behavior such as complex oscillations may also occur in uniform systems far removed from a steady state. There is no assurance that we have yet identified all mechanistic or phenomenological features of temporal order in such systems, but it is a plausible hypothesis that some kind of delayed feedback will be associated with all such systems.

The final section of the manuscript noted that for certain boundary conditions gradients in composition and temperature may spontaneously develop in initially uniform systems despite the tendency of diffusive processes to destroy such gradients. These gradients may then couple to generate very complex types of temporal order. No general methods exist to classify the types of behavior which may or may not be possible in such systems.

Acknowledgment- The development of these ideas was supported in part by a Grant from the united States National Science Foundation.

References

1. R.M. Noyes: in Nonlinear Phenomena in Chemical Dynamics, C. Vidal and A. Paucault, editors (Springer-Verlag 1981) pp. 201-206.
2. J.J. Tyson: J. Chem. Phys. 62, 1010-1015 (1975).
3. A.J. Lotka: J. Amer. Chem. Soc. 42, 1595-1599 (1920).
4. K. Showalter, R.M. Noyes, K. Bar-Eli: J. Chem. Phys. 69, 2514-2524 (1978).
5. R.J. Field, R.M. Noyes: J. Chem. Phys. 60, 1877-1884 (1974).
6. R.M. Noyes: J. Phys. Chem. 88, 2827-2833 (1984).
7. J.J. Tyson: Lecture Notes in Biomathematics. 10. The Belousov-Zhatotinskii Reaction (Springer-Verlag 1976).
8. M. Volmer: Kinetik der Phasenbildung (Steinkopf, Leipzig 1939).
9. K.W. Smith, R.M. Noyes: J. Phys. Chem. 87, 1520-1524 (1983).
10. N. Ganapathisubramanian, R.M. Noyes: J. Chem. Phys. 76, 1770-1774 (1982).
11. O.E. Rössler: Z. Naturforsch. Teil A 31, 259-264 (1976).
12. H.G. Busse: J. Phys. Chem. 73, 750 (1969).
13. J.R. Bond, P. Gray, J.F. Griffiths: Proc. Toy. Soc. London A375, 43-64 (1981).

Mechanisms and Chemically Consistent Models for Isothermal Oscillations in the CSTR

Peter Gray

University of Leeds, Leeds LS2 9JT, United Kingdom

Stephen Scott

Macquarie University, N.S.W. 2113, Australia

Autocatalysis lies at the heart of nearly all isothermal, oscillatory chemical systems. When autocatalysis is strong, the rate of chemical change passes through a pronounced maximum as reaction progresses towards completion. Despite the diversity of real systems, two simple skeleton schemes, sharing the stoichiometry $A \to B$, cover almost the whole range of autocatalytic behaviour encountered in practice. They may be written:

$$A + B \to 2B \quad \text{rate} \propto ab \quad ; \quad A + 2B \to 3B \quad \text{rate} \propto ab^2.$$

Skeleton schemes of this kind have a long history, and even in the present context of the continuous-flow, well-stirred tank-reactor (cstr), their investigation goes back to work by Zel'dovich and Zysin in 1941. When the inflow contains only the species A (or A plus a fixed amount of B), they essentially describe 1-variable systems. Multiplicity is attainable, but not oscillations. Where should we turn for the prototype of oscillatory schemes? This question has received many answers. Some are too simple - Lotka's answers, for example, do not describe stable oscillations - but most are unnecessarily complex, so that their properties are only accessible after much numerical computation, and often have doubtful generality.

Another factor often present, and a frequent source of obscurity, is the "neglect of reactant consumption". In its crudest form, the conservation of matter is set aside; in its more popular form, it is supposed that a single product can be continuously extracted at a rate exactly matching the inflow of a single reactant.

At Bordeaux in 1981 (see also [1]-[5]) we introduced a new model for autocatalysis in a realizable open system, in which the concentrations of A and B were decoupled in as simple a way as possible. Autocatalysis is now accompanied by catalyst consumption or decay:

$$A + 2B \to 3B \qquad k_1ab^2 \qquad (1)$$

$$B \to C \qquad k_2b \qquad (2)$$

This scheme gave remarkably varied dependences of yield (amount of A consumed or amount of B formed in a steady state) on residence time in a continuously flowing, well-stirred, tank reactor. These curves could have anvil shapes (mushrooms) and show isolated regions (isolas). The scheme also gives rise to stable oscillatory behaviour, and it seems to be by far the simplest of all isothermal, chemically reasonable schemes to do so. It can, moreover, be developed and dissected with mathematical rigour and clarity to a degree not possible with any other scheme.

Our purpose here is: (1) to sketch the types of multiplicity and (2) the origin of the different patterns of local stability; (3) more particularly to describe the birth, growth and modes of extinction of stable oscillations; (4) to indicate the relationships of the present scheme to others (and to other circumstances) and finally (5) to address questions of chemical and thermodynamic reasonableness. We end with a brief chronology.

1 Mass-balance Equations and Convenient Dimensionless Forms

Consider the simple, irreversible cubic autocatalysis (1) coupled with a first-order decay or poisoning reaction (2) taking place in a c.s.t.r. The mass-balance equations can be readily written [2] in the following dimensionless form

$$\frac{d\alpha}{d\tau} = \frac{1}{\tau_{res}} (1 - \alpha) - \alpha\beta^2 \tag{3a}$$

$$\frac{d\beta}{d\tau} = \frac{1}{\tau_{res}} (\beta_0 - \beta) - \frac{\beta}{\tau_2} + \alpha\beta^2 \tag{3b}$$

where

$$\tau = k_1 a_0^2 t \qquad \tau_2 = k_1 a_0^2 / k_2 \qquad \tau_{res} = k_1 a_0^2 t_{res} \tag{4}$$

$$\alpha = a/a_0 \qquad \beta = b/a_0 \qquad \beta_0 = b_0/a_0.$$

Here a_0 and b_0 are the concentrations of A and B that would be established by the inflow in the absence of chemical reaction, and t_{res} is the mean residence time.

2 Conditions in a Stationary State

In a stationary state, $d\alpha/d\tau = d\beta/d\tau = 0$, and

$$(1 + \beta_0 - \alpha_{ss}) = \beta_{ss} [1 + (\tau_{res}/\tau_2)].$$

The dependence of α_{ss} on τ_{res} is thus given by:

$$\alpha_{ss} (1 + \beta_0 - \alpha_{ss})^2 = \frac{1}{\tau_{res}} \{1 + \frac{\tau_{res}}{\tau_2}\}^2 (1 - \alpha_{ss}). \tag{5}$$

This is a cubic equation in α_{ss} and a quadratic in the residence time τ_{res}.

2.1 Special Case ($\beta_0 = 0$)

In the special case of no B in the inflow ($\beta_0 = 0$), one solution of this equation is $(1 - \alpha_{ss}) = 0$. This factor may then be cancelled from each side to leave a quadratic equation in α_{ss} for the two remaining stationary states

$$\alpha_{ss} (1 - \alpha_{ss}) = \frac{1}{\tau_{res}} (1 + \frac{\tau_{res}}{\tau_2})^2. \tag{6}$$

Because the left-hand side has a maximum value of $\frac{1}{4}$ (when $\alpha_{ss} = \frac{1}{2}$) and the righthand side a minimum of $4/\tau_2$ (when $\tau_{res} = \tau_2$) this equation can have real solutions only if $\tau_2 \geq 16$. With this condition satisfied, the resulting dependence of the stationary-state extent of conversion $(1 - \alpha_{ss})$ on residence time portrays an isola pattern as shown in Fig. 1a. The range of dimensionless residence time over which the isola extends is given by:

$$\tau_{res}^{\pm} = \frac{1}{8} \tau_2^2 \{(1 - \frac{8}{\tau_2}) \pm (1 - \frac{16}{\tau_2})^{\frac{1}{2}}\}. \tag{7}$$

2.2 General Case

For the general case, with $\beta_0 \neq 0$, the cubic equation (5) may conveniently be solved graphically. Plotting the left and right hand sides of (5) separately as functions of $(1 - \alpha_{ss})$ leads to a cubic curve R_a and a straight line L_a respectively. The intersections of R_a and L_a on such a flow-diagram give the stationary states. The reaction curve R_a is independent of the residence time, but changes in τ_{res} affect the gradient of the flow-line L_a.

At the shortest residence times, L_a is steep and its gradient decreases as τ_{res} is increased. At long residence times, the gradient increases with τ_{res} and L_a

becomes steep again. In between there is a minimum slope for $\tau_{res} = \tau_2$. There are two possible tangencies between R_a and L_a which correspond to the appearance or merging of two stationary-state solutions.

These changes in R_a and L_a give rise to varying patterns for the dependence of the stationary-state extent of conversion of residence time as shown in Fig. 1. The pattern observed for a given system depends upon the particular values for β_0 and τ_2 as shown in Fig. 2. (An S-shape pattern arises when $k_2 = 0$ or $\tau_2 \to \infty$.)

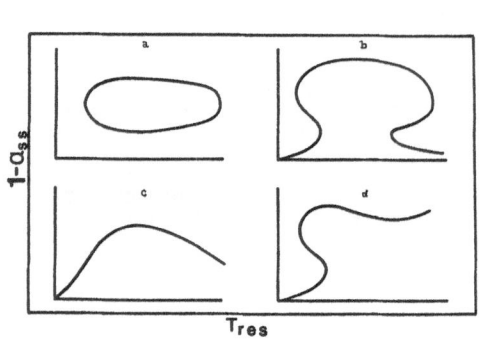

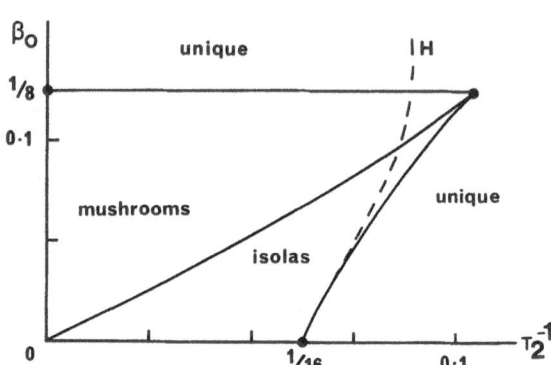

FIGURE 1   Different dependence of stationary-state extent of conversion found for cubic autocatalysis with catalyst decay

FIGURE 2 $\beta_0 - \tau_2$ regions for mushroom, isola and unique dependences of stationary-state on residence time. The line H divides systems which may show Hopf biburcation (high τ_2) from those which may not (low τ_2)

A fifth pattern emerges if the reverse reactions of steps (1) and (2) are accompanied by an uncatalyzed process, $A \rightleftharpoons B$. This may give rise to an isola and an S-shape.

3   Local Stability and Character of Stationary-States

The character of the stationary-states are evaluated in the usual manner [7] from the partial differential coefficients with $\alpha = \alpha_{ss}$ and $\beta = \beta_{ss}$.

$$\frac{\partial}{\partial \alpha}\left(\frac{d\alpha}{d\tau}\right) = -\beta^2 - \frac{1}{\tau_{res}} \qquad \frac{\partial}{\partial \beta}\left(\frac{d\alpha}{d\tau}\right) = -2\alpha\beta$$

$$\frac{\partial}{\partial \alpha}\left(\frac{d\beta}{d\tau}\right) = \beta^2 \qquad \frac{\partial}{\partial \beta}\left(\frac{d\beta}{d\tau}\right) = 2\alpha\beta - \left(\frac{1}{\tau_2} + \frac{1}{\tau_{res}}\right).$$

For zero catalyst inflow, the solution corresponding to no conversion is always a stable node (sn). The middle solution is always an unstable saddle point (sp). The nature of the highest extent of conversion varies with the residence time: it may be a stable node, stable focus (sf), unstable focus (uf) or unstable node (un). For non-zero catalyst inflow there is a further complication as the character along the lowest branch, which now corresponds to non-zero extents of conversion, varies between stable and unstable node or focus. There are two patterns with unique solutions, three isolas and four mushrooms. Some of the different sequences found are indicated on the mushrooms in Fig. 4.

4   Hopf Bifurcation, Limit Cycles and Sustained Oscillations

Points at which a stable focus becomes unstable correspond to points of Hopf bifurcation. These are the conditions at which closed orbits or limit cycles appear

26

around the stationary-state in the α - β phase plane. Limit cycles may be stable or unstable. The former lead to sustained oscillations in the reactant and catalyst concentrations. Unstable cycles do not correspond to observable oscillations but act in a similar way to the separatrices of a saddle point. Trajectories starting within an unstable limit cycle tend to stable stationary-state lying inside. Other trajectories move across the phase-plane to other stable states or cycles. The size of an unstable limit cycle is thus very important for global behaviour.

The line H in Fig. 2 separates those systems (characterized by β_0 and long catalyst lifetimes τ_2) which exhibit one or more Hopf bifurcation from those with lower values of τ_2 which do not possess such a point. For circumstances with no catalyst in the inflow ($\beta_0 = 0$), limit cycles occur only if $\tau_2 > 16$;

$$\tau_{res} \text{ (Hopf)} = \frac{1}{4} \tau_2^{3/2} \{[1 + (4/\tau_2^{\frac{1}{2}})]^{\frac{1}{2}}\}^2 \cong \tau_2^{3/2} [1 - 2 \tau_2^{-\frac{1}{2}} - \ldots] \qquad (8)$$

Stable limit cycles require larger values still, and if $\tau_2 = 16$, the first stable limit cycle is not found until $\tau_{res} \cong 28.5$.

5 Phase Portraits: Arrangements of Singular Points and Limit Cycles in α - β Phase Plane

Different combinations of stable and unstable singularities and stable and unstable limit cycles give rise to the nine different portraits in the phase-plane portraits displayed in Fig. 3. Three display a unique stationary-state, but differ in the number of limit cycles. Of the six portraits with three singularities, two have no cycles and four have either one or two.

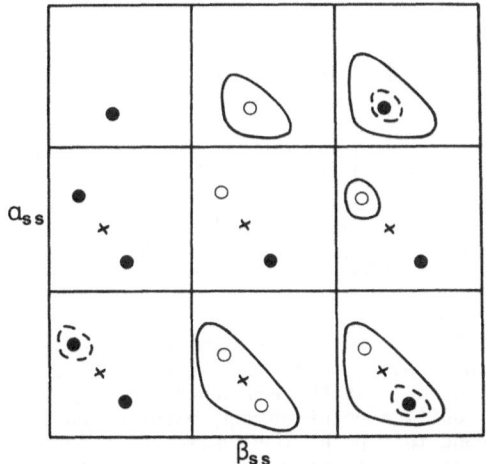

α_{ss}

β_{ss}

FIGURE 3 Phase portraits: •, stable node or focus; o, unstable node or focus; x, saddle point. Solid curves denote stable limit cycles, broken lines give unstable cycles

These phase-portraits for isothermal autocatalysis are exactly the same as the nine patterns discovered by Uppal *et al.* for single, irreversible exothermic reaction. Subsequently Kwong and Tsotsis located three more possible portraits for the non-isothermal reactions with the Arrhenius temperature-dependence ($\varepsilon > 0$). Although Vaganov *et al.* present a total of 35 patterns (from various combinations of one or three stationary-states and up to three limit cycles) they do not prove that all these occur for the simplest case studied here. It appears that twelve portraits are the only ones possible in the non-isothermal case.

6 The Varied Patterns of Oscillatory Behaviour

So far as oscillations are concerned, "supercritical" Hopf bifurcation corresponds to the simplest development. From such a point, oscillations grow in amplitude from nothing (and diminish continuously to nothing on reversal) according to a square-root law:

$$\text{amplitude} \propto [t_{res} - t_{res} \, (\text{Hopf})]^{\frac{1}{2}}.$$

This, however, is not the only possible behaviour. At some particular flow rate, an experimenter may encounter a sudden transition from no oscillations at all to fully fledged spiky patterns: reversing the procedure will then show marked hysteresis. Similarly a period of growth of oscillatory amplitude may end very abruptly when the stable limit cycle in phase space 'bursts' as a result either of collision with the separatrices of a saddle point or of coalescence with an unstable limit cycle. Examples of these types of behaviour are set out below and illustrated in Fig. 4 in terms of the observations which would be recorded by an experimenter slowly decreasing the flow-rate in each case.

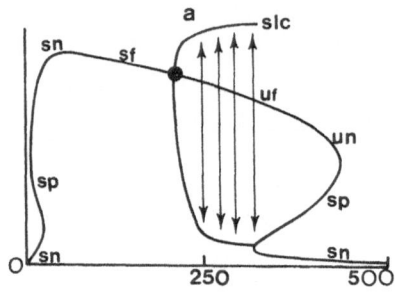

FIGURE 4   Mushroom dependences of $1 - \alpha_{ss}$ on τ_{res} showing different behaviour at long residence time:
(a) $\tau_2 = 40$, $\beta_0 = {}^1/15$;
(b) $\tau_2 = 50$, $\beta_0 = 0.1$;
(c) $\tau_2 = 50$, $\beta_0 = 0.12$.
Also shown are envelopes of limit cycles in $\alpha - \beta$ plane

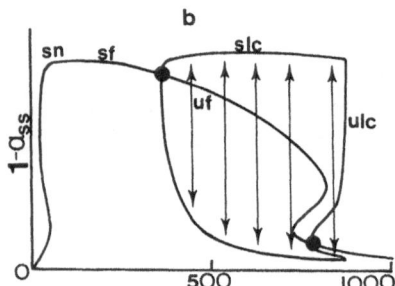

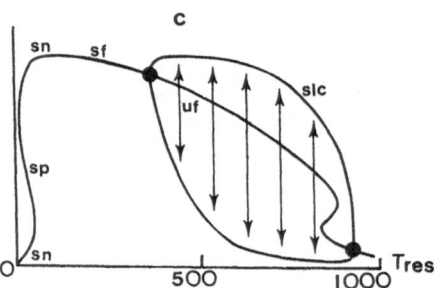

Figure 4(a)   For an experiment set up with an initial residence time $\tau_{res} = 100$, there is a unique stationary-state solution which has stable focal character. Small perturbations decay in a damped oscillatory fashion. As the residence time increases, the degree of damping becomes less until a point of Hopf bifurcation is encountered, in this case at $\tau_{res}^* = 100$. Beyond this point the system moves into a series of self-excited, sustained oscillations. These grow from a vanishingly small amplitude at τ_{res}^*: initially the amplitude is proportional to $(\tau_{res} - \tau_{res}^*)^{\frac{1}{2}}$. The oscillations have a non-zero period at the point of Hopf bifurcation which is determined by the imaginary eigenvalues of the Jacobian matrix $\lambda = \pm i\omega_0$, $p = 2\pi/\omega_0$. There is no hysteresis at this "supercritical" bifurcation [8]; decreasing the residence time simply causes the amplitude of the oscillations to shrink back to zero at τ_{res}^*. As the residence time is increased further, the oscillations become larger and separated by longer periods. The envelope surrounding the upper branch of stationary-state in Fig. 4(a) shows the highest and lowest values of $1 - \alpha$ attained during the pulses as a function of residence time.

At $\tau_{res} = 312.5$, a saddle point and a stable node appear in the phase-plane, corresponding to the two other stationary-states. Shortly after this, the limit cycle grows sufficiently large to touch the separatrix of the saddle point and form a homoclinic orbit. The period of motion around this cycle becomes infinite and

28

beyond this point oscillations are not found. There is thus a sudden quenching of
large amplitude oscillations. For all residence times greater than τ_{res} = 313 the
system sits at the lowest, stable state. There is also hysteresis at this limit.
As the residence time is decreased, we pass the point at τ_{res} = 313 at which the
limit cycle bursts on the upward journey, but do not now move from the lower branch.
Only when there is a saddle-node bifurcation of the lower two stationary states at
τ_{res} = 312.5 does the system jump back into large oscillations.

Figure 4(b) Again,if an experiment is started at τ_{res} = 150 there is a unique
stable node. On increasing the residence time, a supercritical Hopf bifurcation of
the above type is encountered at τ^*_{res} = 326 and approximately sinusoidal, small-
amplitude oscillations set in. These become larger and have longer periods as τ_{res}
increases. At τ_{res} = 782 there is a saddle-node bifurcation, but the two emerging
stationary-states now lie within the limit cycle. The lowest extent of conversion
is at first an unstable node and then becomes an unstable focus. For τ^*_{res} = 786
there is a point of Hopf bifurcation along the lower branch. Here an unstable limit
cycle emerges to surround the stable focus, still within the original stable limit
cycle. As the residence time is further increased, the upper two solutions merge
and disappear. Finally, the unstable limit cycle becomes large enough to touch the
stable cycle and oscillations are abruptly quenched. Beyond this point the system
sits on the lowest branch.

Again there is hysteresis at this long-residence-time end of the range of
oscillatory behaviour. If τ_{res} is now reduced, the system stays on the stable lower
branch until the Hopf bifurcation point at τ^*_{res} = 786, whence it jumps back into
large amplitude oscillations about all three stationary-states.

Figure 4(c) A further increase of the catalyst inlet concentration leads to the
pattern in Fig. 4c. The behaviour at long residence-times is now quite different.

Once the oscillations have set in at the first point of Hopf bifurcation (τ^*_{res}
= 349), there are two saddle-node bifurcations at τ_{res} = 921 and τ_{res} = 927 within
the limit cycle. The first gives rise to the emergence of an unstable lower branch
and the saddle point corresponding to the middle branch. At the second,the middle
branch merges with the highest extent of conversion. The latter occurs before the
lowest stationary-state has regained its stability, i.e. before the second point of
Hopf bifurcation. When the unstable focus finally becomes stable, at τ^*_{res} = 939,
the limit cycle shrinks to a point and the oscillations vanish at this second
supercritical bifurcation. There is now no hysteresis at either end of the
oscillatory range.

7 Comments on the Status of the Simple Skeleton Scheme

The cubic rate-law model presented here affords a prototype for many systems of
practical interest, including solution-phase kinetics and enzyme reactions. It
should be stressed that steps 1(a) and (b) do not have to be regarded as elementary
steps: they may reflect combinations of such steps. Thus the reaction between
arsenite and iodate ions, which is autocatalytic [9,10] in the production of I^-,
is well approximated at constant pH by the cubic rate expression

$$\text{rate} \propto \{a + [I^-]\}[I^-][IO_3^-]$$

where $[I^-] + [IO_3^-]$ = constant, over some range of concentration.

The present model differs significantly from the so-called "pool chemical"
approximation approach, which assumes that the concentration of a precursor to
reactant A would somehow be held constant, in two important ways. First there is
a finite maximum value which may be attained by the reaction rate, $viz.$

$$R_{max} = \frac{4}{27} k_1 a_o^3,$$

whereas the "pool chemical" rate shows hyperexponential growth to infinitely large values [11,12]. Secondly, multistability is a real feature of the cstr equations. Karmann and Hinze [13] have shown that the multiple solutions for the "pool chemical" approach are not stable to the inclusion of lower-order steps and their reverse reactions.

In this context it is worth mentioning that the "pool" scheme:

$$\text{precursor} \rightarrow A \text{ (constant rate)}$$

$$A + 2B \rightarrow 3B$$

$$B \rightarrow C \text{ (leaves as fast as A enters)}$$

does generate stable oscillations [14]. In essentially this form, it was studied by Sel'kov in 1968, and rediscovered by Tyson and Light in 1973. Schlögl's rather different form, with reversible reactions but with A kept constant (!), does not seem to have been such a fruitful growth point.

We may also note that the cubic dependence on concentration embodied in step (1b) is not a necessary feature for sustained oscillations. If the decay reaction for the catalyst has a less simple form, for example

$$B \rightarrow C \qquad \text{rate} = k_2 b/(1 + rb), \qquad (9)$$

the quadratic non-linearity of step (1a) is sufficient [14].

A rate-law of the form (9) may arise in enzyme reactions, or in heterogeneous catalysis obeying a Langmuir-Hinshelwood law. The oxidation of carbon monoxide at low pressures displays sustained oscillations under isothermal conditions in an open system [15], and results indicate that an important role is played by the reactor surface.

8 Augmented Schemes: Questions of Chemical Realism, Thermodynamic Self-consistency and Detailed Balance

8.1 Augmented Irreversible Schemes

It has been argued that it is chemically unreasonable to ignore completely the uncatalyzed step. Whether this argument is accepted or not is debatable, but augmented schemes including either $A \rightarrow B$ or $A + B \rightarrow 2B$ or both these steps in addition to reactions (1) and (2) are interesting in themselves and deserve study, though they involve more algebra. We have now examined them in detail [14]. Their most important property in the present context is that, in a cstr, they can still show the types of behaviour highlighted above. They also show three very important differences. First, at long residence-times (zero flow-rates), all species A disappears; second, the range of conditions over which 'exotic' behaviour occurs is diminished; and third, two new bifurcation diagrams ($1-\alpha$ vs t_{res}) appear, one showing S-shapes and the other showing an isola plus an S-shape.

8.2 Incorporating Reversibility of Reaction Steps

All chemical changes are in principle reversible, and it was thought until quite recently that the unusual properties of this kind of autocatalysis depended on its reversibility. Clearly this is not the case here, although in other circumstances reversibility may be necessary to produce exotic behaviour. The effect of reversibility is simply to bring reaction to a halt before conversion is complete; so long as $K_e = k_1/k_{-1}$ is large, no other qualitative changes are important. As K_e gets less, so interesting behaviour is lost.

8.3 Satisfying the Principle of Detailed Balance

It is well known that if all three steps $A \rightarrow B$, $A + B \rightarrow 2B$ and $A + 2B \rightarrow 3B$ are reversible, then for each one $(k_f/k_r) = K_e$. Can the present scheme still show multiplicity and oscillations under this restriction of detailed balance? The answer [14] is yes, although over a narrower range of parameters. The scheme is kinetically robust and thermodynamically self-consistent. This is all the more remarkable in view of the fact that the Brusselator fails this test [6]: in order to Brusselate, the principle of detailed balance must be suspended.

9 Some Comments on Antecedents and Chronology

(a) The present system in a cstr was originally proposed [4] as a counter example to a spurious generalization about stationary state patterns. Showalter [10] illustrated the value of the predictions in an imaginative cstr study of the iodate plus arsenite reaction catalyzed by iodide ion. Previous cstr work with the single step $A + 2B \rightarrow 3B$ (but with $k_2 = 0$) goes back to Zel'dovich (1941). He was followed by M.C. Lin (1981) who looks at $mA + nB$. A catastrophe-theory study by D'Anna and Lignola is currently (1984) in progress in Naples.

(b) The third order step $A + 2B \rightarrow 3B$ lies at the heart of more complex early models such as Turing's work and the Brusselator. From those times on, complexity reigns and the (false) beliefs that models of oscillatory reactions must involve reversibility and more than one autocatalytic or autoinhibitory step are widespread.

(c) Sel'kov (1968) studies and Tyson and Light (1973) mention the present reaction scheme in the context of the 'chemostat' or 'pool chemical' approach. A strange form (constant A) is invoked by Schlögl (1971-2) in a comparison with first-order phase transitions. Oscillations in the reversible Sel'kov model and pool chemical context are also studied by Richter, Rehmus and Ross (1981, 1984); in 1984, Escher and Ross also extend this to cstr circumstances.

(d) In the biochemical field, the Sel'kov model and pool chemical context are exploited by other workers, notably Hess (1965+). In this field there is still the double question: (i) how close are the elementary schemes to the actual complex kinetics and (ii) how close is the "chemostat" model to real conditions of inflow and (especially) outflow?

Acknowledgments

We are grateful for research support from the EEC and the SERC and for the award of a Queen Elizabeth II Postdoctoral Fellowship to SKS. We also thank Dr D.G. Knapp and Dr J.H. Merkin of Leeds and Prof. B.F. Gray of Sydney for helpful discussions.

References

1 S.K. Scott: Ph.D. dissertation, University of Leeds (1982).
2 P. Gray and S.K. Scott: Chem. Engng. Sci. 38, 29-43 (1983): 39 (1984).
3 S.K. Scott: Chem. Engng. Sci. 38, 1701-1708 (1983).
4 P. Gray and S.K. Scott: J. phys. Chem. 87, 1835-38 (1983): J. Chem. Phys. 79 6421-3 (1983): Ber. Dtsch. Bunsenges 87, 379-382 (1983).
5 P. Gray and S.K. Scott: Chemical Instabilities (Eds. G. Nicolis and F. Baras); Reidel (1984).
6 B.F. Gray, P. Gray and S.K. Scott: J. Chem. Soc. Faraday Trans. 2 (1984) in the press.
7 A.A. Andronov, A.A. Vitt and S.E. Khakin: Theory of Oscillators, Pergamon Press (1966).

8 B.D. Hassard, N.D. Kazarinoff and Y.-H. Wan: Theory and applications of Hopf
       bifurcation Cambridge University Press (1981).
9 H.-G. Lintz and W. Weber: Chem. Engng. Sci. 35, 203-208 (1980): F.A. Papsin,
       A. Hanna and K. Showalter: J. phys. Chem. 85, 2575-82 (1981).
10 N. Ganapathisubramanian and K. Showalter: J. phys. Chem. 87, 1098-99 (1983):
       J. chem. Phys. 80, 4177 (1984).
11 G.A.M. King: J. chem. Soc. Faraday Trans. 1 79, 750-8 (1983).
12 R.M. Noyes: J. phys. Chem. (1984) in the press.
13 K.-P. Karmann and J. Hinze: J. chem. Phys. 72, 5476 (1980).
14 S.K. Scott: J. chem. Soc. Faraday Trans. 2 (1984) in the press: J.H. Merkin,
   D.J. Needham and S.K. Scott: Proc. R. Soc. Lond. A (1984) in the press.
15 P. Gray, J.F. Griffiths and S.K. Scott: Proc. R. Soc. Lond. A (1984): 20th
       Symp. (Int.) Combust. Ann Arbor (1984).
16 B.F. Gray and T. Morley Buchanan: in press (1984).

Analysis and Qualitative Modelling of Experimentally Observed Dynamic Features

Moshe Sheintuch and Dan Luss

Department of Chemical Engineering, University of Houston, University Park
Houston, TX 77004, USA

1. INTRODUCTION

Chemically reacting systems may exhibit a variety of dynamic and steady-state multiplicity features. The parameter space is separated by hypersurfaces called bifurcation sets into regions with qualitatively different types of behavior. These sets intersect or coalesce at singular points next to which several states with qualitatively different behavior exist [1-4]. We define a singular point to be a physical center if three or more regions with different observed behavior intersect transversely or one at which two regions intersect tangentially.

Experimental bifurcation diagrams (maps) which describe transitions (bifurcations) caused by the change of one (two) operating variable(s) can be used to predict new types of behavior (and the corresponding operating conditions) for systems for which no mathematical model exists as illustrated by EPSTEIN et al. [5] and HAROLD and LUSS [6] using the theoretical background presented in [7-10]. Knowledge of the physical centers enables construction of the simplest model, which accounts for all the observed features, rational organization of experimental or theoretical knowledge and a systematic finding of parameters corresponding to certain phase-plane behavior.

Consider a system characterized by two dynamic variables which is described by the two autonomous equations

$$\epsilon \dot{w} = f(w, v) \tag{1}$$

$$\dot{v} = g(w, v) . \tag{2}$$

The steady-state solutions satisfy the relation

$$f = g = 0 . \tag{3}$$

In some cases v may be eliminated from these two eqns. to give a single steady-state eqn. of the form

$$F(w) = 0 . \tag{4}$$

The bifurcations occurring in this system are of five different types [1, 3]. It is relatively easy to predict bifurcations in the number of steady-state solutions (saddle-node bifurcation) or the emergence of oscillatory solutions at a Hopf bifurcation. A much more difficult task is the prediction of a global bifurcation to oscillatory solutions at a global-saddle node (a homoclinic curve passing through a saddle node), generalized Hopf bifurcation (coalescence of an unstable and stable limit cycle to form a semi-stable one) or a saddle-loop bifurcation (a homoclinic curve passing through a saddle point).

The dynamic analysis of the system is significantly simplified when the time scales of the two variables are widely separated, say $\epsilon \to 0$. Use of this limiting case yields important physical insight into the structure of the dynamic behavior, which unfortunately cannot be obtained by analyzing a model in which both time scales are of the same order.

Heterogeneous catalytic systems (single pellets, wires or foils) exhibit relaxation oscillations with a wide difference between the characteristic times of the oscillations and the chemical reaction (inverse turn-over number). This dynamic behavior indicates that widely different time scales exist in the system. In some cases multipeak and chaotic oscillations are observed, indicating that more than two dynamic variables need to be included in any lumped model of the system. We ignore here these observations and assume that the system may be described by a model having two variables with widely different time scales.

We discuss in this work the identification of experimentally observed bifurcations and the use of at least two different bifurcation diagrams to deduce the existence of physical centers and high order singular points. We then illustrate how to design a limiting dynamic model (one with $\varepsilon \to 0$) which has the same center(s) and present a systematic approach to the analysis of the dynamic features of the model. The existence of the center does not guarantee that the sequence of observed transitions agrees with the observations. We propose a scheme for modifying the model so that it predicts the proper order of transitions.

2. IDENTIFICATION OF EXPERIMENTALLY OBSERVED TRANSITIONS

Identification of the type of experimentally observed transitions may be useful for predicting operating conditions corresponding to different or new dynamic features and in the development of mathematical models of the system. Unfortunately, the identification of observed bifurcation is, in general, a difficult task, and a certain ambiguity often complicates this process. Some of the difficulties associated with this identification may be overcome by examining a family of transitions, i.e. construction of a bifurcation map. The parameter sets describing a specific transition lie on continuous curves and this reduces the uncertainty associated with the analysis of only one bifurcation diagram.

We examine here some of the possible bifurcations which can exist in a system characterized by two dynamic variables, having no more than three steady-state solutions and up to one stable oscillatory state. Six possible bifurcation diagrams which describe the branches of both the stable and unstable states are shown in Fig. 1. The diagrams show the observed (stable) and unobservable (unstable) behavior next to each of the bifurcations.

The most common transition to oscillatory behavior is at a Hopf-bifurcation. Four different types of Hopf bifurcations can occur in this system. The first, denoted by H_1, is a (supercritical) Hopf bifurcation, which leads to a transition from a branch of stable unique states to stable oscillatory states with a smooth (sometimes steep) dependence of the amplitude on the distance from the bifurcation point. H_2 is the same transition as H_1,

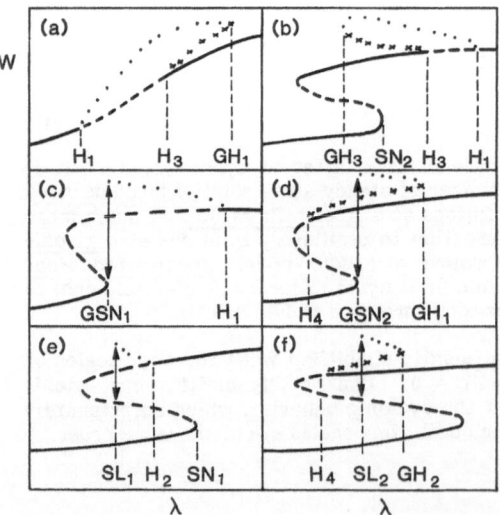

Fig. 1. Possible bifurcations in a system of two differential equations with up to two stable steady states and one stable limit cycle. Solid (broken) lines represent stable (unstable) steady state solutions. Dots (x) represent maximal value of stable (unstable) oscillatory solutions

but an additional branch of either stable or oscillatory states exists in the transition region (see Fig. 1.e, 2). H_3 is (subcritical) Hopf bifurcation in which a stable (observable) limit cycle surrounds the unstable limit cycle. (This transition causes a hysteresis between the stationary and oscillatory states (Figs. 1.b and 2)). H_4 is a subcritical Hopf bifurcation in which no limit cycle surrounds the unstable oscillatory state next to the transition point (Figs. 1.d, 1.f, 2). H_4 bifurcation leads to a jump from a stable branch of stationary states to another branch of either stable or oscillatory states. The transition has the same observable features as a <u>saddle-node</u> (SN) bifurcation. This coalescence of a node and a saddle is observable only if the node is stable at the SN point (SN_1 in Fig. 1.e). When the three states are surrounded by a stable and unstable limit cycle, an SN_2 transition occurs within the observed oscillatory state (Fig. 2).

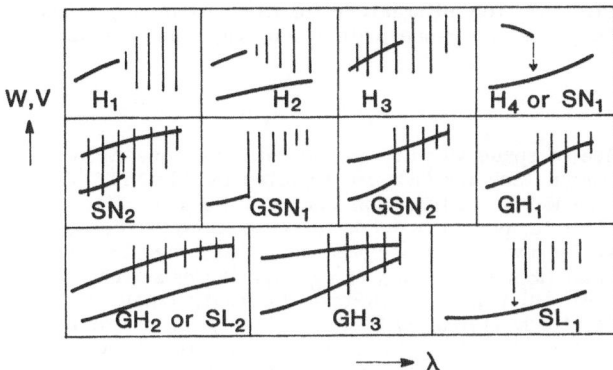

Fig. 2. Observable features of the bifurcations (symbols correspond to bifurcations shown in Fig. 1). Solid line denotes stable steady states, vertical line–amplitude of limit-cycles

A <u>global saddle-node</u> bifurcation (GSN) describes a hard bifurcation from a saddle node point to a branch of oscillatory state. No hysteresis occurs in this case. The case in which the node, surrounded by the limit cycle, is unstable is denoted as GSN_1 (Fig. 1.c), while if it surrounds an unstable cycle and a stable state it is of type GSN_2 (Figs. 1.f and 2).

At a <u>global-Hopf</u> bifurcation (GH) a semi-stable cycle appears and splits into a stable and an unstable oscillatory states. This bifurcation leads to a hysteresis between the oscillatory and stationary branches of observable states. We denote by GH_1 the situation that the semi-stable limit cycle surrounds unique stationary states (Fig. 1.a), by GH_2 when it encircles one of two stable states (Fig. 1.f), and by GH_3 when it surrounds both observable stationary states (Fig. 1.b). A stable branch of oscillatory solutions emerges or disappears at a <u>saddle-loop</u> bifurcation (SL). There exists in this case another branch of stable solutions on both sides of the SL point. When the emerging limit cycle surrounds an unstable state, the transition is of type SL_1 (Fig. 1.e), while if it surrounds a stable stationary state and an unstable limit cycle it is of type SL_2 (Fig. 1.f). Figure 2 indicates that in some cases (H_1, GSN_1, H_2, H_4, SN, SL_1) no overlap exists between the stationary states and the amplitude of the oscillatory states in the vicinity of the transitions. All these transitions have been observed in heterogeneous catalytic systems. Overlapping states occur when at least one stationary state is within the range of amplitude of an oscillatory state next to the bifurcation point (H_3, SN_2, GSN_2, GH, SL_2). The literature does not contain any clear example of overlapping states near transition points in a heterogeneous catalytic system.

Figure 2 indicates that even within that limited class some transitions have similar observable features. For example, the observed features of H_4 and SN_1 or transitions GH_2 and SL_2 are the same. The ambiguity associated with the identification of observed transitions increases when several transitions occur close to each other. For example, the observed behavior next to a global saddle-node (GSN_1) is very similar to that of a Hopf bifurcation of type H_3 which is very close to a GH_2.

When ε is very small (widely different time scales) the amplitude of the oscillatory solutions grows rapidly with the distance from the Hopf bifurcation point and the experimental distinction between hard and soft bifurcation disappears. In the limiting case of ε = 0 a Hopf bifurcation is hard, and a generalized Hopf bifurcation can occur only if five solutions of f = 0 exist for some v.

When two unstable solutions coalesce the bifurcation is not observable. Non-observable bifurcations may lead to pitfalls in the analysis of observed data. For example, when two adjacent non-observable saddle-node points are within a region of stable oscillatory states one assumes usually that the branch of steady-states is single valued even though it is a multivalued function.

We illustrate now the identification of the bifurcation for two examples. This information will be used later to develop a simple model with ε → 0, which has the same bifurcations.

Example 1

An experimental study of the influence of space velocity on the catalytic conversion of carbon monoxide in a continuously stirred tank (CSTR) was reported by PLICHTA and SCHMITZ [11]. Figures 3 and 4 describe the bifurcation diagrams for two different reactant feed concentrations and surface temperatures. For the case shown in Fig. 3 a single high conversion state is observed for low space velocities. This phase plane behavior is denoted by U. At point A supercritical Hopf bifurcation of type H_1 causes a transition to a unique oscillatory state (O). At point B a saddle-node bifurcation causes the emergence of another branch of stable steady states, so that the phase plane contains a non-overlapping stable limit cycle and a stationary state (OS). At point C a hard bifurcation causes the disappearance of an oscillatory state, most probably by a saddle-loop transition. This implies that to the right of C the phase plane contains three solutions, only one of which is stable and observable (U_3). Eventually, an unobserved SN transition may cause a transition to a phase plane with a single stable state (U).

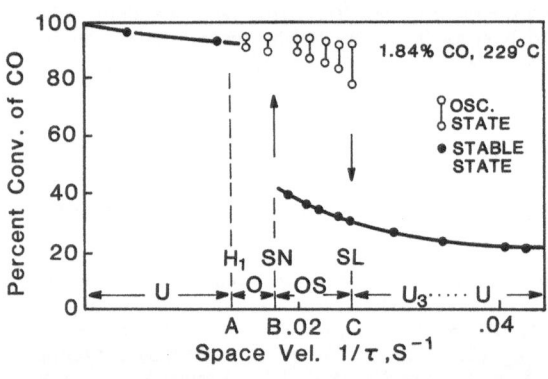

Fig. 3. Bifurcation diagram observed by PLICHTA and SCHMITZ [11] during CO oxidation on Pt

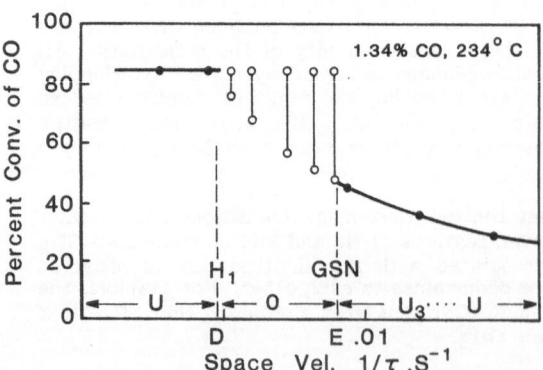

Fig. 4. A different bifurcation diagram observed in CO oxidation [11]

36

Figure 4 describes a different bifurcation diagram. At point D a Hopf bifurcation of type H_1 causes a transition to an oscillatory state. At point E apparently, a global saddle node causes a shift from an oscillatory to a stable state. Another possible explanation is that at point E a very steep Hopf bifurcation (either type H_1 or H_3 + GH) occurred. Assuming that a smooth transition exists between the two diagrams (Figs. 3 and 4) we search for a center that yields no other types of bifurcation diagrams between the two. A simple explanation is that the transition is a GSN. This implies that to the right of point E the phase plane is of type U_3 similar to that to the left of C in Fig. 3.

Example 2

The isothermal oxidation of propylene over a Pt wire was studied by SHEINTUCH and LUSS [12]. At high temperatures a continuous high activity branch exists and the low activity branch is observed only for concentratios larger than a critical value (see Fig. 5, $T_S = 202°$). Along the upper branch bifurcations of type H_1, SN_1 and H_2 exist. Note that the first transition was not actually observed. At higher temperatures the oscillations disappeared.

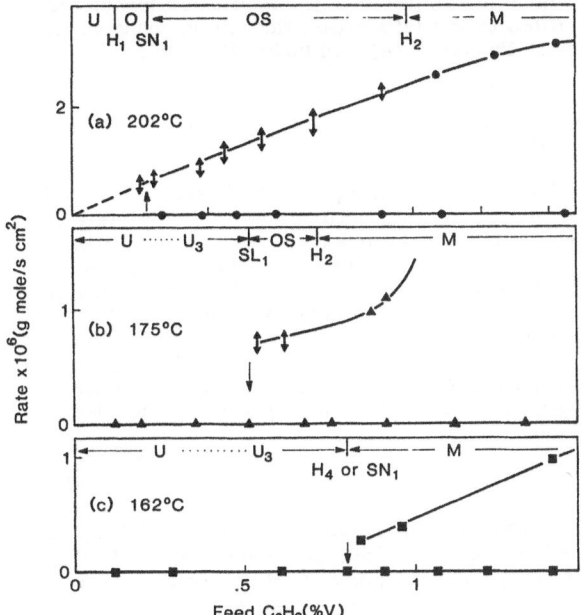

Fig. 5. Sequence of bifurcation diagrams observed during C_3H_6 oxidation on an isothermal Pt wire [12]

Decreasing the temperature caused a low reaction rate branch to exist for all reactant concentrations (see Fig. 5, 175° and 162°) while for 202°C a high rate branch existed for all reactant concentrations. At 175°C that branch was extinguished upon decreasing propylene concentration by an SL bifurcation. The extinction at 162°C may be due to either a H_4 or SN_1 bifurcation. Thus, between 175°C and 162°C a smooth transition from SL_1 to either H_4 or SN_1 must occur. The H_2 bifurcation that existed at 175°C disappeared at 162°C.

3. MATHEMATICAL MODELLING

A mathematical model needs to predict observed dynamic features such as phase planes and bifurcation diagrams, sets, maps and centers. Proper identification of the observed features is essential for the development and testing of a model. Local features, such as bifurcation sets and centers, provide guidance in determining the functional dependence on the state variables. Once this dependence has been fixed, global features, such as the organization of centers in the parameter plane and the orientation of bifurcation sets, depend on the functional dependence of f and g in (1-2) on the parameters.

Models containing only saddle-node and Hopf bifurcations can be readily developed. When the time scales of the two dynamic variables are similar no simple procedure exists for the development of models predicting global transitions such as saddle-loop, global Hopf and global saddle-node. However, when the time scales are widely separated, a simple graph of the two functions f = 0 and g = 0 can predict the dynamic behavior, and it is possible to derive simple analytical conditions describing any of the five bifurcations. A systematic approach for determining all the possible phase planes and bifurcation sets and centers in the global parameter space was described in [13] for this case, and is reviewed briefly in the next section.

We describe first a procedure for developing a <u>qualitative</u> <u>model</u> which predicts all the bifurcation sets when $\varepsilon = 0$. A <u>quantitative</u> <u>model</u> can then be developed, by use of mathematical functions which incorporate knowledge of the rate processes, constraints on the values of the state variables, and knowledge of the qualitative features f and g need to satisfy. There exists no simple scheme which guarantees that a quantitative model will predict also the proper sequence of transitions as the bifurcation variable is changed. A scheme for modifying the model to accomplish this task [14] will be briefly described.

4. DYNAMIC FEATURES OF LIMITING CASE

When $\varepsilon = 0$ the dynamic motion consists of a slow motion along the stable section of f = 0 and a rapid motion elsewhere. The slow motion persists until either a steady state is reached or a fast transition to another stable section of f = 0 occurs. The stable (unstable) section of f = 0 is one with $f_w < 0$ ($f_w > 0$). Any intersection of f = 0 and g = 0 is a steady-state solution. The dynamic behavior of (1-2) depends on the location of the steady states on the graph of f = 0. The stability of a state can be determined by inspection of the two null curves f = 0 and g = 0 in the (v, w) phase plane.

These graphs enable derivation of simple conditions defining the various bifurcations. The stability of a node changes by crossing a limit (turning) point of f = 0. At this (Hopf) bifurcation

$$f(w^\ell, v^\ell) = f_w = g(w^\ell, v^\ell) = 0 . \tag{5}$$

A saddle-node (SN) bifurcation occurs when the two null curves are tangent. When the tangency point is on a limit cycle a global saddle node (GSN) is formed. A saddle loop bifurcation (SL) occurs when a saddle point crosses the boundary of a limit cycle at a point S defined by

$$v^S = v^\ell \qquad w^S \neq w^\ell. \tag{6}$$

A global Hopf bifurcation (GH) can occur with $\varepsilon = 0$ only if at least five steady-state solutions are feasible [13]. We do not consider such cases here.

We define a bifurcation set to be the set of all parameters for which a specific bifurcation occurs. For example, a SNS consists of all the parameters satisfying

$$F(w, \underline{p}) = F_w(w, \underline{p}) = 0 , \tag{7}$$

while the LPS (limit-point set) consists of those satisfying condition (5). A major advantage of the use of a model with $\varepsilon = 0$ is the ability to divide the global parameter space into regions with different behavior by use of simple algebraic expressions [13]. To illustrate this consider the system

$$\varepsilon \dot{w} = -w^3 + p_0 w - v = f(w, v, p_0) \tag{8}$$

$$\dot{v} = p_2 w^2 + p_1 w - v - p_3 = f(w, v, \underline{p}) . \tag{9}$$

Simple analysis [13] indicates that for $p_0 < 0$ the only possible behavior is U or M, while for $p_0 = 0$ $f = 0$ has a cusp point. For $p_0 > 0$ the four parameter (8-9) can be reduced by a simple transformation [13] into a similar set but with $p_0 = 1$. To simplify the analysis we consider that special case.

We select p_1 to be the bifurcation variable, so that a bifurcation diagram describes the sequence of phase plane obtained as p_1 is changed for fixed values of p_2 and p_3. The qualitative features of a bifurcation diagram can change only at p_2 and p_3 values, for which two bifurcation sets intersect either tangentially or transversely, or if a cross-section of the bifurcation set for a fixed p_3 has an extremum point, i.e. $\partial p_2 / \partial p_1 = 0$. Every bifurcation set is a two-dimensional surface in the three-dimensional parameter space. An intersection between any two bifurcation sets or any transition (either hysteresis or isola) variety (see [8] for a definition) is a line in the parameter space. The projection of these lines into the (p_2, p_3) plane separates parameter regions with different bifurcation diagrams. Knowledge of the behavior in one region is often sufficient to predict those in other regions based on knowledge of the transitions occurring at each line.

Simple analytical expressions defining the various bifurcation sets and boundaries of regions with different diagrams can be obtained readily for (8-9). Figure 6 describes the various boundaries which represent the intersection of the LPS with itself (C), with the SNS either tangentially (D) or transversely (E) and with the SLS (F). The intersection of the SNS with the SLS is shown as line G, with itself at a cusp point as H_F and the local extremum of the SNS is line I_F.

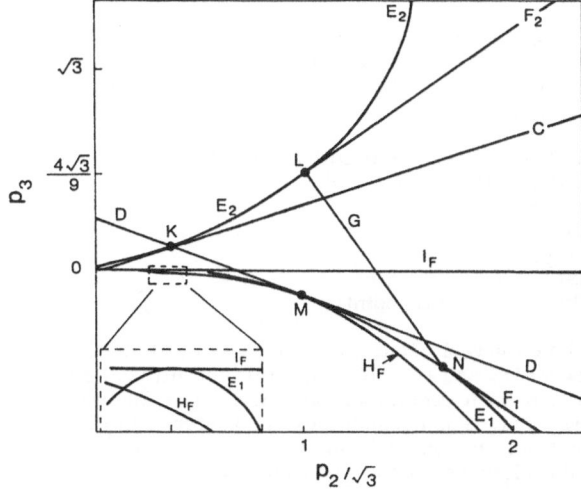

Fig. 6. Projections in the p_1 direction of the intersections of the various bifurcation sets of (8, 9)

We define a bifurcation map to be a cross-section of the various bifurcation sets in the (p_1, p_3) plane. New phase plane are created in these maps only upon crossing a p_2 value at which either three bifurcation sets intersect or two bifurcation sets intersect with an hysteresis or isola variety. Nine intersections of this type exist in the (p_2, p_3) plane. Not all of these represent transitions which are observable. Furthermore, only five of them are actual intersections, i.e. the three sets have the same p_1 value. These are points O, K, L, M and N in Fig. 6. Thus, four different type of structurally stable bifurcation (p_1, p_3) maps exist for $p_2 > 0$.

A detailed description and analysis of all the bifurcation maps is presented in [13]. We describe here only two cases (Fig. 7). A thick line in the map describes a transition in an observable feature of the system. The phase plane in each region is denoted by a letter. For example, U^a and U^b represent a unique steady state on the upper or lower branch of $f = 0$, respectively. For $1 < \sqrt{3}p_2 < 3$ four distinct regions are organized around center E_2 at the intersection of the LPS (Hopf bifurcation type H_4 to the left of E_2 and H_1 to the

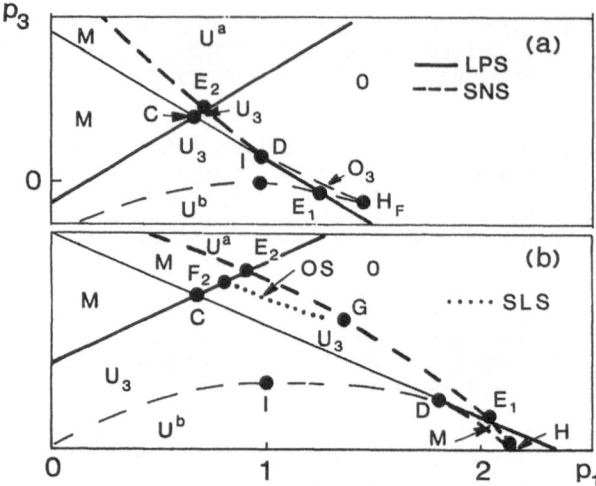

Fig. 7. Bifurcation maps in which the bifurcation sets separate regions with different dynamic behavior ($1 < \sqrt{3}p_2 < 3$ in a, $3 < \sqrt{3}p_2 < 5$ in b)

right of it) and the SNS and GSNS (section E_2D). Next to this center phase planes U, O, U_3, and M exist.

At $p_2 = \sqrt{3}$ (points L and M in Fig. 6) points D, H_F and E_1 coalesce and then separate in a different order, creating a new observable region ($\bar{M}$). Moreover, at this p_2 centers G and F_2 emerge from center E_2, which changes its nature due to the formation of an OS region. Figure (7.b) exhibits six centers (C is not observable) −

C − transversal intersection of two LPS. Observable only if both are Hopf bifurcations.
D − tangential coalescence of the SNS and LPS (H_4).
E_1 − intersection of LPS (H_1 or H_4) with SNS and GSNS.
E_2 − intersection of SNS and LPS (H_1 or H_2).
F_2 − intersection of SLS and LPS (H_2 or H_4).
G − intersection of SLS with SNS and GSNS (GE_1 in Fig. 7.b).
H_F − tangential coalescence of two SNS branches (a cusp point).

The arrangement of the regions next to an observable center C is of type U, M, U and O. This behavior was first found by BOISSONADE and De KEPPER [10] and utilized by EPSTEIN et al. [5] to find parameter regions with oscillatory behavior. The analysis clearly indicates that other types of centers exist. For example, next to center G in Fig. 7.b regions with U, OS and O exist. Moreover, an oscillatory region is not always obtained at the extension of the boundaries of the multiplicity region as shown by point D and H in Fig. 7.b.

5. DEVELOPMENT OF QUALITATIVE MODELS

Schematic diagrams of $f = 0$ and $g = 0$ which describe the simultaneous occurrence of two or more bifurcation sets at a physical center are very useful for the development of qualitative models for $\varepsilon = 0$. Small perturbations (unfolding) of this case describe the various bifurcation sets and phase plane diagrams in the vicinity of the center. The figure should be drawn so that a smooth transition from one center to the other is possible.

To illustrate this procedure, consider the two bifurcation diagrams for carbon monoxide oxidation (Figs. 3-4). They include three different types of observed phase plane behavior, a single stable state (U or U_3), a globally attracting oscillatory solution (O or O_3) and a non-overlapping single steady state and a limit cycle (OS). U_3, O and OS phase-planes exist in a vicinity of a center at which an SL and SN point coalesce. For $\varepsilon = 0$ such a center may be explained by the schematic graphs of $f = 0$ and $g = 0$ shown in Fig. 8. A saddle-node bifurcation explains the transition between phase-planes O to OS, a saddle-loop bifurcation

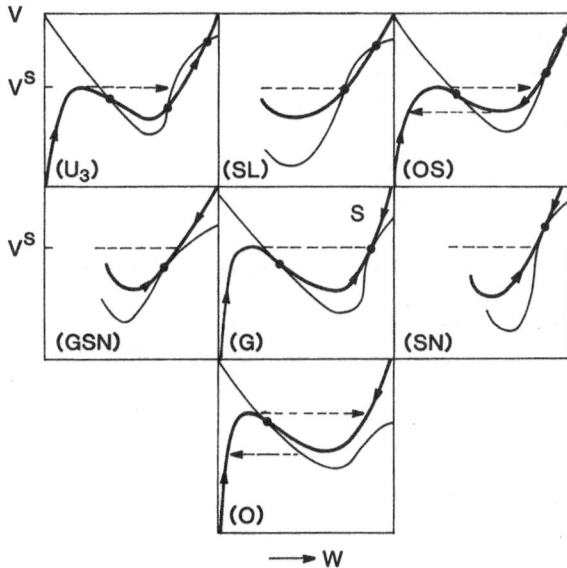

Fig. 8. Phase plane diagrams obtained by small perturbations of center G. (Only the vicinity of point S shown for GSN, SN and SL). Thick line f = 0, thin line g = 0. Steady-state solutions are intersections between the two graphs

that of OS to U_3, and a global saddle-node bifurcation that of O to U_3. These three bifurcations are obtained by small changes in the position of the two null curves next to point S in center G.

Figure 8.G indicates that a simple model describing this center can be obtained if the following conditions are satisfied –

a. f(v, w) = 0 is a single-valued continuous function of w having up to three solutions for some v, i.e. it has two feasible limit points. The change of sign of f_w along f = 0 requires a rate expression describing either reactant inhibition or product acceleration.
b. g(v, w) = 0 is a single-valued continuous function of w having up to two solutions for some v. The sign change of g_w implies that a positive feedback step must be incorporated into this expression. One branch of g should attain an asymptotic value for large w to prevent solutions from moving outside the feasible region.
c. The functions f = 0 and g = 0 are tangent for some parameter at point S while the third node is unstable.

Functions different than those shown in Fig. 8 can also create a center at which a SN, GSN and a SL coalesce. Figure 9.a describes such a center for a case that f = 0 is a multivalued function of w and v. Since both f_w and f_v change their sign along f = 0 the rate expression needs to account for a positive feedback dependence on w and on v. The equation for the slow variable (g) may be monotonic in this case. The organization of the three regions around this center (Fig. 9.b) is identical to that around center G (Fig. 8).

It should be noted that other types of transitions may lead to a similar organization of regions in the parameter space. When I_f, the isola variety of f = 0, is crossed, two branches of the function coalesce and separate in a different order. This can account for an OS to U_3 transition as seen in Fig. 10. This bifurcation which satisfies the conditions f = f_w = f_v = 0 is equivalent to a SL transition. When g = 0 passes through the double sloped point (Fig. 10.S) one obtains a center at which a tangential coalescene of the I_f, SN and LP occurs (Fig. 11). The perturbations (unfoldings) of this case give the phase-plane diagrams shown in Fig. 10. Note that the transition between O to O_3 is by an unobserved SN, while

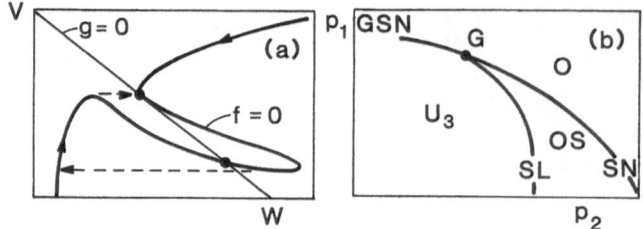

Fig. 9. Two null curves that form a type G center (a) and the corresponding bifurcation map (b)

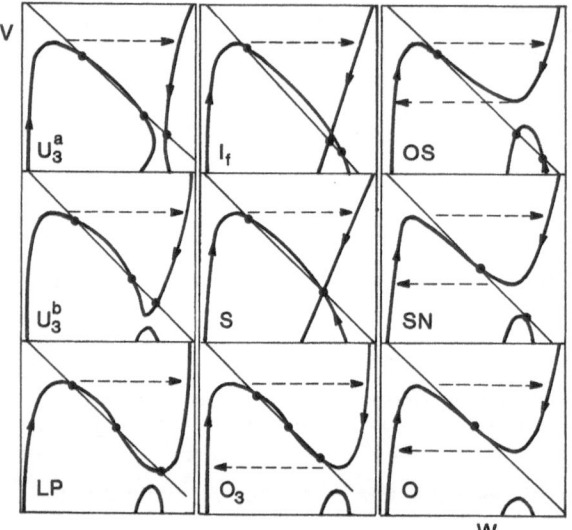

Fig. 10. Phase-plane diagrams obtained by small perturbations of center S.

the transition from U_3^a to U_3^a is by an unobserved I_f. The above examples point out that the graphs of the two null curves of the qualitative model should have qualitative features similar to those of the experimental system.

A qualitative model should describe all the experimentally observed centers. In the m dimensional parameter space the bifurcation sets are m-1 dimensional hypersurfaces and the centers are m-2 dimensional surfaces. When more than one center is observed, higher order singular points should be found at which several of the centers coalesce. The corresponding diagram of the two null curves can be perturbed to give the various centers. The higher the order of the singular point, the larger is the number of defining conditions and the smaller is the class of functions which can be selected for f and g. Thus, knowledge of the highest order singular point is very useful for model development. The order of these points determines also the minimal number of parameters which need to be in the model to realize all the possible perturbations.

Figure 12 illustrates the formation of singular points of different orders. The center F in that figure is formed by the intersection of a LPS and a SLS, i.e. coalescence of a Hopf and saddle-loop bifurcations. Perturbations of this center give the H_2, H_4 and SL transitions, which can explain the behavior observed in the bifurcation diagrams of propylene oxidation (Fig. 5 175°C and 162°C). Figure 12.L illustrates a case corresponding to the coalescence of centers F and G, i.e. the coalescence of a Hopf SN, GSN and SL points. Point L in Fig. 6 is such a singular point. An even higher order singular point is obtained upon the coalescence of the cusp points of f = 0 and F = 0 (Fig. 12.O).

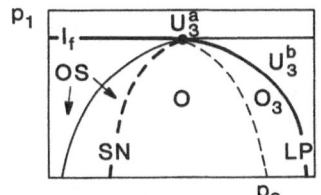

Fig. 11. A local bifurcation map of the center shown in Fig. 10.S. A thick line denotes an observable transition, thin line an unobservable bifurcation set

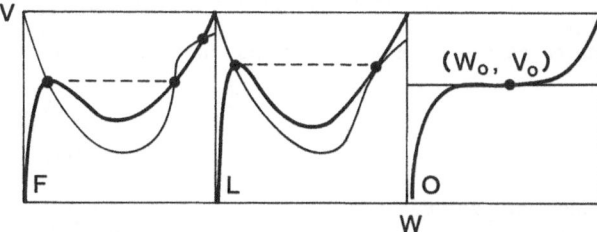

Fig. 12. Diagrams of singular points created by two (F), three (L), or four (O) intersecting bifurcation sets

In the vicinity of a singular point the behavior depends only on the nature (defining conditions) of the singularity and not on the global functional dependence of f and g. Thus, we can use the simple qualitative polynomial model defined by (8-9) as an efficient tool for finding the features of a much more complex system having centers F and G. In developing a qualitative model, one should attempt to use functions f and g which have qualitative features similar to those of the actual system, which is usually rather intricate, and should contain the minimal number of parameters needed for the unfolding of the highest order singular point.

6. GLOBAL ANALYSIS

Models containing the proper centers will predict the observed phase-planes and bifurcations but may not predict the correct sequence of transitions upon changes of the bifurcation variable. It is essential to check whether the model predicts the transitions in the correct order, and if necessary to modify it.

To illustrate this point, consider the qualitative model described by (8-9) with $p_0 = 1$. It predicts all the phase-plane diagrams and bifurcations observed in the CO oxidation experiments (Figs. 3-4). The experimental bifurcation diagrams are qualitatively identical to those obtained by the model by changing p_3 at fixed p_1 next to point G in Fig. 7.b. The following sequence of phase-planes (and bifurcations) are obtained with descreasing p_3

U^a (LP) O (SN) OS (SL) U_3 (unobserved SN) U^b
U^a (LP) O (GSN) U_3 (unobserved SN) U^b .

Both sequences agree with the experimental data. We developed recently [14] a model for this case requiring it to have a G type center. That model did also predict the correct sequence of transitions.

The phase-plane behavior observed during the oxidation of propylene (Fig. 5) are identical to those observed during CO oxidation (U, M, O, OS). However, the sequence of observed transitions differs from those predicted by the qualitative model (8-9) using either p_1 or p_3 as the bifurcation variable. The three observed transitions may be accounted for by the following sequences

at 202°C: U^b (LPS) O (SN) OS (LP) M
at 175°C: U^b (unobserved SN) U_3 (SL) OS (LP) M
at 162°C: U^b (unobserved SN) U_3 (LP) M .

An interesting and important question is whether it is possible, by modifying the functional dependence of the model on some of the parameters, to change the order in which the various transitions occur, i.e. the global organization of the regions in the parameter space. To illustrate this, consider the bifurcation maps shown in Fig. 7. Suppose we vary an experimental variable (say λ) that affects both p_1 and p_3. Then, the sequence of bifurcation depends on the order in which the graph of $\underline{p}(\lambda)$ intersects the different bifurcations as λ is changed. Clearly, this scheme does not introduce any new phase-plane behavior or bifurcations, but may affect the order of the bifurcations. The transformation of $\underline{p}(\lambda)$ may give any of the three observed sequences as any adjacent phase-plane behavior exists in the bifurcation map (Fig. 7). The general discussion of parameter transformations that give a desired sequence of transitions is rather intricate and abstract. We consider here only local changes next to a center caused by a single valued, invertible transformation of the form

$$p_1 = h(s_1, s_2, \lambda) \tag{10}$$

$$p_2 = \lambda \tag{11}$$

where all the other p_i ($i \geq 3$) are independent of $\underline{s}$ and λ.

For a fixed s_1 and s_2 (10-11) define a curve in the (p_1, λ) space, which can be used to map the bifurcation sets into the (s_1, λ) space (for a fixed s_2 and p_i ($i \geq 3$)).

To illustrate the scheme, consider a center created by transversal intersection of two bifurcation sets in the (p_1, λ) plane (Fig. 13). When no trajectory is tangent to any of the bifurcation sets, the mapping may either (a) not produce any qualitative change, or (b) may produce new bifurcation sequences due to inversion of the slopes of both sets (Fig. 13.a) or just one set (Fig. 13.b). When one trajectory is tangent to a bifurcation set (say at point K) the transformation produces a local extremum point in the set (Fig. 13.c). A more singular case occurs when the trajectory is tangent at the center (Fig. 13.d). In these two cases the transformation leads to new sequences of transitions which cannot occur in the original system, such as repeated intersections of the same set.

These simple illustrations show that the observed sequence of bifurcations may be modified by changing the functional dependence of the parameters. A more detailed discussion of this approach is presented elsewhere [14].

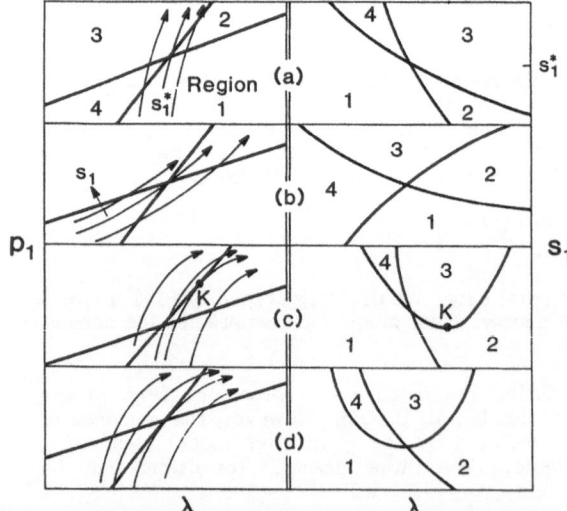

Fig. 13. Transformations of a local bifurcation map (solid lines) by modifying the dependence of f and g on a parameter. Thin lines are $h(s_1, s_2, \lambda)$. In all four cases $\partial h / \partial s_1 > 0$.

7. CONCLUSIONS and REMARKS

We presented here a method of identifying experimentally observed bifurcations for systems characterized by two dynamic variables. The uncertainty associated with the analysis of a single bifurcation diagram may be reduced by finding a sequence of these diagrams and of physical centers at which a transition in the bifurcation sets occurs.

The number of possible centers is rather large. We described some of these centers for systems with widely different time scales ($\epsilon \to 0$). A much larger variety of centers exists when ϵ is finite. BOISSONADE and De KEPPER [7] compared the center formed by the intersection of two Hopf bifurcation sets for $\epsilon = 1$ and $\epsilon = 0$. The additional behavioral features found for $\epsilon = 1$ existed only in a very narrow domain of parameters and would probably escape experimental detection. It is essential to develop analytical criteria or guidelines about the difference between the structure next to a center with $\epsilon = 0$ and that with finite ϵ.

We have used the information about an identified center to construct qualitative model when $\epsilon = 0$. This assumption enabled a rapid construction of null curves which define such a center and gave important insight into the features that f and g need to have. When $\epsilon \neq 0$ defining conditions can be still derived for local bifurcations, i.e. ones which can be studies in the neighborhood of steady-state solutions. These conditions define relations among various partial derivatives of f and g [1, 3, 15], and may be used to construct qualitative models. Unfortunately, no simple conditions are available for defining centers formed by global bifurcations, which cannot be studied next to a steady state. An example of such a case is a center of type G at which a SL, SN and GSN coalesce. It is possible in such cases to search for higher-order singularities, at which the oscillatory states emerge from a steady state. While the defining conditions of these points can be expressed analytically [1, 3, 15], their complex form usually yields no physical insight into the key features of the system.

We have shown how a modification of the dependence of the functions f and g on the parameters may affect the organization of the center and sequence of observed bifurcations. The scheme should be very useful for introducing exotic features into theoretical models, such as isolated branches of oscillatory solutions, and in modifying models describing experimental data. A similar approach for modifying the parameters may be used for systems with $\epsilon \neq 0$ when a bifurcation map is available. However, in many cases the construction of this map may be a very difficult task.

This work illustrates the benefits and advantages of incorporating knowledge of the behavior of dynamical systems to the analysis and design of experimental studies.

ACKNOWLEDGEMENT

We are thankful to the Welch Foundation for support of this work.

REFERENCES

1. A. A. Andronov, E. A. Leontovich, I. I. Gordon and A. G. Maier: Theory of Bifurcations of Dynamic Systems on a Plane (John Wiley, New York 1973), pp. 253.

2. Chow S. N. and J. K. Hale: Methods of Bifurcation Theory (Springer-Verlag, New York 1982).

3. J. Gukenheimer and P. Holmes: Nonlinear Oscillations, Dynamical Systems, and Bifurcation of Vector Fields (Springer-Verlag, New York 1983).

4. M. Golubitsky and D. Schaeffer: Singularities and Groups in Bifurcation Theory (Springer-Verlag, New York 1985).

5. I. R. Epstein, K. Kustin, P. De Kepper and M. Orban: Scientific American 248, 112 (1983).

6. M. P. Harold and D. Luss: Chem. Eng. Sci., to appear (1985).

7. O. J. Boissonade and P. De Kepper: J. Phys. Chem. 84, 501 (1980).

8. V. Balakotaiah and D. Luss: Chem. Eng. Sci., 37, 1611 (1982).

9. V. Balakotaiah and D. Luss: Chem. Eng. Sci., 38, 1709 (1983).

10. V. Balakotaiah and D. Luss: Chem. Eng. Sci., 39, 685 (1984).

11. R. Plichta and R. A. Schmitz: Chem. Eng. Commun. 3, 387 (1979).

12. M. Sheintuch and D. Luss: Ind. Eng. Chem. Fundls. 22, 109 (1983).

13. M. Sheintuch and D. Luss: Chem. Eng. Sci., to appear (1985).

14. M. Sheintuch and D. Luss: in preparation.

15. L. Pismen: Chem. Eng. Sci., 39, 1063 (1984).

A Physicist's Description of Chemical Oscillations

Fritz Jähnig

Max-Planck-Institut für Biologie, Corrensstraße 38
D-7400 Tübingen, Fed. Rep. of Germany

1. Introduction

Many examples of oscillations are known in physics: Rotations of
planets, oscillations of a harmonic oscillator, rotation of an
electron in a magnetic field, or the variety of eigenmodes in
many-particle systems: sound waves, spin waves, etc. To study these
oscillations theoretically, one usually starts from the energy of the
system, derives the equations of motion which contain the reactive
forces, and solves them for the eigenmodes, which result as undamped
oscillations. Damping is introduced in a second step by adding
appropriate dissipative forces in the equations of motion. Thus,
there is a clear-cut distinction between reactive forces which
follow from the energy and give rise to oscillations, and
dissipative forces which follow from the entropy and lead to
damping. In describing chemical oscillations, a different approach is
usually applied. Energy is not considered, the equations of motion
(the chemical kinetic equations) are immediately written down
without any distinction between reactive and dissipative terms.
Therefore, the driving force for chemical oscillations is not
obvious.

This difference in the description of physical and chemical
oscillations was the starting point for the present work. Does this
difference in description reflect a fundamental difference between
physical and chemical oscillations, or is it merely a matter of
different formulations of the theory? In the latter case, it should
be possible to write down the energy of a chemical reaction system
from which the undamped oscillations would follow in the same way as
for physical systems.

The first tentative answer to this question would be in favor of
a fundamental difference: Chemical oscillating systems are in a
non-equilibrium state, a flux of particles through the system
keeping them far from equilibrium, whereas physical systems are
usually in an equilibrium state /1/. Expressed in other words, the
chemical systems are not invariant under time-reversal, whereas
physical systems are. That a system 's behavior under time-reversal
is of fundamental importance, was pointed out already by Onsager
/2/. According to his rules, for systems which are time-reversal
invariant and described by variables which are even under time-
reversal, the matrix of transport coefficients is symmetric and even
under time reversal, leading to pure dissipation. This applies to
chemical systems (whose variables are the particle concentrations)
in the absence of external fluxes. Only if time-reversal invariance
is broken by external fluxes, the transport matrix may contain an
antisymmetric and odd part, which gives rise to oscillations /3-5/.
Physical oscillating systems are usually described by a set of
variables, some of which are odd under time-reversal, such as the
momentum of a particle. Then, according to Onsager's rules, for time-

reversal invariant systems the transport coefficients relating even and odd variables are antisymmetric and odd. This permits oscillations in equilibrium systems.

There are, however, two well-known exceptions of physical oscillating systems which are not time-reversal invariant: A particle in a rotating coordinate system and a charged particle in a magnetic field /6/. They must be classified as non-equilibrium systems in analogy to chemical oscillating systems. Nonetheless, their energy can be specified as for physical equilibrium systems, the equations of motion be derived and solved for the undamped oscillation. For an electron in a magnetic field these are the cyclotron oscillations. Thus, a rotating particle and an electron in a magnetic field might provide the link between physical and chemical oscillating systems. We will therefore start with a description of these two systems,and then try to apply this description to chemical oscillations.

2. Theory of Non-Equilibrium Systems

We consider a particle in two spatial dimensions. The variables describing its state are the positions x and y, and the momenta p_x and p_y. The energy H of the system is a function of these variables and for small deviations from the steady state x=y=0 may be derived as an expansion in powers of x, y, p_x, p_y. For simplicity, we assume rotational symmetry in the xy plane,and restrict ourselves to the quadratic approximation,obtaining

$$ H = \frac{1}{2m}\left(p_x^2 + p_y^2\right) + \frac{\beta}{2}\left(x^2 + y^2\right) + \frac{\gamma}{2}\left(p_x y - p_y x\right) + \frac{\delta}{2}\left(p_x x + p_y y\right). \quad (1) $$

The first term represents the kinetic energy, m denoting the mass of the particle, the second term the potential energy, ß denoting the strength of the harmonic oscillator potential. The third and fourth term are characteristic for a system in non-equilibrium, because the coefficients γ and δ must be odd under time-reversal to yield an energy which is even. Thus, γ and δ reflect the existence of external fluxes through the system,which lead to a broken time-reversal symmetry of the state of the system. Equilibrium, i.e. time-reversal symmetry, would require $\gamma = \delta = 0$. The fourth term in eq.(1) is unimportant, because it can be removed by an appropriate transformation of the variables /7/ , and will be omitted in the following [1].

The Hamilton equations of motion, $\dot{p}_i = -\partial H/\partial x_i$ and $\dot{x}_i = \partial H/\partial p_i$, yield

---

[1] An energy formally identical to eq.(1) has been used to describe a detuned laser, which also represents a system in non-equilibrium /7/.

48

$$\dot{x} = \frac{1}{m} p_x + \frac{\gamma}{2} y \qquad\qquad \dot{p}_x = -\beta x + \frac{\gamma}{2} p_y$$

$$\dot{y} = \frac{1}{m} p_y - \frac{\gamma}{2} x \qquad\qquad \dot{p}_y = -\beta y - \frac{\gamma}{2} p_x \, . \tag{2}$$

Elimination of p_x and p_y leads to the Newtonian equations of motion

$$m\ddot{x} = -\beta x + \frac{1}{4} m\gamma^2 x + m\gamma \dot{y}$$

$$m\ddot{y} = -\beta y + \frac{1}{4} m\gamma^2 y - m\gamma \dot{x} \tag{3}$$

or in vector notation with $\vec{x} = (x,y,0)$ and $\vec{\gamma} = (0,0,\gamma)$

$$m\ddot{\vec{x}} = -\beta\vec{x} + \frac{1}{4} m\gamma^2\vec{x} + m\dot{\vec{x}} \times \vec{\gamma} \, . \tag{4}$$

To solve these equations, one introduces the complex variable $u = x+iy$ and makes the ansatz $u = u_o e^{i\omega t}$ obtaining the eigenfrequencies as

$$\omega = -\frac{\gamma}{2} \pm \left(\frac{\beta}{m}\right)^{1/2} . \tag{5}$$

Let us now consider several special cases.

(a) Harmonic oscillator: $\gamma = 0$
The equation of motion, eq.(4), becomes

$$m\ddot{\vec{x}} = -\beta\vec{x} \tag{4a}$$

expressing the balance of the inertial force $m\ddot{\vec{x}}$ and the oscillator force $-\beta\vec{x}$. The eigenfrequencies result as

$$\omega = \pm \left(\frac{\beta}{m}\right)^{1/2} . \tag{5a}$$

Depending on the initial conditions x_o, y_o the particle undergoes an oscillatory motion in the xy plane, e.g. a rotation. The sense of the rotation may be positive or negative, as typical for a system without externally imposed fluxes ($\gamma = 0$).

(b) Particle in a rotating coordinate system: $\beta = 0$, $\gamma = 2\Omega$
The parameter Ω denotes the angular frequency of the rotation of the coordinate system, and is odd under time-reversal as required. The equation of motion, eq.(4), and the eigenfrequency, eq.(5), result as

$$m\ddot{\vec{x}} = m\Omega^2 \vec{x} + 2m\dot{\vec{x}} \times \vec{\Omega} \tag{4b}$$

$$\omega = -\Omega . \tag{5b}$$

The particle rotates in the xy plane, the sense of rotation being opposite to the sense of rotation of the coordinate system (in a fixed coordinate system the particle does not move). The forces in the equation of motion are the centrifugal force $m\Omega^2\vec{x}$ and the Coriolis force $2m\dot{\vec{x}} \times \vec{\Omega}$. The centrifugal force acts in radial direction, irrespective of the sense of rotation of the coordinate system (the sign of Ω). The Coriolis force acts perpendicularly to the velocity $\dot{\vec{x}}$ of the particle (Fig.1), hence is the driving force for rotation of the particle. Its sign depends on the sign of Ω , so that the external field $\vec{\Omega}$ determines the sense of rotation of the particle. Thus, the external field via the Coriolis force is responsible for the particle rotating with a distinct sense of rotation.

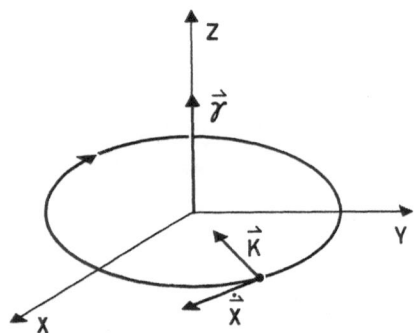

Fig.1. Rotation of a physical or chemical system in the presence of an external field $\vec{\gamma}$ which gives rise to the force $\vec{K}$ acting perpendicularly to the velocity $\dot{\vec{x}}$. Examples for $\vec{K}$ are the Coriolis force, the Lorentz force, or the "chemical oscillation force"

(c) Charged particle in a magnetic field: $\gamma = qB/m$, $\beta = m\gamma^2/4$
Here q denotes the charge of the particle and B the strength of the magnetic field, which is odd under time-reversal, so that again γ is odd as required. The equation of motion, eq.(4), and the eigenfrequency, eq.(5), become

$$m\ddot{\vec{x}} = q\dot{\vec{x}} \times \vec{B} \tag{4c}$$

$$\omega = -\frac{q}{m}B . \tag{5c}$$

The Lorentz force $q\dot{\vec{x}} \times \vec{B}$ in the equation of motion is analogous to the Coriolis force (Fig.1). Correspondingly, it is the driving force for rotation of the particle with a distinct sense of rotation. Since the Lorentz force is determined by the magnetic field $\vec{B}$, the external magnetic field or the externally imposed magnetic flux through the system are responsible for the rotation of the particle with a distinct sense of rotation. Thus, these oscillations are basically different from the oscillations of a harmonic oscillator. The difference originates in the existence of externally imposed fluxes, which break the time-reversal symmetry and define a time direction for the system's motion.

(d) Chemical reaction system: $\gamma \neq 0$

If the spatial coordinates x_i and the momenta p_i are replaced by the concentrations X_i of the chemical species and their conjugate momenta P_i, respectively, the framework of eqs.(1)-(5) describes a chemical reaction system with externally imposed fluxes. The Hamilton equation $\dot{P}_i = - \partial H / \partial x_i$, e.g., correlates the temporal change of the momenta P_i to the forces $\partial H / \partial x_i = \mu_i$, the chemical potentials. Under the simplifying assumption $\beta = m \gamma^2/4$, thereby eliminating any harmonic oscillator-like force, the equations of motion, eq.(3), can be integrated once to yield

$$\dot{X} = \gamma Y \qquad \dot{Y} = - \gamma X$$

or in vector notation

$$\dot{\vec{X}} = - \Lambda \vec{X} \qquad \text{with} \tag{6}$$

$$\Lambda = - \gamma \begin{pmatrix} 0 & 1 \\ -1 & 0 \end{pmatrix}. \tag{7}$$

Eq.(6) is identical with the linearized form of the chemical kinetic equations commonly used in the analysis of reaction systems. The relaxation matrix Λ, eq.(7), is of extremely simple form, but nevertheless has two important properties: It is antisymmetric and odd under time reversal. It has been shown that these properties are characteristic for chemical non-equilibrium systems and necessary for chemical oscillations /5/.[2] Indeed, the relaxation

---

[2]Actually, these symmetry properties are required for the transport matrix $L = \Lambda \Gamma$, where $\Gamma_{ij} = \langle x_i x_j \rangle$ denotes the fluctuations /5/. In the case $\beta = m \gamma^2/4$, the fluctuations diverge, as known for an electron in a magnetic field, but releasing this condition L is easily shown to fulfil the required symmetry properties.

matrix of eq.(7), apart from being isotropic in the XY plane, is
identical with the one for the Lotka-Volterra model /8/, which leads
to oscillations.

Having made contact with the conventional description of chemical
reaction systems, we now demonstrate the analogy between chemical
and physical oscillations. The equation of motion, eq.(4), and the
eigenfrequency, eq.(5), become

$$\ddot{\vec{X}} = \dot{\vec{X}} \times \vec{\gamma} \tag{4d}$$

$$\omega = \gamma . \tag{5d}$$

The system rotates in concentration space with a sense of rotation
determined by γ.[3] The term $\dot{\vec{X}} \times \vec{\gamma}$ in the equation of motion is the
analogue of the Coriolis or Lorentz force. It may be called
"chemical oscillation force", because in analogy to these physical
forces it is responsible for the rotation of the chemical system in
concentration space, with a distinct sense of rotation (Fig.1). The
chemical oscillation force depends on $\vec{\gamma}$, which, in analogy to the
magnetic field $\vec{B}$ may be interpreted as a chemical external field
acting on the concentrations. Since this field is odd under time
reversal, it is connected with a flux of particles through the
system, in analogy to the magnetic flux. Hence, the externally
imposed flux of particles gives rise to the chemical oscillation
force which drives the reaction system to rotate in concentration
space with a distinct sense of rotation.

We are not familiar with the notion of a chemical oscillation
force, and one may ask for the origin of this force. Again the
analogy to the physical forces may help to find an answer. The
Coriolis and the Lorentz force are sometimes called "apparent"
forces, because they are derived from more fundamental forces - the
Coriolis force from the inertial force upon rotation of the
coordinate system, and the Lorentz force from the electrostatic
Coulomb interaction upon a Lorentz transformation. In analogy, the
chemical oscillation force may be derived from the law of mass
action upon rotation of the concentration space, the details of this
transformation depending on the reaction system under consideration.

---

[3] In the literature, the eigenfrequencies often are given as $\omega = \pm \gamma$
/8/. However, upon specification of the amplitudes of the
oscillation, one can easily verify that both solutions lead to the
same motion, i.e. a distinct sense of rotation.

For the special cases of the Lotka-Volterra or the Brusselator model, the chemical field $\vec{\gamma}$ can be determined from the result of the conventional analysis (e.g. ref.8). Since $\omega = \gamma$ and ω has been obtained as $\sqrt{k_3/k_4}\,k_1 A$ for the Brusselator model, $A \xrightarrow{k_1} X$ representing the first step of the reaction, one obtains $\vec{\gamma} = (0, 0, \sqrt{k_3/k_4}\,k_1 A)$. Hence, $\vec{\gamma}$ is given qualitatively by $k_1 A$, the flux of particles leaving reservoir A and entering the system.

Up to now we have discussed physical and chemical systems which can oscillate under the influence of external fields, but they do not show limit cycle behavior, i.e. upon inclusion of dissipation they relax to $x = y = 0$ instead of being stable at finite amplitudes x_o, y_o of the oscillations. To include limit cycle behavior into our description, one must release the assumption of a quadratic energy. The energy H, eq.(1), in terms of the variables $U = X + iY$ and $\dot{U}$, is expressed as

$$H = \frac{1}{2\alpha}\,|\dot{U}|^2 + \frac{1}{2}\left(\beta - \frac{\gamma^2}{4\alpha}\right)|U|^2$$

with $1/m$ replaced by the coefficient α. For the expansion up to higher order terms, the fast oscillating part of U is split off, $U = \Psi\,e^{-i\gamma t}$, and the energy expanded in the slowly varying variable Ψ. In lowest order approximation the time-dependence of Ψ is neglected, so that keeping terms up to fourth order in Ψ one obtains for the energy associated with the variable Ψ

$$H' = \frac{1}{2}\left(\beta - \frac{\gamma^2}{4\alpha}\right)|\Psi|^2 + \frac{1}{4}\beta_2\,|\Psi|^4 \tag{8}$$

with $\beta_2 > 0$. This is the usual Landau-type expression for phase-transitions, Ψ representing the order parameter. The steady state solution follows from the condition of minimal energy, $\partial H'/\partial \Psi^* = 0$ (equivalent to the Hamilton equation $\dot{\Pi} = -2\partial H'/\partial \Psi^* = 0$), and results as

$$\Psi_o = \begin{cases} 0 & \text{for } \beta > \gamma^2/(4\alpha) \\[2ex] \left(\dfrac{\gamma^2/(4\alpha) - \beta}{\beta_2}\right)^{1/2} & \text{for } \beta < \gamma^2/(4\alpha) . \end{cases} \tag{9}$$

Thus, by increasing the particle flux γ one reaches a critical point or phase-transition of second order at $\gamma_c = (4\alpha\beta)^{1/2}$, above which the amplitude of the oscillations remains finite (Fig.2). Such a phase-transition is often called hard mode instability.

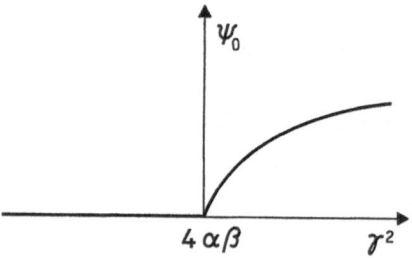

Fig.2. Variation of the order parameter Ψ_0 representing the oscillation amplitude with the particle flux γ through the system

If we allow for a slow time dependence of Ψ , one must keep in the expression for H' the lowest order term in the conjugate momentum π , which is $1/2\gamma_2 i \, (\pi \Psi^* - \pi^* \Psi)|\Psi|^2$. The Hamilton equation $\dot{\Psi} = 2 \, \partial H'/\partial \pi^*$ then yields the equation of motion for Ψ

$$\dot{\Psi} = - i\gamma_2 \, \Psi_0^2 \, \Psi \qquad (10)$$

with the solution

$$\Psi = \Psi_0 \, e^{-i\gamma_2 \Psi_0^2 t} \, . \qquad (11)$$

Indeed, Ψ varies slowly in time, since the frequence of this critical mode increases proportional to Ψ_0^2 above the phase-transition.

The complete solution describing a chemical reaction system with possible limit cycle behavior result as

$$U = \Psi_0 \, e^{-i(\gamma + \gamma_2 \Psi_0^2)t} \qquad (12)$$

with Ψ_0 given by eq.(9). Actually, this solution has the same form as the one for the Brusselator model, derived from the chemical kinetic equations /8,9/. In this conventional solution, the parameters $\alpha, \beta, \gamma, \ldots$ are replaced by the rate constants and reservoir concentrations of the specific reaction scheme. The solution obtained by us, on the other hand, is of more general validity, since it was derived solely on the basis of the symmetries in space and time of the system. The restricting assumption of rotational invariance in space can easily be generalized. An open point within our description remains the expansion of the energy in powers of the variables. Within the Landau theory, there is no way to decide whether β_2 in eq.(8) is positive or not. If $\beta_2 < 0$, in order to keep the energy finite at large amplitudes, a sixth order term with a coefficient $\beta_3 > 0$ must be taken into account. In this case, the phase-transition to a limit cycle turns out to be of first

order.[4] As a further generalization, the Landau-type description
can be renormalized, if necessary, by inserting the appropriate
critical exponents. As a final remark, it should be mentioned that
the hard mode instability has already been treated within the
framework of the Landau theory by others /11,12/, however, solely on
the level of kinetic equations such as eq.(10) without considering
an energy.

3. Conclusion

The aim of the present work was the introduction of an energy for
chemical oscillating systems.in order to describe their reactive
behavior, i.e. undamped oscillations. This was achieved by
establishing a complete analogy between chemical oscillating systems
and certain physical systems.such as a particle in a rotating
coordinate frame or an electron in a magnetic field. An electron in
a magnetic field rotates in space, because it is driven by the
external magnetic field or flux which acts on the electron via the
Lorentz force. Equivalently, the particle flux through a chemical
reaction system gives rise to a "chemical oscillation force" which
drives the system to rotate in concentration space with a distinct
sense of rotation. The universal feature of these systems is their
broken time-reversal symmetry, broken by the external fluxes. This
symmetry property distinguishes them from systems such as a harmonic
oscillator, which are time-reversal invariant. Usually, this
distinction is made by the expressions "equilibrium" and
"non-equilibrium". However "flow-equilibrium" may be a better
characterization of systems with externally imposed fluxes, because
apart from their different behavior under time-reversal.they can be
treated in the same way as equilibrium systems.

Dissipation has been neglected completely in the present work,
but as usual in physics this can be incorporated into the theory
following conventional procedures (e.g. ref.13).[5] Evidently, this
neglect is more serious in flow-equilibrium than in equilibrium
systems, because the first ones can exhibit limit cycle behavior and
thus be in permanent motion connected with dissipation.
Nevertheless, an electron in a magnetic field is often treated in
physics without taking account of dissipation. One might argue that
the electron could be put in vacuum or in a superconducting solid to
approach the dissipation-free case, but even there,the rotating
electron is subject to radiation damping. Thus, neglect of
dissipation always describes an idealized case. Actually, the
electron is studied without dissipation to understand the reason why
it rotates! Hopefully, the analogy between an electron in a
magnetic field and chemical oscillating systems.which has been
discussed in the present paper, helps to improve the understanding
of the origin of chemical oscillations.

---

[4]The analysis of instabilities on the basis of the energy is closely
related to the analysis of kinetic equations within the framework of
bifurcation theory /10/.

[5]The influence of spatial symmetries on dissipation in flow-
equilibrium has been investigated recently /14/.

Acknowledgement

I wish to thank Peter Richter for many stimulating discussions during the course of this work.

References

1. G. Nicolis and I. Prigogine: Self-Organization in Nonequilibrium Systems (Wiley, New York, 1977).
2. L. Onsager: Phys.Rev. 37, 405-426 (1931).
3. M. Lax: Rev.Mod.Phys. 32, 25-64 (1960).
4. K. Tomita, T. Ohta and H. Tomita: Progr.Theor.Phys. 52, 1744-1765 (1974).
5. F. Jähnig and P.H. Richter: J.Chem.Phys. 64, 4645-4656 (1976).
6. P.C. Martin: Measurements and Correlation Functions (Gordon and Breach, New York, 1968).
7. P.H.Richter and S. Grossmann: Z.Phys. 248, 244-253 (1971).
8. P.H. Richter, I. Procaccia and J. Ross: Adv.Chem.Phys. 43, 217-268 (1980).
9. Y. Kuramoto and T. Tsuzuki: Progr.Theor.Phys. 52, 1399-1401 (1974).
10. V. Balakotaiah and D. Luss: Chem.Engineering Science 38, 1709-1721 (1983).
11. H. Haken: Z.Phys. B21, 105-114 (1975).
12. A. Nitzan, P. Ortoleva, J. Deutch and J. Ross: J.Chem.Phys. 61, 1056-1074 (1974).
13. F. Jähnig and H. Schmidt: Ann.Phys. (N.Y.) 71, 129-166 (1972).
14. G. Nicolis and I. Prigogine: Proc.Natl.Acad.Sci. USA 78, 659-663 (1981).

Multiplicity-Induced Oscillations in Porous Catalysts

Bernold Fiedler

SFB 123, Institute of Applied Mathematics, Im Neuenheimer Feld 294
D-6900 Heidelberg, Fed. Rep. of Germany

Applying an abstract global bifurcation result for reaction diffusion systems with two parameters, we investigate how multiplicity of steady states leads to time-periodic oscillations. For illustration purposes we consider the exothermic reaction $A \to B$ in a (bounded) non-isothermal porous catalyst pellet Ω

$$u_t = \Delta u - \phi^2 r(u,v)$$
$$L v_t = \Delta v + \beta \phi^2 r(u,v)$$

(1)

with Arrhenius kinetics $r = u \exp(-\gamma/v)$ and Dirichlet boundary conditions $u = v = 1$ for the concentration u of A and the temperature v (see [1]). As parameters, we choose the size ϕ of the pellet and the Lewis number L, keeping $\beta, \gamma > 0$ fixed.

The stationary solutions (U,V) of (1) do not depend on L. We assume that the stationary solutions form an S-shaped curve as ϕ is varied (fig. 1); see [1, § 4] for more details. Let (U_0, V_0) denote the solution at a saddle node bifurcation $\phi = \phi_0$. Then the linearization of (1) at (U_0, V_0) for $L = 1$, $\phi = \phi_0$ has a one-dimensional kernel generated by (u,v) and similarly the kernel of the L^2-adjoint of the linearization is one-dimensional, generated by (u^*, v^*). Let $<,>$ denote

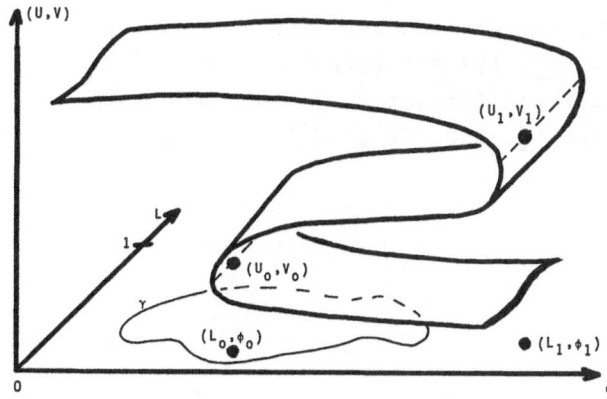

Fig. 1.   An S-shaped curve of steady states with oscillations along the curve γ

the L^2 scalar product on Ω ; then for

$$L_0 := - \frac{\langle u, u^* \rangle}{\langle v, v^* \rangle} \tag{2}$$

an algebraically double eigenvalue zero occurs for the linearization of (1). An algebraically double eigenvalue zero is a hybrid between steady state bifurcation (real eigenvalues) and Hopf bifurcation to periodic oscillations (imaginary eigenvalues). Some easy analysis shows that $0 < L_0 < 1$. A similar situation occurs along the fold (ϕ_1, U_1, V_1) at $L = L_1 \in (0,1)$.

From the results of [2] , we can now draw the following conclusion. Let γ be a closed analytic Jordan curve in the positive (ϕ, L)-plane containing only one of the points $\{(L_0, \phi_0) \ , \ (L_1, \phi_1)\}$ in its interior. Then along this curve (1) has a continuum Z of periodic orbits with (virtual) periods getting arbitrarily large on Z ; see [2] for a definition of virtual periods. Various mechanisms can lead to large periods on Z , e.g. homoclinic orbits or infinite sequences of period-doublings (chaos).

The same approach applies to the reaction $\sum n_i A_i \rightarrow B$ and to the branching steps $A \rightarrow B_1, \ldots, A \rightarrow B_n$. It also provides a deeper understanding of computational results for tubular reactors [3] and allows an analysis of oscillations in the limit of large activation energy. Oscillations in two diffusively coupled CSTRs with Brusselator kinetics [4] or Oregonator kinetics can be treated as well in this spirit.

References

1   R. Aris: The Mathematical Theory of Diffusion and Reaction in Permeable Catalysts, Clarendon Press, Oxford, 1975.

2   B. Fiedler: Global Hopf Bifurcation of Two Parameter Flows, preprint.

3   K.F. Jensen, W.H. Ray: The Bifurcation Behavior of Tubular Reactors, Chem.Eng.Sci. 37, 199-222 (1982).

4   I. Schreiber, M. Marek: Strange Attractors in Coupled Reaction-Diffusion Cells, Physica 5D, 258-272 (1982).

Temporal Oscillations Induced by Boundary Conditions in Systems with and Without Spatial Patterns

Hermann Riecke and Lorenz Kramer

Institut für Theoretische Physik, Universität Bayreuth
D-8580 Bayreuth, Fed. Rep. of Germany

We investigate the influence of boundary conditions on a semiinfinite 2-component reaction-diffusion model. These boundary conditions are thought of as arising from a membrane that may be chemically active. Thus we consider the set of equations

$$\partial_t \underline{n} = \underline{\underline{D}}\partial_x^2\underline{n} + \underline{f}(\underline{n}) \qquad x < 0 \qquad (1)$$
$$\underline{\underline{D}}\partial_x\underline{n} = \underline{g}(\underline{n}) \qquad x = 0 \qquad (2)$$

where $\underline{n}$ is a concentration vector, $\underline{\underline{D}}$ a (diagonal) diffusion matrix, and $\underline{f}$ and $\underline{g}$ are the reaction rates in the bulk and in the membrane, respectively. For this model, the influence of boundary solutions on bulk patterns - with emphasis on wavelength selection - have been discussed in [1]. Here we concentrate on oscillatory boundary solutions arising from the _interplay_ of bulk and membrane (for details see [2]). We find that they can still occur when neither oscillations in the bulk nor in the membrane alone are possible as is the case when autocatalysis is confined to the bulk and coupling of the two reactants to the membrane or vice versa.

The calculation is based on the stability analysis of a steady state $\underline{\tilde{n}}$. Linearization around $\underline{\tilde{n}}=\underline{n}-\underline{u}$ yields

$$\partial_t \underline{u} = \underline{\underline{D}}\partial_x^2\underline{u} + \underline{\underline{M}}\,\underline{u} \qquad x < 0 \qquad (3)$$
$$\partial_x \underline{u} = \underline{\underline{S}}\,\underline{u} \qquad x = 0 \qquad (4)$$

with $M_{ij}=\partial f_i/\partial\tilde{n}_j$, $S_{ij}=D_i^{-1}\partial g_i/\partial\tilde{n}_j$. For simplicity we assume $\underline{\tilde{n}}$ to be homogeneous, which requires $\underline{f}(\underline{\tilde{n}})=0=\underline{g}(\underline{\tilde{n}})$. This is possible in a general manner if $f_1 \propto f_2$, $g_1 \propto g_2$ (implying $|\underline{\underline{M}}|=0=|\underline{\underline{S}}|$) or $f_1 \propto g_2$, $f_2 \propto g_1$ or $\underline{f}=0$. Using the ansatz $\underline{u} \propto \exp(\omega t + iqx)$ one finds three kinds of instabilities with i) $q=0$, $\omega \in R$, ii) $q \in R$, $\omega \in R$, iii) $q=0$, $\omega \in C$. With simple nonlinearities $(-u_1^3, -u_2^3)$ added to (3) they lead to the following solutions in numerical simulations: i) homogeneous static state, ii) space-periodic static state, iii) homogeneous oscillation with associated phase waves.

For the surface instabilities there are three interesting limiting cases. In Fig. 1 the stability diagram is shown for a case where coupling occurs only in the bulk ($\underline{\underline{S}}$ diagonal, $|\underline{\underline{S}}|=|\underline{\underline{M}}|=0$). Here, autocatalysis in the bulk is not necessary if $|\underline{\underline{M}}| \neq 0$. If the coupling is confined to the membrane and autocatalysis to the bulk, the condition for oscillatory instability reads (for $M_{11}=M_{22}=M$)

$$4\sqrt{D_1 D_2}\,|\underline{\underline{S}}| > (\sqrt{D_1}S_{11}+\sqrt{D_2}S_{22})^2 > 2\sqrt{D_1 D_2}\,|\underline{\underline{S}}| -2M \ . \qquad (5)$$

Taking $M=0$ the same condition describes the third case where no reactions take place in the bulk. Diffusion in the bulk, however, still plays a crucial role as one can see from $\underline{\underline{D}}$ entering (5).

Now we turn to the combination of bulk and surface instabilities. As shown in Fig.1 surface oscillations can occur simultaneously with

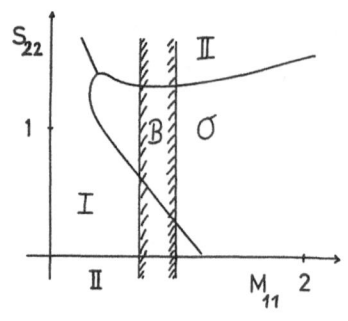

Fig.1 Stability diagram for $D_1=1$ $D_2=2$ $M_{12}=1=-M_{21}$ $M_{22}=-1/M_{11}$ $S_{11}=S_{12}=S_{21}=0$
I : homogeneous state is stable
II: monotonous surface instability
   ($\omega \in \mathbb{R}$)
O : oscillatory surface instability
   ($\omega \in \mathbb{C}$)
B : bulk pattern (between hatched lines)

static bulk patterns. Beyond threshold,these patterns are generally stable in a finite band of wavevectors,which is limited by the Eckhaus-instability [3] and is further restricted by boundary conditions that lead to static solutions with reduced amplitude near the boundary [1],[4],[5]. Oscillatory boundary solutions can similarly restrict the band as shown in Fig.2. Combining surface oscillations with those in the bulk we find two regimes. If the surface oscillations have lower frequency than the bulk oscillations then they usually become locked-in at a rational multiple of the bulk frequency. If, on the other hand, the surface frequency is higher,then the membrane acts as a pacemaker and a growing target pattern emerges.

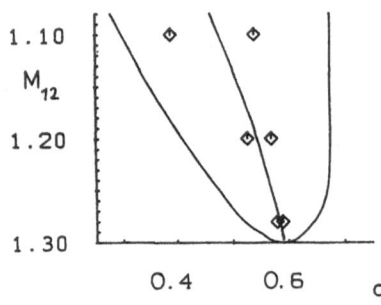

Fig.2 Wavelength selection by oscillating boundary solution
$M_{11}=1$ $M_{12}=-M_{21}$ $M_{22}=-1.2$ $D_1=1$ $D_2=4$
$S_{11}=S_{12}=S_{21}=0$ $S_{22}=0.8$
E : band edge due to Eckhaus-instability
◆ : band edge due to oscillating boundary solution

1. P.C. Hohenberg, L. Kramer, H. Riecke (Physica D, submitted)
2. H. Riecke, L. Kramer (to be published)
3. L. Kramer, W. Zimmermann (Physica D, submitted)
4. M.C. Cross, P.G. Daniels, P.C. Hohenberg, E.D. Siggia
   J. Fluid Mech. 127, 155, 1983
5. L. Kramer, P.C. Hohenberg (Physica D, in press)

On the Classification of Local Disorder in Globally Regular Spatial Patterns

Andreas W.M. Dress

Fakultät für Mathematik, Universität Bielefeld
D-4800 Bielefeld 1, Fed. Rep. of Germany

Abstract

It is shown in terms of simple examples how local perturbations of a globally regular spatial pattern can be classified by associating in a canonical way to each local perturbation an element of the symmetry group of the unperturbed pattern. Various applications towards the existence and compatibility of various combinations of local irregularities are discussed.

In the study of metal surfaces, one often finds the atoms arranged in regular patterns, e.g. in the popular hexagonal pattern (cf. Fig. 1). If, for one reason or another, the surface gets perturbed locally, local disorder may arise (cf. Fig. 2) where in Fig. 2a we find two pentagons and one octagon embedded into an otherwise regular hexagonal pattern. Counting the numbers V, E and F of vertices, edges and faces, respectively, in such a pattern with local irregularities, which globally fit into a regular hexagonal frame and, hence, into an otherwise hexagonally tessellated torus (cf. Fig. 3), one finds - using Euler's formula on the torus -

(1) $\qquad V - E + F = 0.$

So, as long as there are precisely three edges (and three faces) which meet at every vertex, which implies

(2) $\qquad 3V = 2E$

one has for the average number of edges of the various faces, i.e. the number $\sum_{n\geq 3} n \cdot F_n / \sum_{n\geq 3} F_n$, if F_n denotes the number of n-gons, the relation

(3) $\qquad \sum_{n\geq 3} n \cdot F_n / \sum_{n\geq 3} F_n = 2E/F = 6E/3F = 6E/(3E-3V) = 6E/E = 6,$

i.e. this number equals 6.

Hence, it is not a surprise that in Fig. 2a the existence of two pentagons is balanced by the existence of one octagon. But one is also led to expect that an even more simple deviation from regularity should be possible, i.e. a deviation which produces precisely one pentagon and one heptagon inside the regular hexagonal frame (cf. Fig. 4), and therefore it seems to be a remarkable fact that in the study of metal surfaces, such configurations have never been observed. So it seems to be justified to ask whether the obstruction to the occurrence of such irregularities is of a physico-chemical or of a geometric nature.

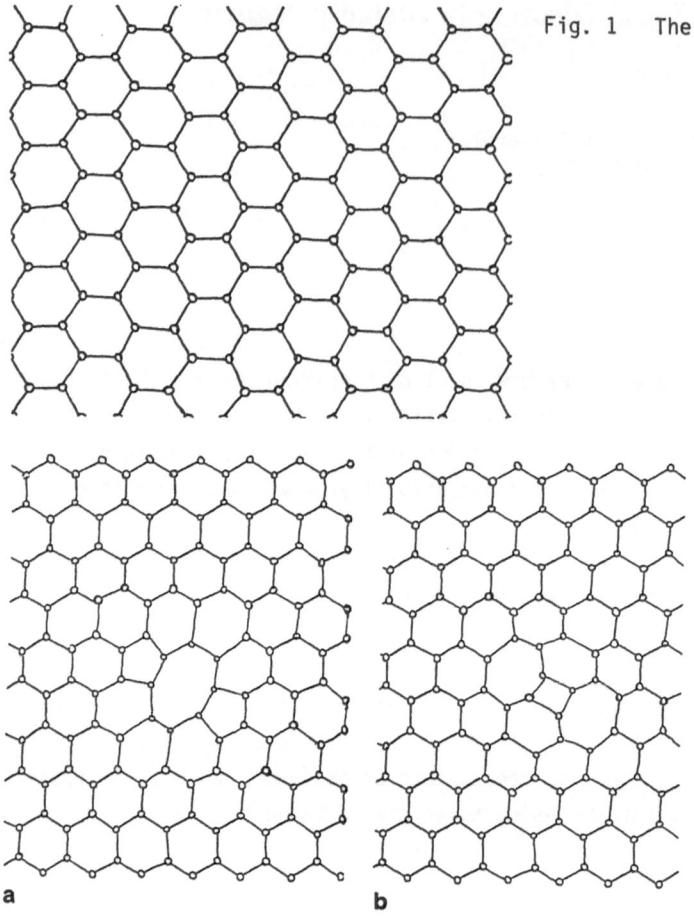

Fig. 1   The regular hexagonal pattern

Fig. 2   Locally perturbed, but globally regular hexagonal patterns, depicting one octagon and two pentagons, and two heptagons and one 4-gon, respectively

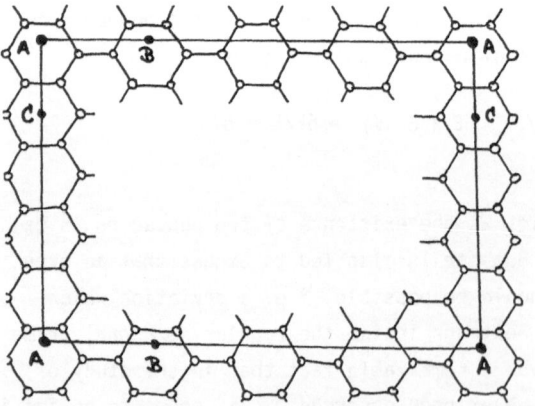

Fig. 3   A pattern, which fits into the above hexagonal frame can be used to tessellate the torus by identifying opposite edges of the above rectangle as usual

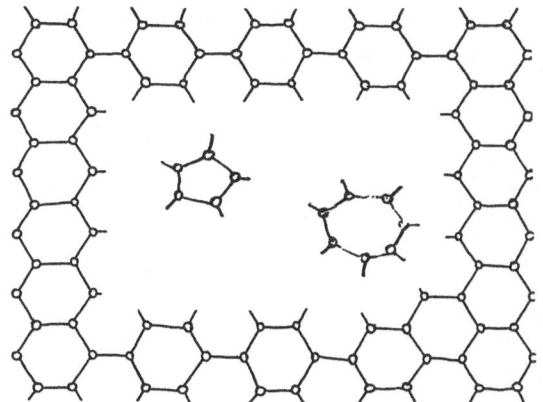

Fig. 4 Can the above picture be
completed by inserting hexagons only,
so that at each vertex in the inside
there meet precisely three edges ?

It is the purpose of this note to describe a simple mathematical method which not
only allows to conclude that for geometric reasons such configurations cannot occur,
but also allows to deduce all possible configurations for other combinations of ex-
ceptional n-gons, and which can also be modified easily to apply to other types of
regular patterns and to one-dimensional irregularities in 3-space,whose classification is an important task in the theory of imperfect crystals. In spite of these
possible generalizations and their importance in crystallography, for the sake of
clarity I will restrict myself in the rest of this note to hexagonal patterns and
their local perturbations.

So assume that inside the infinite regular tessellation of the euclidean plane
by regular hexagons a few local perturbations have occured. Remove from the euclidean plane the set X of barycenters of all non-hexagons (cf. Fig. 5) and compare
the remaining punctured plane with a copy of the regularly tessellated euclidean
plane. Choose an oriented edge $\vec{e}$ and $\vec{e}'$ as an edge of reference in both. Around each
cut out barycenter choose a closed path of oriented edges, starting and ending with
$\vec{e}$ (cf. Fig. 6). Finally, follow up the "same" path in the regularly tessellated
plane, starting with $\vec{e}'$, and note that in general, the last edge $\vec{e}''$ in this path

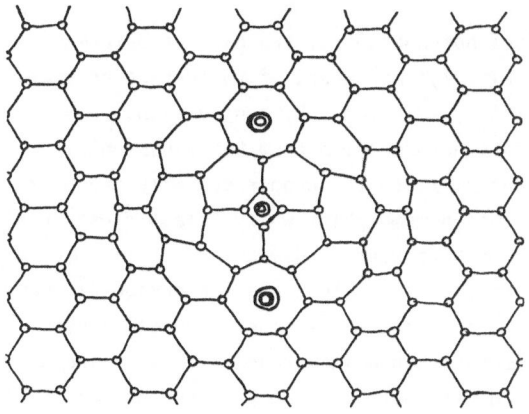

Fig. 5 A locally perturbed pattern,
with barycenters of all non-hexagons
indicated by circles.

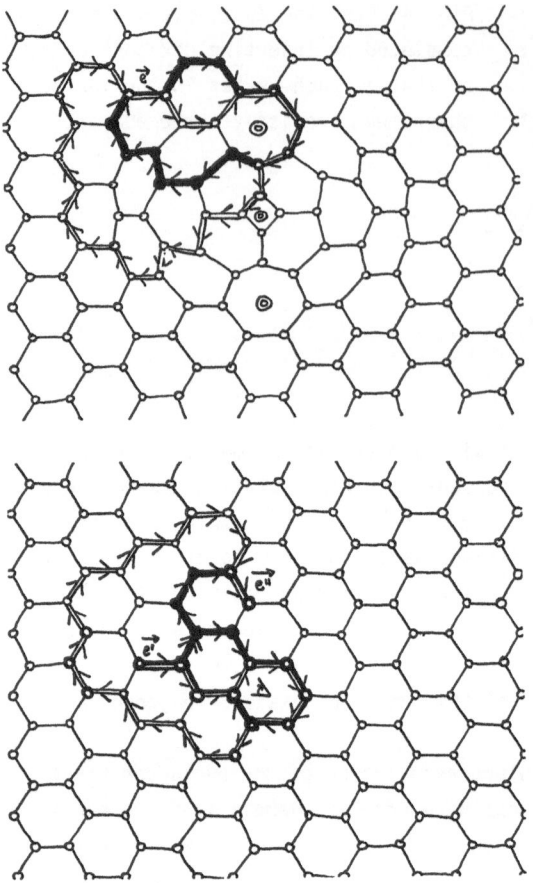

Fig. 6 Two homotopic paths around one of the two heptagons

Fig. 7 The corresponding paths in the regularly tessellated plane

will not coincide with $\vec{e}'$ (cf. Fig. 7). Hence, there exists a unique orientation-preserving symmetry of the regularly tessellated plane which maps $\vec{e}'$ onto $\vec{e}''$. More-over, if the polygon whose barycenter was surrounded by the path was an (6+k)-gon, then the resulting symmetry is a rotation around the center of a hexagon in the "position" of the (6+k)-gon by $k \cdot \frac{2\P}{6}$.

It is easily seen that this symmetry does not change if the path is deformed homotopically in the punctured plane (cf. Fig. 7); moreover, if the reference edge $\vec{e}$ is replaced by another one, then the symmetry is changed by conjugation, only.

More precisely, our construction defines a homomorphism from the fundamental groupoid $\P_1 = \P_1(\mathbb{E} \smallsetminus X)$ of the punctured plane $\mathbb{E} \smallsetminus X$ into the proper symmetry group of the regularly tessellated plane, and it is this homomorphism which "classifies" in a rather neat way the type of local disorder we have been studying.

Moreover, once we assume our irregularities to fit into a regular hexagonal frame, the symmetry associated to the path going along this frame, including all irregu-larities altogether (so we may also call it the path going around ∞) must coincide with the identity, whereas on the other hand, it must coincide with the product of

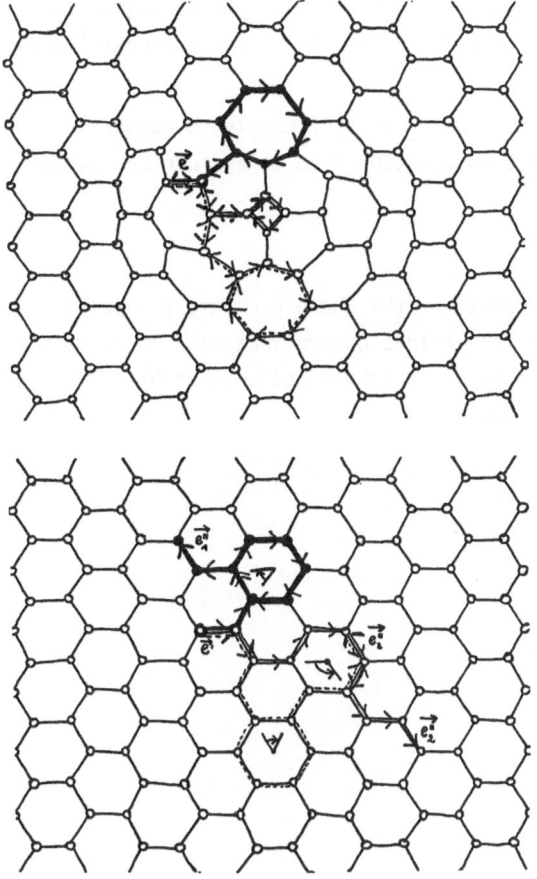

Fig. 8 The product of the symmetries, associated to the paths around the three barycenters, coincides with the symmetry associated to their product, which is homotopic to p_∞ whose associated summetry is the identity

the symmetries, associated to the individual irregularities, taken in an appropriate order (cf. Fig. 8).

Hence, the Eulerian condition (3) which only implies that the symmetry associated with the path around ∞ is a translation, is obviously not a sufficient condition for global regularity. Moreover, it follows that it cannot happen that a globally regular pattern has just two irregularities, since the product of two rotations at different centers never coincides with the identity - so in particular a local disorder giving rise to just one pentagon and one heptagon is excluded. And it shows that in a locally irregular pattern with, say, 2 pentagons and one octagon, the irregularities have to be placed in such a way that the centers of the corresponding rotations by $-\frac{2\pi}{6}$, $-\frac{2\pi}{6}$ and $+4\cdot\frac{2\pi}{6} = -2\cdot\frac{2\pi}{6}$ form a triangle with angles $\frac{\pi}{6}$, $\frac{\pi}{6}$ and $\frac{2\pi}{6}$. Hence, if for instance one of the two pentagons has a common edge with the octagon, the other pentagon must have in common with the octagon the opposite edge, i.e. the only possibility is the one depicted in Fig. 2.

More generally, if we have precisely k irregularities, then the relative position of any k-1 irregularities with respect to each other determines the position of the last irregularity. E.g. if our local perturbation gives rise to two pentagons and two heptagons, then the centers of the corresponding rotations form a parallelogram with angles $\frac{\pi}{6}$, $\frac{5\pi}{6}$, $\frac{\pi}{6}$ and $\frac{5\pi}{6}$.

It is also easy to work out a procedure by which it is possible to construct local perturbations of globally regular patterns starting from an appropriate system of rotations.

Acknowledgement

I am grateful to Rothschild for pointing out to me the above problem. I hope that a more thorough study of this problem and its various generalizations, based on the above method, will appear in a joint paper with Rothschild and some of his collaborators, who have also worked on this problem.

Some Proposals Concerning the Mathematical Modelling of Oscillating Heterogeneous Catalytic Reactions on Metal Surfaces

A.W.M. Dress, M. Gerhardt, N.I. Jaeger[*], P.J. Plath[*], and H. Schuster

Fakultät für Mathematik, Universität Bielefeld
D-4800 Bielefeld 1, Fed. Rep. of Germany

Introduction

The experimental study and the interpretation of complex temporal phenomena such as oscillations, multistability and chaotic behavior in heterogeneous catalytic reactions (mostly oxidation reactions) has drawn increasing attention in recent years. The mathematical modelling was found to provide new insights into the fundamental processes of heterogeneous catalysis on metals and supported metal catalysts [1-13].

The variety of observed phenomena seems to exclude the possibility of a unique mathematical model to treat all these cases under one heading - particularly since theoretical work on chaotic behavior [14-16] has shed new light on these phenomena, focussing attention on formerly neglected parts of the observations.

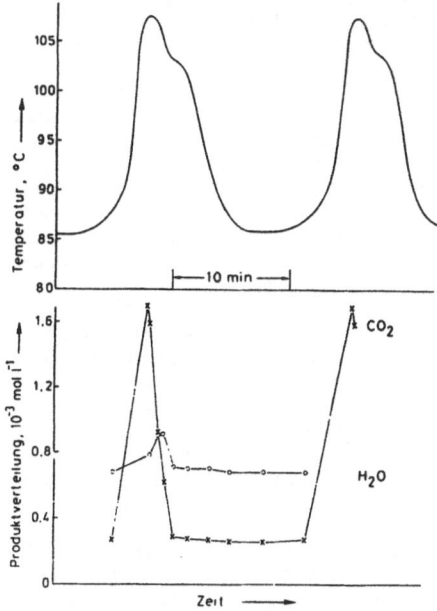

Fig. 1 Typical temperature oscillation and corresponding product distribution for the catalytic oxidation of methanol at a lower level of the system temperature

Therefore we have developed mathematical models only with regard to the experiments concerning methanol oxidation [17] and CO-oxidation [18] on palladium-supported catalysts. Both reactions exhibit rather complex behavior, especially over extended periods of time (Fig. 1 and 2). In the following, we will present some basic considerations concerning the mathematical modelling of both reaction systems.

---

[*] Institut für Angewandte und Physikalische Chemie, Forschungsgruppe Angewandte Katalyse, Universität Bremen, D-2800 Bremen 33, Fed. Rep. of Germany

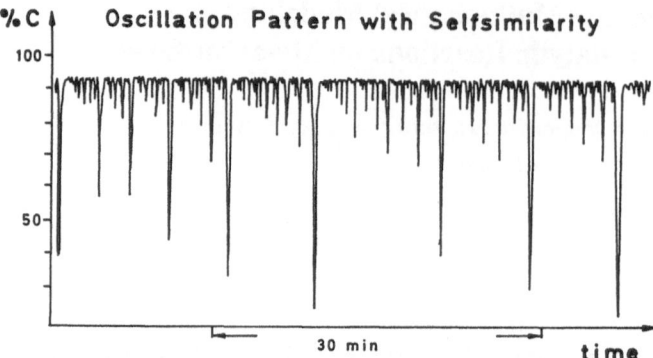

%C

Oscillation Pattern with Selfsimilarity

100—

50—

|← 30 min →|

time

Fig. 2 Oscillations exhibiting selfsimilarity in the catalytic oxidation of carbon monoxide on Pd crystallites. C : conversion of CO to CO_2

Methanol Oxidation

While discussing models of methanol oxidation, it became apparent that the observed oscillations could not be explained by analyzing only the possible periodic reactions on the surface of an individual Pd-crystallite - as discussed for instance in the rather attractive form of surface-games ("Oberflächenspiele") in [1]. Even if we understood perfectly well how a single crystallite could autonomously start to catalyse periodic reactions, this could not explain the observed phenomena, since autonomous oscillations of millions of independent crystallites would only result in almost constant noise.

So we had to find a cause, which coerces those crystallites - at least after some time - to oscillate in synchrony. The experiments, which were carried out under shallow bed conditions with the catalyst resting on a silver plate, suggested heat transport across the catalyst bed as the reason for synchronization. More precisely, we have adopted the hypothesis that once the strongly exothermic CO_2-production surpasses a certain critical threshold-value on one crystallite, it incites increased CO_2-production in the vicinity of this crystallite, and thus may start a chain reaction, which "flares off" the adsorbed carbon monoxide in the whole system.

This hypothesis is supported by the fact that raising the basic temperature of the system, which in general was kept at 80°C facilitated or even provoked the onset of phases of highly increased CO_2-production, while lowering the temperature below 80°C inhibited the onset of such phases or even prevented them completely. Though this hypothesis might seem rather natural, it still has an important consequence: it implies that the analysis of isothermic models will be of marginal value with regard to our experiments, and that the reaction temperature will be an essential variable affecting the whole process and not to be eliminated.

But let us go back to our hypothesis that heat transfer synchronizes the system. In order to make use of this assumption, it is first necessary to give precise chemical and physical meanings to the purposely rather vague terms "critical threshold value", "inciting CO_2-production", "flaring off adsorbed CO", etc. Second, a simple mathematical model must be designed, which at least qualitatively allows us to mathematically reexamine the intuitive ideas concerning the onset of oscillating or bistable behavior. Precise meanings of the above vague terms may be found in [17]: a temperature dependent phase-transition of the Pd-crystallites and its reversal appears to be a plausible mechanism to explain the observed phenomena from the physico-chemical point of view.

On a more detailed level, we want to discuss our proposals concerning the mathematical description of the basic mechanism. According to the philosophy and the techniques of "adiabatic approximation" as discussed in a rather general form by H. Haken in his excellent book on synergetics [19] (see also [20,21]), the essential features of the equilibrium and nonequilibrium behavior even of rather complicated systems involving many variables can be described quite often in terms of the inter- actions of just a few dominant modes. In general, these dominant modes represent appropiate c o m b i n a t i o n s of the original variables,and are related to them via a particular coordinate transformation. So, the dominant modes cannot be iden- tified with any one of the original variables. Still, very often they are found to involve relatively few of the variables in an essential way[+]. Moreover, one hopes that once the dominant modes have been isolated by an appropriate mathematical ana- lysis of a complete description of the system, the interactions among these modes will be in basic agreement with a chemist's intuitive perception of the phenomena.

In our case,we have therefore taken the liberty to begin by trying to set up a rather simple-minded mathematical model conforming with our intuitive perception of the basic driving forces of the reaction, and only later to develop a "realistic" system of equations involving all of the physically and chemically relevant vari- ables we were aware of.

So we have started by defining what we have come to call the "ideal storage model", a system involving just two variables A and B and two equations. The vari- able B is thought to represent the amount of whatever has been stored away at a certain moment, while the variable A, which is confined between 0 and 1, represents the fraction of B being consumed per unit time. In view of the fact that the com- putations will ultimately be done on a digital calculator, we have described the interaction between those two variables by difference rather then by differential equations,using a discrete time variable. Since the difference $B(n+1)-B(n)$ between what has been stored at time "n+1" and time "n" coincides with the difference $C-A(n)\cdot B(n)$ between the amount C, which is continuously added to the storage,and the amount $A(n)\cdot B(n)$ of what has been used up at time n, our first equation reads

$$B(n+1) = B(n) + C - A(n) \cdot B(n).$$

Moreover, we assume that the "rate" $A(n+1)$ by which the stored quantity is used up at time "n+1" depends monotonely on the amount $A(n)\cdot B(n)$, which just has been "burned away", i.e. we suppose the existence of a monotonely increasing function $F(x)$ $(x \geq 0)$ with values between 0 and 1 such that the following equation holds:

$$A(n+1) = F(A(n)\cdot B(n))$$

The shape of $F (...)$ can be chosen quite arbitrarily - as long as $F (...)$ is mono- tonely increasing. The idea is that sharp increases of F occur at certain threshold values.

The stability analysis of the unique stationary state $A = F(C)$ and $B = C/F(C)$, which has been carried out in [24] shows that the stability of this state can in- deed easily be read off from the "graph" of the function $y = F(x)$ in the (x,y)-plane (Fig. 3): if the tangent of the curve $y = F(x)$ at $x = C$ has a positive intercept $F(C) - C \cdot F'(C)$, the stationary state $A = F(C)$, $B = C/F(C)$ is stable. It is un- stable if the intercept is negative, in which case the system oscillates around the stationary state,or - in case of a rather steep ascent or discontinuous jump in $F(C)$ - the system may also behave in an almost bistable fashion, jumping from a state with a comparatively high rate A and a rapidly diminishing stored supply B to a state with a rather low reaction rate A and a slowly increasing store B.

If we consider a system, whose function $y = F(x)$ is like that depicted in Fig. 3, and if we start with a small rate C by which the system is being refilled, the system will be in a steady state with a small "reaction rate" $F(C)$. As we increase C the steady state will become unstable,and the system will start to oscillate a- round it in a way which is quite analogous to the well-known Hopf bifurcation [25],

---

[+]Compare for instance the concept of a "quasi-species" in M. Eigen and P. Schuster [22] and W. Ebeling [23]

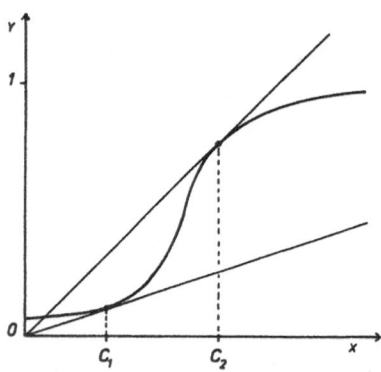

Fig. 3. Graph of a function F of interest in the model of ideal storage and tangents from the origin. The increase of C (rate of influx) from below C_1 leads to oscillations at $C = C_1$ and to a new stable steady state at $C = C_2$

until a state of "maximal" oscillations is reached. After that, further increasing C will lead to smaller and smaller oscillations, until finally a new stable steady state is reached, which is characterized by a high reaction rate A = F(C). Note that the stored amount B = C/F(C) is large in general for $F \ll 1$ and small for $F(C) \approx 1$.

Since this corresponds to the experimental observations, we were quite confident that our simple "ideal storage" model had indeed caught the essential features of the dynamics behind the experimentally observed data. So, based on our experience with the ideal storage model, we went on to develop a rather detailed "realistic" system of equations, including any of the individual simple reaction steps we supposed to take place in the experiment, and describing the complete scheme of chemical and physical interactions we assumed to be responsible for the observed phenomena. Unfortunately, this model is much too complicated to be studied qualitatively from the purely mathematical point of view, but at least the numerical tests we have run so far have been quite successful (Fig. 4,[26]).

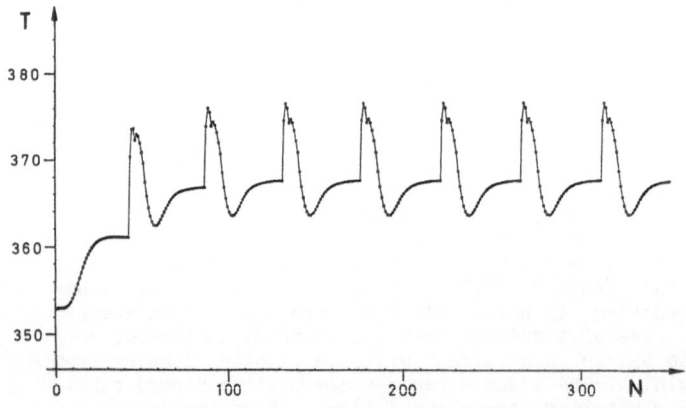

Fig. 4 Computer simulation of a detailed mathematical model of the oscillating methanol oxidation [26]. N: number of discrete time intervals, T: temperature of a palladium crystallite

We hope to be able to present a thorough study of this model in the near future. From the point of view of pure mathematics, it may also be worthwhile to remark that the relatively weak requirement of monotonicity of F (...) can be fulfilled in many different ways. Depending on the choice of F (...), the "limit cycles" which occur may be either a "continuous" orbit or a finite periodic orbit (Fig. 5).

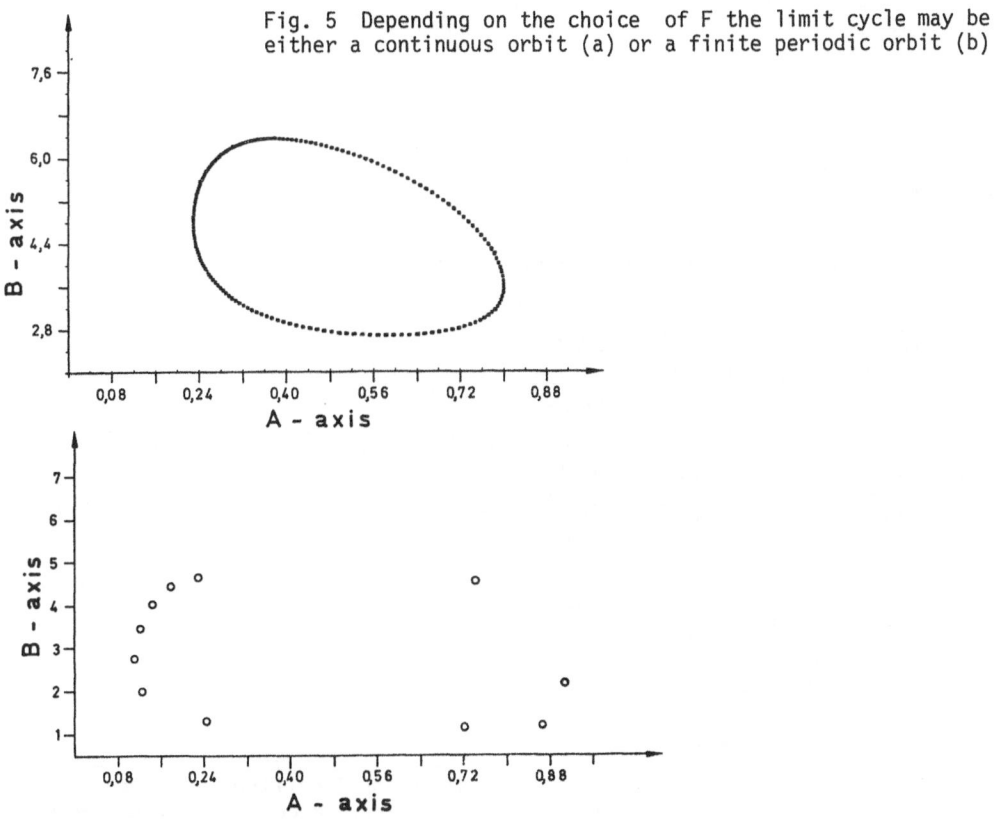

Fig. 5 Depending on the choice of F the limit cycle may be either a continuous orbit (a) or a finite periodic orbit (b)

CO-Oxidation

Finally, we would like to discuss the rather peculiar CO-oxidation curves shown in Fig. 2. These curves depict a sequence of increasing sharp peaks of reduced CO_2-production, which culminate in an almost total absence of CO_2-production, after which the whole process starts again. Moreover, interspersed between the continuously growing peaks of reduced CO_2-production, one finds a series of smaller peaks, which seem to mimic the global pattern on a smaller scale, giving rise to a temporal pattern of selfsimilarity. Thus, according to Mandelbrot [27], it may also be considered as some sort of a fractal pattern.

   At first, we were quite surprised when we discovered that a rather similar pattern occurs in a seemingly completely different context: if for any natural number n we denote by K(n), the number of integers k with $0 \le k \le n$ for which the binomial expression $\binom{n}{k}$ is odd, then the function K(...) seems to mimic the behavior of our chemical system (Fig. 6)[+]. Trying to understand this coincidence, we first reinterpreted the numbers K(n) in terms of a one-dimensional "cellular automaton" [28]: if one considers a system Σ, whose states have been identified with the finite subsets of the set $\mathbb{N} = \{0,1,2,3,...\}$ of natural numbers, in such a way that its dynamics are defined via the following state transition rule: if the subset $A \subseteq \mathbb{N}$ represents the state $\Sigma(n)$ of the system Σ at time t = n, then the subset $A' \subseteq \mathbb{N}$, which represents the state $\Sigma(n+1)$ of Σ at time t = n+1 is given by : $A' = \{a \in \mathbb{N} | a \in A \text{ and } a-1 \notin A\} \cup \{a \in \mathbb{N} | a \notin A \text{ and } a-1 \in A\}$. It follows easily from the structure of the Pascal triangle that if we start with $\Sigma(o) = \{0\}$, then the car-

---

[+]Actually, it is well known and not difficult to prove that $\ln_2(K(n))$ equals the number L(n) of non-zero digits in the 2-adic expansion of n, i.e. one has $K(n) = 2^{L(n)}$.

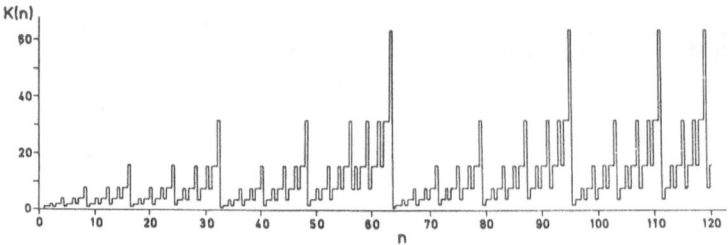

Fig. 6 Representation of the number sequence defined by K(n)

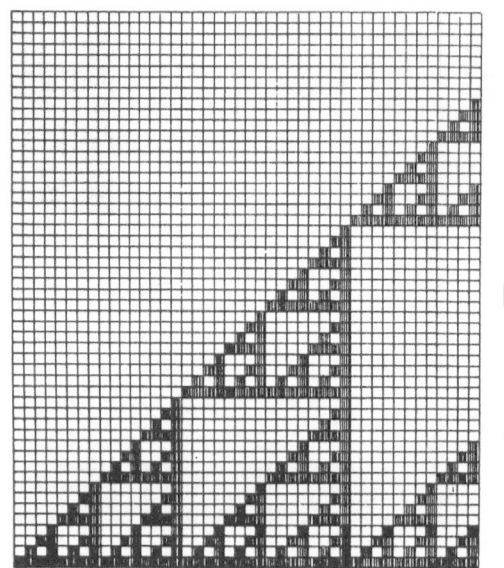

Fig. 7 Time evolution of the one-dimensional cellular automaton. The hatched areas mark the elements of the subset A ⊂ N representing the state Σ (n) for each time t=n

dinality of $\Sigma(n)$ equals K(n) (Fig. 7). Moreover, if we restrict the states of our system to subsets of a finite section of IN, say $\{0,1,2,...,2^k-1\}$ for some k, then - starting again with $\Sigma(o) = \{o\}$ - our system becomes periodic with period 2^k and for $0 \le n < 2^k$ we have - as before - $\# \Sigma(n) = K(n)$, so that it resembles our chemical system even more convincingly.

We believe that this resemblance is not just accidental, and that an explanation can be found, which is based on an oxidation-reduction cycle the catalytically active palladium crystallites undergo. An active particle will be at a high temperature due to a high rate of CO_2-production. Due to the high local temperature the Pd-metal will be converted into palladiumoxide, which is not active in the oxidation reaction. As a consequence, the local temperature drops into a range where the reduction of the oxide by CO is favored, and hence the crystallite is returned to its active state again [29]. Thus - idealizing the situation - we suppose every crystallite always to be in one of two possible states: it either catalyzes CO-oxidation at a constant positive rate, or it stops catalyzing CO-oxidation by being transformed into a non-active state. So the state of our system at any given moment is defined by the set A of active crystallites, considered as a subset of the set of all crystallites. Furthermore, we assume that the CO not converted at the blocked site will have to be oxidized at sites in the immediate vicinity. This leads to higher reaction rates at neighboring particles, the transformation of which to the non-active state will be accelerated as a consequence. We hope that by specifying this interaction in an appropriate way we may be able to define a cellular automaton (two- or three-dimensional and perhaps using stochastic rather than deterministic transition rules), which can be considered as a reasonable model for the chemical system, and which still can be shown to behave similarly to the "Pascal automaton", producing larger and larger competing regions of synchronized crystallites until finally the mode of synchronization of one of those regions wins out and blocks the whole system for a short moment. Thereafter, the same game starts a new beginning with rather small areas of synchronized crystallites.

Though the first attempts to define such a stochastic cellular automaton still have some shortcomings, which, we hope, can be eliminated soon, their overall behavior seems to be close enough to what we are looking for to encourage further research in this direction (Fig. 8).

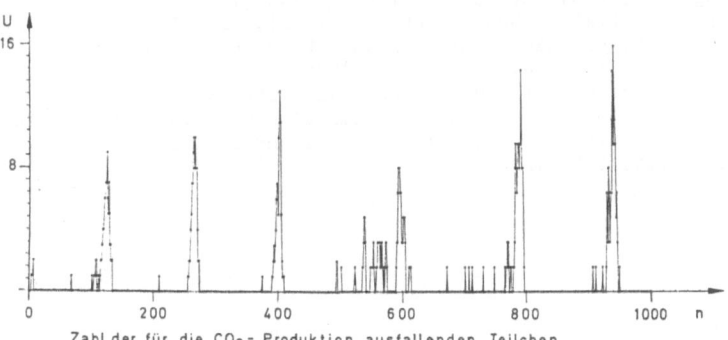

Zahl der für die CO_2- Produktion ausfallenden Teilchen

Fig. 8 Computer simulation of a 2-dimensional cellular automaton as a model for the CO-oxidation. U: number of Pd-crystallites not active in the conversion of CO to CO_2; discrete time $t = n \in \mathbb{N}$

In any case, we are convinced that cellular automata can be used to develop interesting mathematical models to study synchronizations in synergetic networks.

Acknowledgement

Two of the authors (N. Jaeger and P. Plath) gratefully acknowledge financial support from the Stiftung Volkswagenwerk.

References

1 E. Wicke, P. Kummann, W. Keil, J. Schiefler: Ber. Bunsenges. Phys. Chem. 84, 315 (1980)
2 H. Suhl: Surf. Sci. 107, 88 (1981)
3 B.C. Sales, J.E. Turner, M.B. Maple: Surf. Sci. 114, 381 (1982)
4 Z. Kurtanjek, M. Sheintuch, D. Luss: J. Catal. 66,11 (1980)
5 K.F. Jensen, W.H. Ray: Chem. Eng. Sci. 37, 1387 (1982)
6 A. Haberditzl, N. Jaeger, P.J. Plath: Z. Phys. Chem. (Leipzig) 265, 449 (1984)
7 M.P. Cox, G. Ertl, R. Imbihl, J. Rüstig: a) Surf. Sci. 134, 1517 (1983); b) Phys. Rev. Lett. in press
8 M. Greger, B. Ihme, M. Kotter, L. Riekert: Ber. Bunsenges. Phys. Chem. 88, 427 (1984)
9 R.A. Schmitz, G.T. Renola, A.P. Zioudas: Publ. Math. Res. Cent. Univ. Wisconsin, Madison 44, 177 (1980)
10 K. Rajagopalan, D. Luss: Chem. Eng. Sci. 38, 473 (1983)
11 R.A. Schmitz: this volume
12 E.G. Seebauer, J.M. Vohs, R.I. Masel: Ind. Eng. Chem. Fundam. 23,19 (1984)
13 M.G. Slin'ko: Kinetika i Kataliz 24, 803 (1983)

14 O.E. Rössler in: Springer Series in Synergetics (Ed. H. Haken), Vol. 2, p. 184,
    Springer Verlag, Berlin 1977
15 Springer Series in Synergetics (Ed. H. Haken), vol. 11, "Chaos and Order in Na-
    ture", Springer Verlag, Berlin 1984
16 "Evolution of Order and Chaos" (Ed. H. Haken), Springer Verlag, Berlin 1982
17 N.I. Jaeger, P.J. Plath, E. van Raaij: Z. Naturforsch. 36a, 395 (1981)
18 N.I. Jaeger, K. Möller, P.J. Plath: Z. Naturforsch. 36a, 1012 (1981)
19 H. Haken:"Synergetics", Springer Verlag, Berlin 1977
20 B. Aulbach: ZAMM 62, T28 (1982)
21 B. Aulbach in: Mathematics in Industry (H. Neunzert, Ed.), Stuttgart 1984, p. 229
22 M. Eigen, P. Schuster: The Hypercycle: A Principle of Natural Self Organisation
    Springer Verlag, Berlin 1979
23 R. Feistel, W. Ebeling: studia biophysica, Berlin 71, 139 (1978)
24 A. Dress, N.I. Jaeger, P.J. Plath: Theoret. Chim. Acta 61, 437 (1982)
25 E. Hopf: Ber. Math.-Phys. Klasse Sächs. Akad. Wiss., Leipzig, 94, 3 (1942)
26 M. Gerhardt, H. Schuster: Diplomarbeit, Bielefeld 1984
27 B.B. Mandelbrot: "The Fractal Geometry of Nature", New York 1983
28 S. Wolfram, J.D. Farmer, T. Toffoli: Cellular Automata: Proceedings of an Inter-
    disciplinary Workshop, Physica D, Vol. 10 D, No. 1 and 2, 1984
29 K. Möller: Thesis, Universität Bremen 1984

In-Situ IR Study During Oscillations of the Catalytic CO Oxidation

D. Böcker and E. Wicke

Institut für Physikalische Chemie der Universität Münster, Schloßplatz 4
D-4400 Münster, Fed. Rep. of Germany

From earlier investigations it is well known that the oscillations in reaction rate of the CO oxidation on Pt or Pd run parallel with periodic changes of the CO coverage of the metal surface. This could be confirmed by following the periodic changes of the IR absorbance of the chemisorbed CO in an infrared cell with the oscillating reaction in-situ.

Spectra taken as "snap-shots" along the oscillating period revealed that not only the overall absorbance, but also the shape of the spectra, changes. The changes indicate the formation of a densely packed CO chemisorption phase (c (4 x 2) structure with $\Theta_{CO} = 0.5$) besides the "normal" CO domain structure with $\Theta_{CO} = 0.33$, and its removal in the cycle of the oscillations. The removal is accomplished, presumably, by a densely packed chemisorption phase of oxygen atoms (($\sqrt{3}$ x $\sqrt{3}$) R 30° structure), that forms under compression by the dense CO phase, and is known as particularly reactive to chemisorbed CO.

The formation of this aggressive chemisorption structure of oxygen, induced by the dense chemisorption phase of CO, is one autocatalytic step in this mechanism. The second non-linear effect is the ability of CO to built up domains at the metal surface during reaction, by means of a precursor mechanism of adsorption. These two non-linearities are obviously the driving forces of the oscillations.

1. Introduction

Oscillations of reaction rate in heterogeneous catalysis were observed first at the end of the sisties, predominantly with the oxidation of CO and of H_2 on single pellets of Pt carrier catalysts [1 - 3]. Since then, these reactions have been investigated frequently on platinum metals, and became exemplary cases for oscillatory behaviour in catalysis [4 - 8]. Other examples have been found in the oxidation of C_2H_4 on Pt films [9], of C_3H_6 on Pt wires [10], of NH_3 on Pt wires and foils [11], of CH_3OH on Pd carrier catalysts [12] and of H_2 on Ni plates [13], as well as in the oxidation of CO with NO under high - vacuum conditions on Pt ribbons [14]. Only recently, oscillations have been observed by HESSE et al. [15] in our Institute in certain regions of the hydrogenation reaction of C_2H_4 and C_3H_6 on Pt carrier catalysts.

It seems likely that this assortment of catalytic systems is not incidental. Obviously, a platinum metal or Ni as catalyst and reaction components like CO, O_2, NO and olefinic double bonds that interact strongly with the d-band states at the metal surface are

essential for the occurrence of oscillating reaction rates.[*)]
Several mechanisms have been proposed to explain the generation
of such oscillations. No generally valid mechanism can be expected,
but for special cases the most promising conceptions are: i) alternating oxidation and reduction of the catalyst surface [6,8,13];
this seems to be the origin of the oscillations in reaction rate
of the H_2 oxidation on Pt and Ni [6,13]; ii) the generation of
unstable surface phases of the catalyst by reconstruction processes during the reaction; this has been shown by ERTL et al.
[17] to be the reason for oscillations occurring in the CO oxidation on Pt (100) under high vacuum conditions; iii) phase-transitions of chemisorption layers, i.e. formation and disintegration
of chemisorption domains of reaction components at the catalyst
surface. This latter idea has been put forward in former investigations [7], and could be confirmed by the results presented
in the following. Beforehand, however, the main features of the
oscillation regime under normal pressure conditions of the CO
oxidation on carrier catalysts of Pt or Pd will shortly be reviewed.

2. Review of CO Oxidation Oscillations

With increasing CO concentration in air, the reaction rate of CO_2
formation on platinum metal surfaces runs through three regions,
with different kinetic behaviour as shown schematically in Fig. 1.

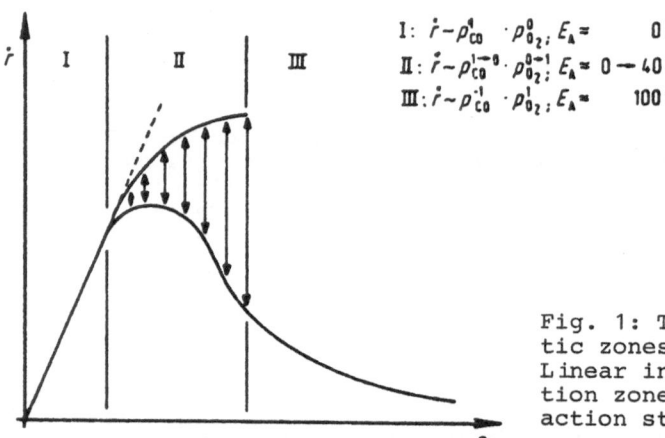

I: $\dot{r} \sim \rho_{CO}^{1} \cdot \rho_{O_2}^{0}$; $E_A =$ 0 kJ/mol

II: $\dot{r} \sim \rho_{CO}^{1-0} \cdot \rho_{O_2}^{0-1}$; $E_A \approx$ 0 — 40 kJ/mol

III: $\dot{r} \sim \rho_{CO}^{-1} \cdot \rho_{O_2}^{1}$; $E_A =$ 100 kJ/mol

Fig. 1: The three characteristic zones of the reaction rate:
Linear increase (I), oscillation zone (II), quenched reaction state (III), after [7]

In region I the reaction is of first order with respect to CO
and of zero order with respect to O_2; obviously, the Pt surface
is covered with a saturated oxygen layer, but still accessible
for CO, the amount of which on the surface is rather small in this
region. The linear increase of the reaction rate turns out to be
independent of temperature, that means the activation energy of
the reaction at the surface and the heat of chemisorption of the
CO cancel one another; the apparent activation energy vanishes.- In

---

*) An exception was reported quite recently by SLINKO and UKHARSKII
[16] who observed thermokinetic oscillations in the oxidation of
cylohexane on zeolite KY at 200 - 250 °C. They explain the phenomenon as alternation between accumulation of oxidation intermediates at the zeolite surface and total oxidation by self-ig-
nition.

region III, on the other hand, the reaction rate is of first
order with respect to O_2, and of minus first order with respect
to CO, thus indicating a dense coverage of the surface with CO
that leaves open only a few positions for oxygen chemisorption. The
apparent activation energy is rather high in this range.

The region II between reveals transition features. The surface
coverage changes from predominantly oxygen to predominantly CO,
whilst the apparent activation energy increases. In this transition range,the oscillations of the reaction rate occur, the
amplitudes of which increase with increasing CO concentration.
The breakdown of the reaction rate at the end of the oscillation
zone is a transition between multiple steady states, the upper one
is an oscillatory-ignited reaction state, the lower one is a
stable quenched state.

The phenomenon that competition and change in surface coverage
between different reaction components leads to a transition zone
with oscillations is not restricted to the case of CO oxidation.
The same behaviour can be demonstrated with the hydrogenation of C_2H_4
or C_3H_6 on Pt catalysts [15]; the hydrogen then replaces the
oxygen in Fig. 1, the olefin the carbon monoxide.

The shape of the oscillations at different temperatures of the
catalyst are represented in Fig. 2. A hydro-reforming catalyst was
applied with 0,3 wt.% Pt on γ-Al_2O_3. The pellets were crushed to
small grains ($\approx$ 0.1 mm), and the pulverized material was exposed
to the gas flow in a shallow bed (0.5 to 1 mm high); by these means
interference of transport processes could be avoided. The reaction
rate was determined by measuring the CO_2 content in the effluent
gas by infrared analysis (URAS). As shown in Fig. 2 the amplitudes
of the oscillations decrease with increasing temperature, the
periods also decrease from 5 s to about 2s. Near the upper limit,
the oscillations degenerate to irregular fluctuations that finally
vanish.

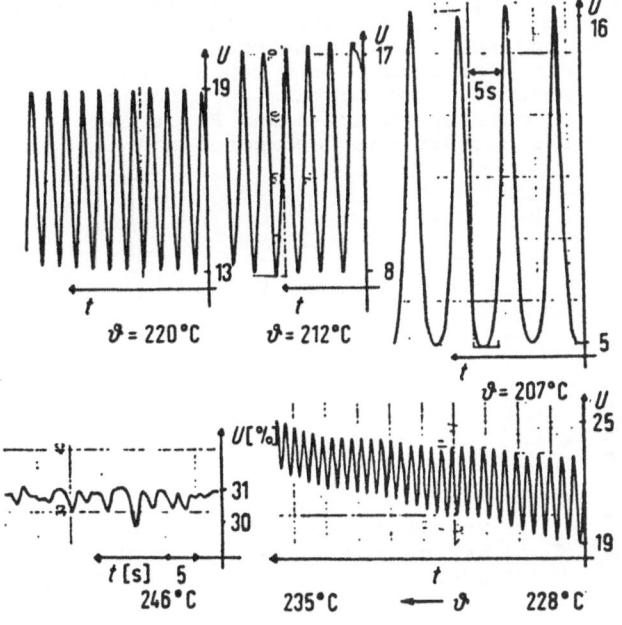

Fig. 2: Reaction rate os-
cillations at different
catalyst temperatures,
after [7]. U = % conver-
sion of CO. (0,5% CO in
air, 10 ml/s, 50 mg ca-
talyst)

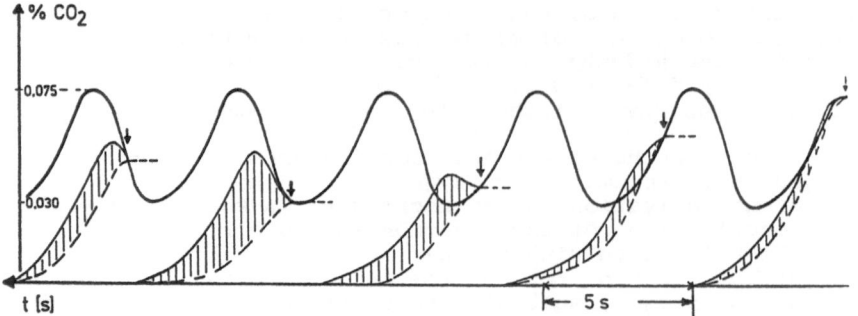

Fig.3: Amount of CO on the catalyst surface in different
stages of the oscillations (hatched areas). The arrows
indicate the stop time of the CO supply. (0,4 % CO in air,
205°C, 50 mg catalyst) [6]

By applying a flushing out method similar to gas chromato-
graphy the CO coverage of the catalyst surface could be determined
in different stages of the oscillation periods. To this end, the CO
supply to the reaction gas mixture was stopped at certain times,
and the amount of CO_2 was measured that left the catalyst layer
still after this stop. The results of this method are shown in
Fig. 3. The hatched areas represent the amount of CO_2 flushed out
after the CO supply was finished (arrows in Fig. 3), and are
therefore equivalent to the amount of CO stored at the catalyst
surface. The CO coverage, obviously, goes up and down in counter-
phase with the reaction rate. These results gave rise to our in-
situ infrared investigations.

3. Experimental and Preliminary Studies

For the in-situ investigations, an infrared cell was constructed
[18] that could be used simultaneously as a flow reactor. Fig. 4
represents a cross-section of the cell, the gas stream flowing
from the right to the left side and the light beam crossing
perpendicular to the picture plain. The square area represents
the catalyst sample, a foil of about 0.1 mm thickness, compressed
from a mixture of finely powdered carrier material and Pt or Pd
nitrate. After the compression, the nitrate was decomposed and

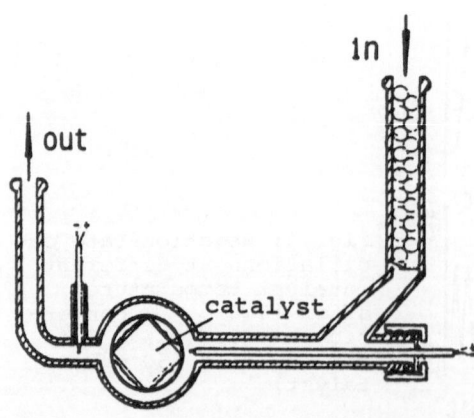

Fig.4: IR absorption cell
as continuous flow reactor,
after [18]

reduced with hydrogen at temperatures of 300 to 500°C. The gas stream, entering the cell from a preheater, passed by both faces of the catalyst foil.

In this cell, the three regions of reaction rate shown in Fig. 1 and their change with temperature could be reproduced in fair agreement with the results obtained in the flow reactor, with the shallow bed of finely-grained technical catalyst. This agreement shows that the oscillation behaviour is rather independent of details in the reactor construction, or in the preparation of the catalyst.

The absorption-reaction cell was used first in a small laboratory IR spectrometer (Beckmann, AccuLab 3). Fig. 5 represents the results obtained with simultaneous registration of the reaction rate and the absorption intensity of the chemisorbed CO in the main peak region at $2080 \, \text{cm}^{-1}$. The reaction rate runs parallel with the IR transmission of the catalyst foil, and thereby confirms the results of the flushing out experiments that the CO coverage of the surface oscillates in counter-phase with the reaction rate.

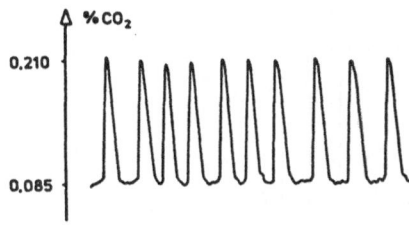

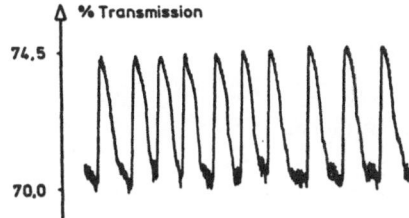

Fig. 5: Reaction rate (above) and transmission of the 2080^{-1} peak of chemisorbed CO in the oscillation regime, after [18]. (0,71 % CO in air, 247°C)

The next step was to try recording of CO spectra at distinct stages of the oscillation periods, for instance in the maxima or the minima of the reaction rate. Those measurements were made with a Fourier-Transform IR Spectrometer (Nicolet 7001; Data Station Nicolet 1180) where a whole CO spectrum with high resolution could be taken within 0.5 s. *) The evaluation of these spectra, however, turned out to be difficult, because in the most interesting spectral region between 1800 and 2000 cm^{-1} the usual carrier materials silica or alumina exhibit too high absorption of their own. After quite a number of tests with different carrier materials in

*) The measurements were performed by D. BÖCKER in the Inst.f. Physikal. und Theoret. Chemie d. Univ. Essen. We gratefully acknowledge the kind permission of Prof.Dr. B. SCHRADER to make use of this highly-efficient and valuable instrument, and the hospitality in his group.

combination with Pt or Pd we finally were successful with finely-grained TiO_2 (Degussa P 25, grain size 10 – 100 nm) and Pd as active component. The titania turned out to have a rather open spectral window in the range mentioned; the measurements, therefore, were made with catalyst foils prepared from Pd (1 wt.%) and TiO_2.

4. Results and Discussion

Oscillations of reaction rate obtained with a Pd/TiO_2 catalyst foil at 230°C and 0,8% of CO in air are shown in Fig. 6 above. At certain stages of the period, as indicated in the figure, spectra were taken which are represented in Fig. 6 below. The comparison of the spectra taken in the rate minima, Nr. 1 and 29, with the spectra near the maximum, Nr. 10 and 11, reveals the expected big difference in overall absorption intensity, but also a remarkable difference in shape. The minima spectra contain a rather sharp peak at 1980 cm^{-1} that is absent in the maxima spectra. Besides this, a broad peak occurs at about 1900 cm^{-1} with a similar shape in the two cases, but very different intensity. According

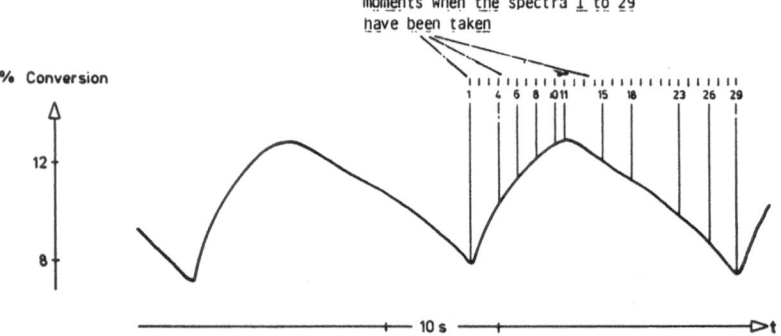

The spectra 1 to 29 taken during an oscillation period show the decrease and subsequent increase of the CO coverage, whilst the reaction rate runs through a maximum.

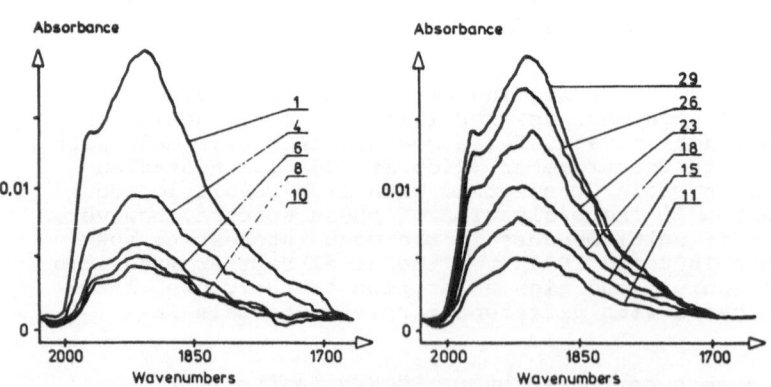

Fig. 6: CO spectra along the ascending branches (1 – 10) and along the descending branches (11 – 29) of the reaction rate

to investigations by BRADSHAW and HOFFMANN [19] the broad peak originates from chemisorbed CO molecules that are coordinated to three or more neighbouring Pd atoms, whereas the sharp peak at 1980^{-1} can be taken as characteristic of a CO species bridged between only two Pd atoms. Accordingly, the two peaks indicate the occurrence of two phases of chemisorbed CO with different structures at the surface: a phase with threefold coordinated CO molecules and a saturation coverage of θ_{CO} = 0.33, and a more dense or "compressed" phase with twofold coordination, leading to a saturation coverage of θ_{CO} = 0.5. Both structures of chemisorption layers have been verified with LEED investigations of CO on Pd (111) planes by ERTL et al. [20-22], the dilute state is a $(\sqrt{3} \times \sqrt{3})$ R 30° structure, the dense state a c(4 x 2) structure. It is to be expected that the polycrystalline Pd particels in the catalyst foil exhibit at their surfaces the densest packed (111) plane most frequently, but there will also be contributions from other crystal planes with different adsorption sites. These will be responsible for the large breadth of the absorption range about 1900 cm^{-1}.

As Fig. 6 below demonstrates, the amount of both chemisorption states at the surface decreases along the ascending branch of the reaction rate (1 to 10 in Fig. 6) and again increases along the descending branch (11 to 29). From measurements with chemisorbed pure CO one would expect a shift of the peaks to higher wave numbers with increasing CO coverage, i.e. increasing packing density of the CO molecules at the surface [19]. However, only minimal shifts ($\leqslant$ 4 cm^{-1}) have been observed here. This behaviour is characteristic of CO chemisorption within a surrounding layer of chemisorbed oxygen [23] where the CO forms islands compressed by the oxygen layer. A

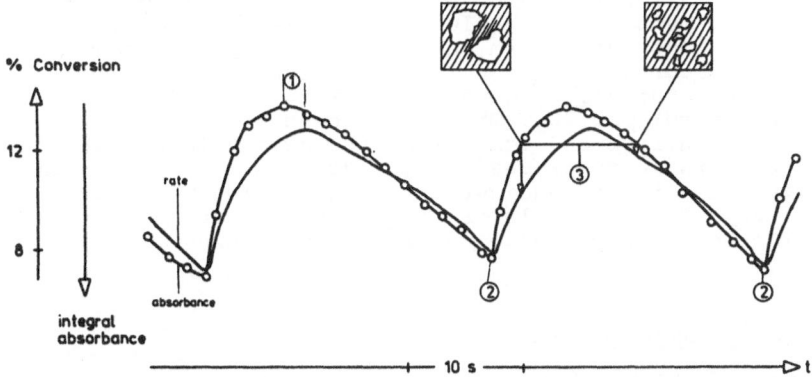

① The integral absorbance, i.e. the CO coverage decreases faster than the reaction rate recovers.
The rate maximum runs behind the coverage minimum (phase shift).

② The minimum reaction rate coincides with the maximum CO coverage (no phase shift).

③ At equal CO coverage the reaction rate on the ascending branch is lower than on the descending branch.
Interpretation: On the ascending branch the CO coverage consists of large domains, on the descending branch, however, of small clusters, providing more O_{ad}-CO_{ad} pairs for reaction.

The ascending branches are steeper than the descending ones: the reaction rate recovers faster than it decreases.

Fig. 7: Oscillations of the reaction rate and of the integral absorbance in the range 2025 cm^{-1} to 1670 cm^{-1}

change of the CO coverage then means a change in the size of the
islands, but no change of the packing density within the CO islands.
Therefore only minimal shifts, if at all, can be expected in this case.

The area below the absorbance curves in Fig. 6 can be taken as a
measure of the overall CO coverage. In Fig. 7 these values have been
plotted vs. time (circles) along with the reaction rate (full line).
The comparison shows that the CO coverage and the reaction rate do
not exactly run parallel, but that the CO coverage reaches its minimum value earlier than the reaction rate its maximum. This unsym-
metric behaviour means that at the same total CO coverage, the reaction rate on the ascending branch is lower than on the descending
branch, as shown at the level (3) in the figure. An obvious and
simple explanation for this effect may be assumed in a different
distribution of the same overall CO coverage in the two cases. If
along the ascending branch the CO is present in the form of rather
large domains or islands within the surrounding oxygen layer, along
the descending branch on the contrary in the form of small clusters
- as is drawn schematically in Fig. 7 - then the length of the border
line where chemisorbed CO molecules and oxygen atoms interact is
different, and this will be true also for the reaction rate.

As a matter of fact, quite a number of investigations during the
last years led to the conclusion that on platinum metals the reaction proceeds predominantly at the border lines between CO and
oxygen domains on the surface. This is an important point for the
development of a mechanism of the oscillations. A second point is
that the interaction of CO molecules and O atoms on the surface
gives rise to domain formation and to discontinuous phase-transitions of the domains that often replaces a continuous variation
of surface coverage. This point gains even more importance by the
observation with LEED methods, on single crystal planes of platinum
metals, that oxygen can not only form the "normal", expanded chemisorption phase with saturation $\theta_O = 0.25$ at the surface - a rather
open 2 x 2 structure - but also a dense phase with $\theta_O = 0.33$
($\sqrt{3}$ x $\sqrt{3}$/R 30° structure) [20,22]. The dense phase is observed when
CO is present at the surface and compresses the oxygen layer to a
denser packing. From these compressed oxygen domains it is known that
they exert a higher reactivity against the neighbouring CO islands
than the normal, expanded oxygen layer [22] (the activation energy
of reaction diminishes as a consequence of the shortening of the CO-O
distance).

5. A Mechanismn of the Oscillations

Starting from the two points mentioned above, a reaction cycle was
developed that combines the spectra in the successive stages of the
oscillation period with the chemisorption structure to be expected.
This reaction cycle is represented with 5 stages in Fig. 8. In the
stage a) on the left side, to begin with, the surface is covered by
the expanded phase of chemisorbed oxygen, and only single CO molecules or pairs are present. Here the CO coverage has its minimum
(spectrum No. 10 in Fig. 6). In the next stage b) the CO nuclei have
grown on to small clusters, the number of CO-O pairs is maximum, and
so is the reaction rate (spectrum No. 11). Proceeding to stage c) the
CO clusters have grown to small islands, and the peak of the dense
phase indicating the C(4 x 2) structure occurs first in the spectrum.
The reaction rate is on the descending branch. Finally in stage d)
the CO islands have grown up to large domains, the reaction rate is
at the minimum. Now the oxygen layer is transformed with compression
by the CO domains into the dense $\sqrt{3}$ structure. This compressed oxygen

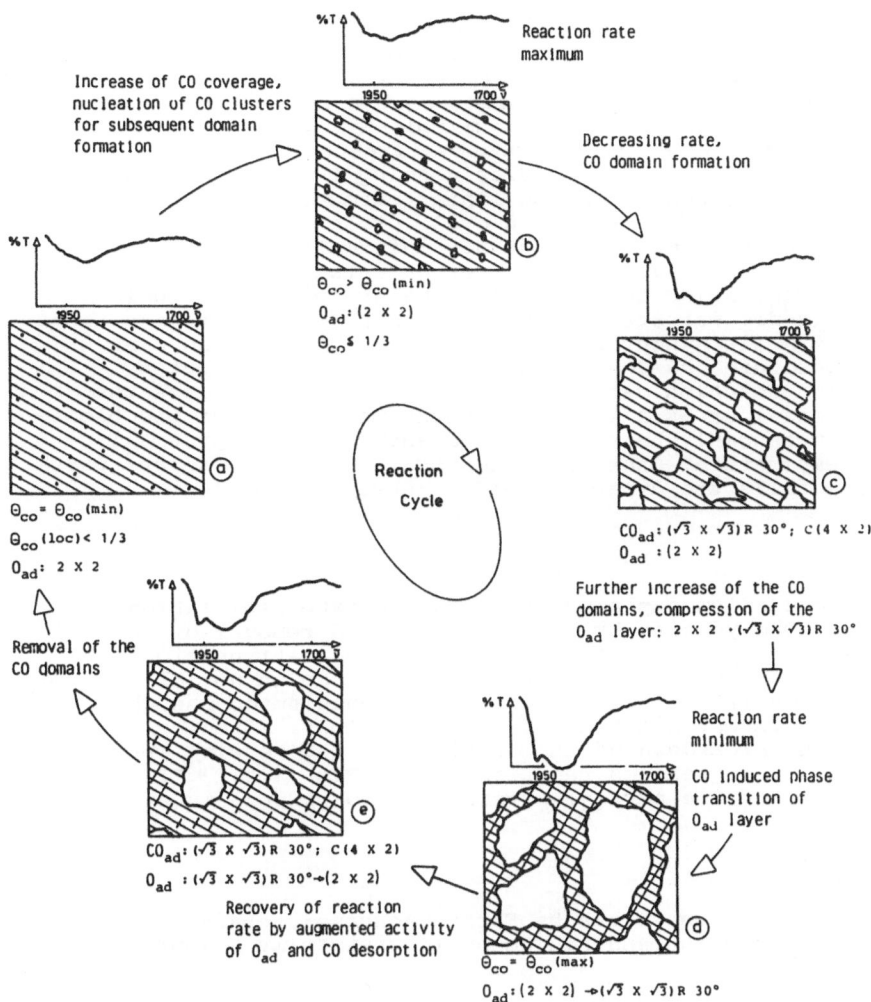

Fig. 8: Reaction cycle of an ascillation period derived from the IR spectra and supposed structures of chemisorption layers

phase with its higher reactivity attacks the CO domains strongly and removes them from the surface (stages d → e → a). As can be seen from the steepness of the curves in Fig. 7 the removal of the CO coverage and the reaction rate in this range exhibit their highest acceleration during the whole cycle. Obviously, one type of ignition process begins here, initiated by the CO induced phase-transition of the oxygen layer to the dense $\sqrt{3}$ structure. Presumably this fast acceleration of the reaction rate generates local superheating that supports the removal of CO from the surface by thermal desorption. Simultaneously, with the CO removal the oxygen phase expands to its normal (2 x 2) structure. Then the cycle starts again.

Besides the non-linearity just described, there is a second non-linearity effective in the reaction cycle. This is the ability of the CO to built up domains on the surface starting from small clusters (stages b → c → d in Fig. 8). This ability can be attributed

to the special adsorption kinetics of the CO. Molecules that strike
the surface on top of a CO island already formed are kept there in
an adsorption precursor state with high lateral mobility [24]. Every
vacant chemisorption site at the border of the island will therefore
be occupied much more likely by a CO molecule from the precursor
state,than by oxygen from the gas phase, irrespective of the additional disadvantage of O_2 molecules needing two neighbouring
adsorption sites. - These two non-linear effects are obviously
the driving forces for the reaction cycle presented in Fig. 8.

Still open is the question about how information is transferred
between neighbouring metal crystallites or catalyst grains,that
provides synchronization in the dynamic behaviour of the whole
system as it is demonstrated by the well shaped oscillation periods
in Fig. 2 or 6. This synchronization seems to be achieved even
by small temperature changes of the catalyst brought about by
the alternating reaction rate. The mechanism that generates the
oscillatory behaviour is a kinetic (isothermal) one, but the oscillations will presumably be synchronized and amplified by small
changes of temperature. Once the synchronization of different parts
of the system fails,the dynamics tend to chaotic behaviour.

6. Conclusions

The proposed mechanism explains the reaction-rate oscillations of
the catalytic CO oxydation by the formation and removal of chemisorption structures, and by two-dimensional phase-transitions of
chemisorbed layers on the catalyst surface. Although not all single
steps are yet confirmed individually, the mechanism on the whole
seems to be consistent and free of contradictions. Moreover, it can
explain also the breakdown of the oscillation regime with abrupt
transition into the quenched reaction state (Fig. 1, II → III), once
a critical limit of the partial pressure ratio p_{CO}/p_{O_2} is exceeded.

7. Acknowledgment

Effective support by the Deutsche Forschungsgemeinschaft providing
the salary for one of us,and equipment,is gratefully acknowledged.

References

1 P. Hugo, Ber.Bunsenges.physik.Chem. 74,121 (1970)
2 E. Wicke, H. Beusch, P. Fieguth, Advances in ACS Symp.Ser.109,
    615 (1972)(paper presented at ISCRE I,1970,Washington,D.C.)
3 H. Beusch, P. Fieguth, E. Wicke,Chem.-Ing.-Tech.44, 445 (1972)
4 M. Sheintuch, R.A. Schmitz, Catal.Rev.-Sci.Eng. 15, 107 (1977)
5 M. G. Slinko, M.M. Slinko, Catal.Rev.-Sci.Eng.17, 119 (1978)
6 E. Wicke, P. Kummann, W. Keil, J. Schiefler, Ber.Bunsenges.
    physik.Chem. 84, 315 (1980)
7 W. Keil, E. Wicke, Ber.Bunsenges.physik.Chem. 84, 377 (1980)
8 J. E. Turner, B.C. Sales, M.B. Maple, Surf.Sci.114, 381 (1982)
9 C. G. Vayenas, B. Lee, J. Michaels, J.Catal. 66 (1980) 36
10 M. Sheintuch, D. Luss, J.Catal. 68, (1981) 245
11 M. Flytzani-Stephanopoulos, L.D. Schmidt, R. Caretts,
    J. Catal. 64, 346 (1980)
12 N. I. Jaeger, P.J. Plath, E. van Raaij, Z.Naturf. 36a (1981)395
13 Z. Kurtanjek, M. Sheintuch, D. Luss, Ber.Bunsenges.Phys.Chem.
    84, 374 (1980)
14 W. Adlhoch, H.G. Lintz, T. Weisker, Surf.Sci. 103, 576 (1981)
15 D. Hesse, private communication (U. Heese, Diploma Thesis
    Münster 1981, W. Bytyn, Diploma Thesis Münster 1982)

16   M. M. Slinko, A. A. Ukharskii, Proc. 8. Int.Congr.Catal.Vol.
      III. p. 243 (1984)
17   G. Ertl, P. R. Norton, J.Rüstig, Phys.Rev.Lett. $\underline{49}$, 177 (1982)
18   D. Böcker, Diploma Thesis, Münster 1981
19   A. M. Bradshaw, F. M. Hoffmann, Surf.Sci.$\underline{72}$, 513 (1978)
20   H. Conrad, G. Ertl, J. Küppers, E. E. Latta, Surf.Sci.$\underline{65}$,
      245 (1977)
21   H. Conrad, G. Ertl, J. Küppers, Surf.Sci.$\underline{76}$ 323 (1978)
22   T. Engel, G. Ertl: "Oxidation of Carbon Monoxide" in "The
      Chemical Physics of Solid Surfaces and Heterogeneous
      Catalysis" Vol. 4, p.73. Ed. D.A.King and D.P. Woodruff
      Elsevier 1982
23   D. M. Haaland, F.C. Williams, J. Catal. $\underline{76}$, 450 (1980)
24   R. J. Behm, K. Christmann, G. Ertl, M.A. van Hove, J.Chem.
      Phys. $\underline{73}$, 2984 (1980)

Spatial Effects and Oscillations in Heterogeneous Catalytic Reactions

J.R. Brown, G.A. D'Netto, and R.A. Schmitz
University of Notre Dame, Department of Chemical Engineering
Notre Dame, IN 46556, USA

1. Introduction

Mathematical models of heterogeneous catalytic reactions are usually based on the assumption of uniformity of the catalyst surface and of the rates of processes which occur thereon. Exceptions are those models which examine the possibility of the formation of islands of chemisorbed species [1] and of symmetry-breaking instabilities [2-7]. Furthermore, experimental observations and measurements of local states on a catalytic surface generally cannot be made. Consequently, most measurements are made of the bulk fluid state, although there have been experimental studies in which in-situ measurements at normal pressures have been reported of surface concentrations, contact potential and temperature. (See, for example, references [9-14]).

We became interested in studying the macroscopic spatial variations in the chemical rate on a catalytic surface during reaction a few years ago for three principal reasons. First, we suspected that certain strange intrinsic dynamics which we had observed in earlier experiments [15] were caused or strongly influenced by local events.

Second, we wanted to investigate experimentally the possibility that spontaneous symmetry-breaking instabilities might occur on catalytic surfaces. Such instabilities, which presumably would lead to the formation of a spatial pattern for the surface state, were described first for reaction-diffusion systems in theoretical studies by TURING [2], SCRIVEN [5,16], PRIGOGINE [3,4] and coworkers and subsequently by many others, some of which have focused directly on heterogeneous catalytic systems [6-8]. If such instabilities were to occur, they would possibly give rise to spatial patterns on the catalytic surface of the type observed for the well-known liquid-phase Belousov-Zhabotinskii reaction [17,18].

Third, we wanted to conduct an experimental inquiry into the behavior of assemblages of interacting particles -- as in a fixed bed reactor. Presumably in such assemblages, any one particle differs slightly at least from any of the others. Further, there are questions about the ability of one particle to communicate and become synchronized with the others. Knowledge about such matters might be important in understanding the behavior of systems of interconnected particles.

It seemed that the best tool for conducting experimental research into these topics was infrared thermography. By means of infrared thermography measurements of thermal infrared radiation taken from points on the catalytic surface during reaction are used to construct a thermogram -- a thermal image of the temperature pattern which, of course, would reflect the spatial pattern of the rate of the reaction for nonisothermal systems. We have been using this thermal imaging approach in recent experiments [19,20] and from the very beginning we were surprised at the spatial non-uniformity of catalyst samples studied. Our observations suggest that the assumptions of uniformity should be made with care.

Here we report further on these studies by showing a sample of thermograms obtained from several forms of catalysts. Our work with thermography is still in an early state, which means that experiments have not necessarily been conducted systematically and results not subjected to detailed scrutiny. In spite of such shortcomings, the results obtained to date serve to illustrate certain behavioral phenomena, such as self-generated oscillations and multiple steady states, which researchers have been reporting for heterogeneous catalytic reactions. But more importantly, they serve the present purpose of illustrating the use of infrared thermography and its potential for studying catalytic processes. Owing to space limitations, we will be able to present only selected illustrations with very brief descriptions here.

2. The Experimental System

The principle components of the experimental system used in the studies reported here were the reactor, an infrared scanner, a TV monitor for display of the thermograms and a product analyzer.

The reactor, constructed of aluminum, was of the continuous-flow gradientless type. Its volume, 415 ml for some experiments and 630 ml for others, was continuously fed with a mixture of H_2, O_2, and N_2 and mechanically agitated so as to be uniformly mixed on a macroscopic scale as verified by independent tracer response tests. Operation was carried out with the feed at room temperature and the reactor walls exposed to room temperature. The pressure was atmospheric. The reaction being studied was the oxidation of hydrogen on platinum. Metallic catalysts in the form of thin foils and gauzes and dispersed catalysts on alumina supports in the form of wafers and spherical pellets were used. Further description of each of these catalysts is given in the section on experimental observations which will contain samples of observations from each. The reactor lid or wall contained a sapphire window, two inches in diameter, which allowed transmission of thermal infrared radiation from the catalyst surface to the detector of the infrared scanner.

The scanner was a Barnes Model RM-50. Its built-in features included an InSb semiconductor which was sensitive to infrared radiation on the wavelength range of 1.8 to 5.5 microns and a scanning mechanism which directed signals from points on the target (catalyst) surface to the detector over a two-second interval. Through appropriate signal conditioning devices, these scanned signals resulted in a colored thermogram updated every four seconds on the TV monitor. With the focusing lens used in these experiments, the scanner had a one-inch square field of view, and it recorded the radiance level from 5376 points in that field. The temperature range of the instrument was adjustable between 0 and 600° C. The temperature resolution was one-eighth of the selected range. The catalysts were calibrated for determination of their effective emissivities so that the radiance level actually measured by the Barnes instrument could be converted to apparent temperature level.

The thermograms displayed on the TV monitor actually contained eight fields of different color. Each color corresponded to a certain temperature range. However, owing to obvious difficulties in publishing such representations, we will be able here to use the thermograms only as indicators of the spatial regions of the color corresponding to the highest temperature on a given thermogram. This is explained in greater detail in a later section.

The concentration of the product water in the bulk gas was measured by an infrared analyzer and recorded continuously. Therefore, the experimental outputs to be shown here are thermograms of the catalyst surface and measurements of water in the bulk gas. The principal experimental parameters were feed concentration and flow rate. Stirring speed was also varied in some instances to examine the effects of gas-solid transport coefficients.

3. Experimental Observations

Selected observations are shown in subsections here for experimental work with four different types of catalyst. Much can be written about catalyst pretreatment procedures and tests of reproducibility. Suffice it to say here that pretreatment generally consisted of heating the catalyst sample in oxygen. The resulting activity was nearly constant through experiments covering a week or two. Results were reproducible from day to day for a particular catalyst sample, but only the general qualitative aspects were similar from sample to sample.

   Platinum Foil. In these studies the catalyst was a square piece of poly-crystalline platinum foil of thickness 0.0076 cm and 2.5 cm on each side. Clamped by its corners in a support frame, it served as the target for the scanning infrared detector.

   Sustained oscillations observed in the experiments have been described in reference [17], which included an abbreviated sequence of thermograms taken during the oscillations. A more complete sequence is shown here in Figs. 1 and 2.[1]

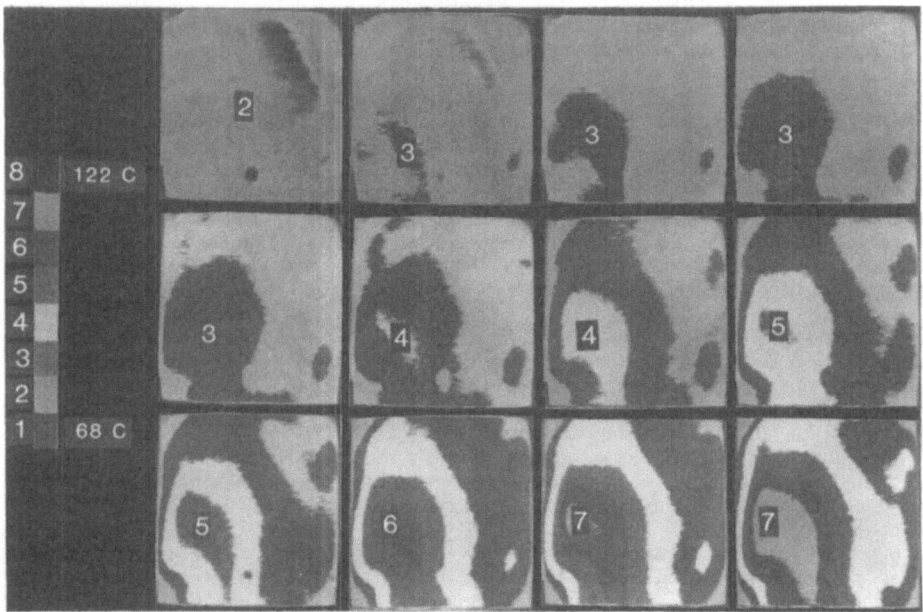

Fig. 1. Sequence of thermograms of a Pt foil during an oscillation (The sequence reads from left to right starting at the top.)

---

[1]The thermograms shown in this paper were originally in color, and the scale shown at the left in each was an eight-color scale with colors from bottom to top corresponding to increasing temperatures. To interpret the thermograms shown here, the reader should notice that there are eight zones including the upper and lower extremes. These are numbered 1 through 8 on the scale, and the total temperature range is approximately linearly divided among the eight. In each thermogram the numbered region defines the principal region of the maximum temperature zone, and the number inserted in the region defines its zone on the scale. For the number 1, temperatures in the shaded region are at or below the minimum on the scale, and for 8, they are at or above the maximum.

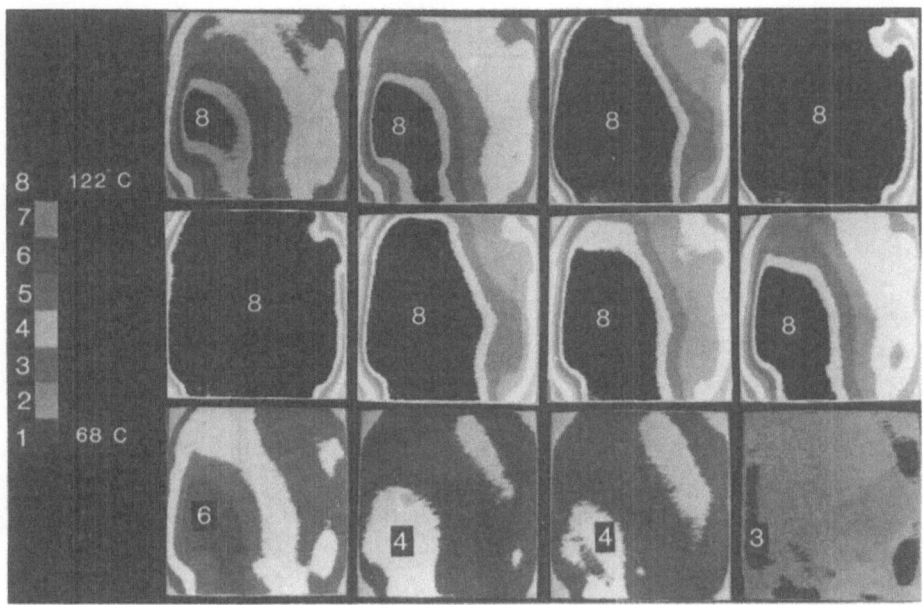

Fig. 2. Continuation of the sequence shown in Fig. 1 (The thermograms in Figs. 1 and 2 were taken from one cycle of an oscillation which had a period of about seven minutes.)

These thermograms, show the development of an oscillation beginning from a region of relatively high activity. From that region the reaction wave spreads over the entire surface. In tests with a number of samples, we found the location of the region of high activity to be unpredictable. On some samples it would be at an edge or corner, but for a given sample it did not move. The maximum temperature on the foil reached about 170°C, and the maximum variation over the surface was about 70°C during the oscillations.

Platinum Gauze. In these experiments, we employed a 2.5 cm square piece of platinum gauze in a manner similar to that described above for the foil. The gauze was 52 mesh with a strand diameter of about 0.01 cm.

Figure 3 shows the experimental response to changes in the concentration of hydrogen in the feed. The data show bistablility of steady states and sustained oscillatory behavior (for conditions on the dashed portion of the curve). The resulting oscillations of the bulk gas for the case of 3.5% hydrogen in the feed are shown in Fig. 4. As with the foil, these oscillations have simple cycles with repeated features.

The thermograms shown in Fig. 5, selected from one cycle of the oscillations, show a more spatially complex spread of the reaction front. As in the case of the foil, there is a dominant active region, but owing to the gauze structure, spatial communication is now more difficult and irregular. Here a close examination of the thermograms shows that the point of maximum temperature moves during a cycle of the oscillations. The maximum temperature observed on the gauze during these oscillations was about 200°C, and there was always an appreciable spatial variation. For example, on the eighth thermogram (all black) of Fig. 5, the variation over the gauze was about 55°C.

In some experiments with the gauze, a stepwise change in a parameter would cause the state to go from a smooth stable steady state through several regular oscillations of large amplitude and then eventually to a "fluttering" mode. This

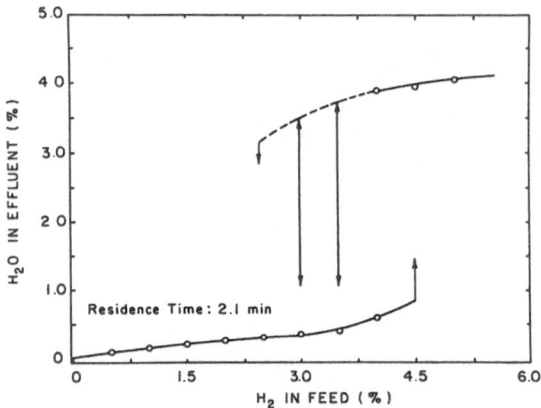

Fig. 3 Steady-state hysteresis loop for a
Pt gauze (The circles represent steady-state
data while the two-sided arrows represent
amplitudes of sustained oscillations.)

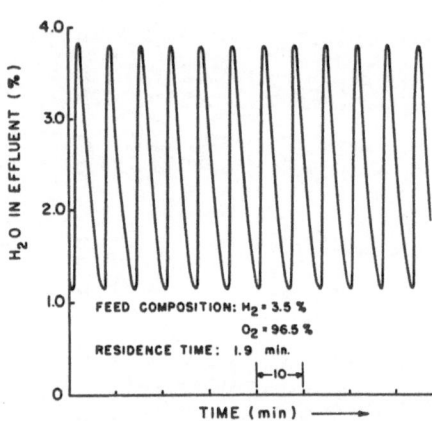

Fig. 4 Experimental oscillations
observed during the oxidation of
hydrogen on a Pt gauze under conditions of Fig. 3 with 3.5% hydrogen in the feed gas

mode, characterized by fast changes of small amplitude in the spatial detail of
the thermograms, seemed to exist over a well defined range of experimental conditions and was distinctly different in behavior from other stable and oscillatory
states. We suspect that it may correspond to a breakdown in synchronization of
local oscillatory regions, but a more extensive experimental study of them is
required.

Supported Wafers. We obtained platinum-on-alumina (1 wt% Pt) supported
catalyst in the form of wafers of 0.15 cm thick and 2 cm diameter as commercial
samples. Extensive physical description of them was provided, including a
description of their preparation in powder form by the incipient wetness technique
and of the pressing of the catalytic powder into wafer form. For the experimental
results shown here, the wafers were placed in the reactor on a small platform so

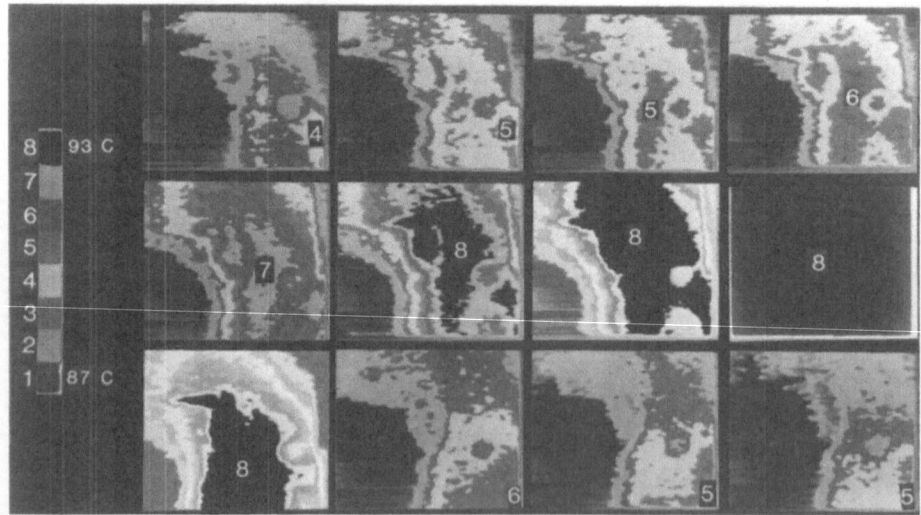

Fig. 5. Selected thermograms of a Pt gauze during one cycle of the oscillations
shown in Fig. 4 (The sequence reads from left to right starting at the top.)

that only their top side was exposed directly to the bulk gas. Other support frames which allowed greater exposure of the catalyst were also tested, and the spatial temperature patterns were affected somewhat. At this time we are somewhat uncertain about the role of the support frame.

The most noteworthy feature of the observations was the spatially spotty activity in all cases with a number of catalyst samples. The thermogram at the upper left in Fig. 6 shows a typical stable steady state. Clearly there is poor spatial communication between various regions of the catalyst. The greatest activity for the case shown in Fig. 6 was on the small spot near the top, where the temperature reached about 140°C. The temperature on the bottom active regions was about 90°C. Thus there was approximately a 100°C temperature variation over these catalyst wafers at steady state.

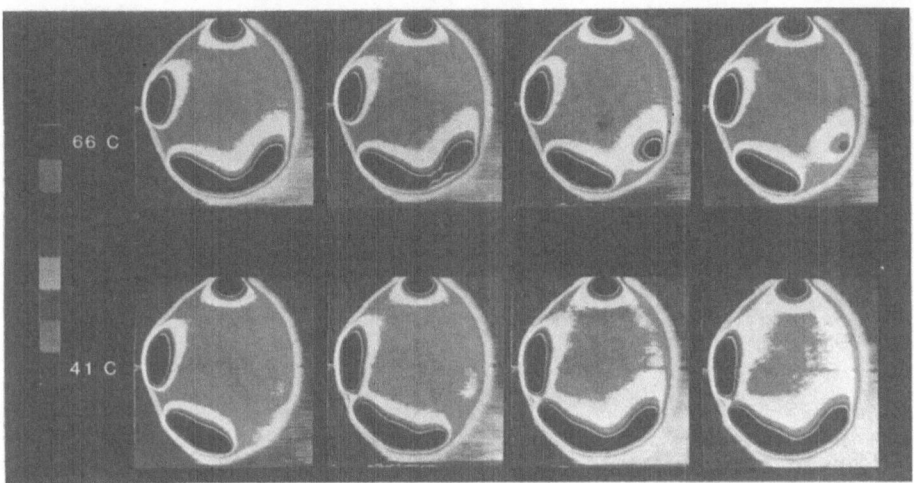

Fig. 6. Thermograms of slow temporal changes of the active regions on a supported Pt wafer for a feed composition of H_2 = 5.7%, O_2 = 94.3% and a residence time of 1.96 minutes (The black region in these thermograms corresponds to zone 8 -- the zone of highest temperature.)

In most cases such active spots were not truly steady. There would usually be some temporal change in size, and sometimes a coalescence and redispersion of spots would occur over a period of several hours. Such changes, ordinarily not evident in the gas phase composition analysis, are shown by the sequence of thermograms in Fig. 6 which covered a period of several hours of observation.

With one wafer sample, over a narrow range of parameters, we observed sustained oscillations of low amplitude. These were described in reference [20]. A set of thermograms, presented here for such an oscillation in Fig. 7, shows that these oscillations were caused by a single oscillating spot. It is noteworthy that the other spots seem to be unaffected. The maximum temperature on the thermograms of Fig. 7 was about 130°C in the region at the top of the particle.

In one accidental case reported elsewhere [19] we found a number of spots oscillating simultaneously, some strongly interacting, but most were apparently not synchronized. The recording of the bulk concentration for that case was noticeably oscillatory but with irregular amplitudes and no apparent periodicity.

Presumably the spotty spatial activity here was caused by a maldistribution of the platinum in the support, by spatial nonuniformity of physical characteristics from the pressing process, or by complicated interactions between the wafer and

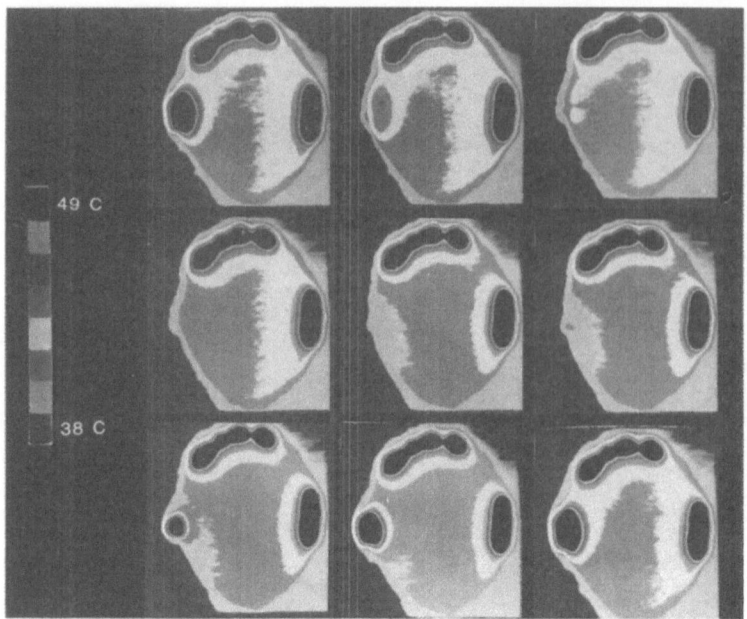

Fig. 7. Thermograms of a supported Pt wafer during one cycle (approximatly 15 minutes) of an oscillation for a feed composition of H_2 = 5%, O_2 = 95% and a residence time of 1.8 minutes (The black region in these thermograms corresponds to zone 8 -- the zone of highest temperature.)

its support platform. We have not yet conducted tests to ascertain the spatial characteristics of the samples.

Spherical Pellets. As with the wafers, we obtained catalysts in the form of spherical pellets as commercial samples, but from a different source. The pellets contained 0.3 wt% platinum contained entirely in a "thin" shell near the outer surface of the alumina support. The average pellet diameter was 0.5 cm.

In our experiments, the pellets were strung on a fine aluminum wire which passed through a hole drilled along a diametrical line. All sides of a particle, or of a string of particles, could be exposed to the infrared scanner by externally rotating the wire.

The commonly observed feature from experiments with seven different particles individually was a single region of relatively high activity. An example is shown by the series of thermograms in Fig. 8. The thermogram on the upper left was obtained with the pellet turned so that maximum activity was facing the imager, and the one on the lower right was obtained after the same particle was turned 180°. (The bands at the outer edges of the projected particle should be ignored because they are caused by the curvature of the spherical surface.) In no case did we observe the spotty activity that was prevalent in the thermograms for wafers. We suspect that the distribution of activity with the pellets was a consequence of the commercial preparation process, but we have not yet made a definite determination.

We have not encountered any sustained oscillatory states with the pellets. Our report here amounts to a brief description of an experimental study of steady states.

We first conducted experiments with pellets individually in the reactor. The results for two pellets of particular interest here are shown in Fig. 9. The

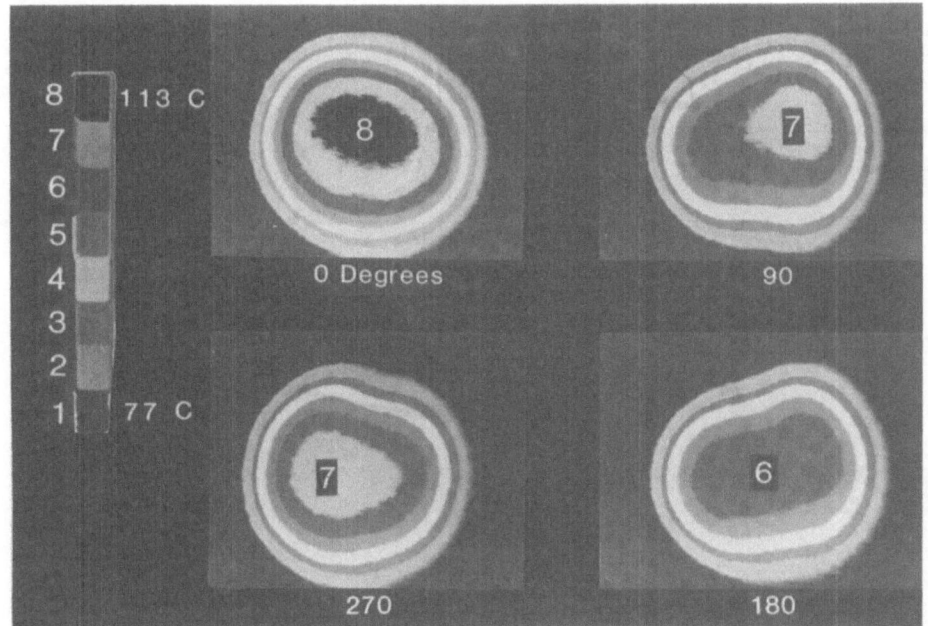

Fig. 8. Four thermograms of a single supported Pt pellet taken on four different rotations with feed conditions of H_2 = 1.95%, O_2 = 50.0% and N_2 = 48.05% with a residence time of 0.5 minutes

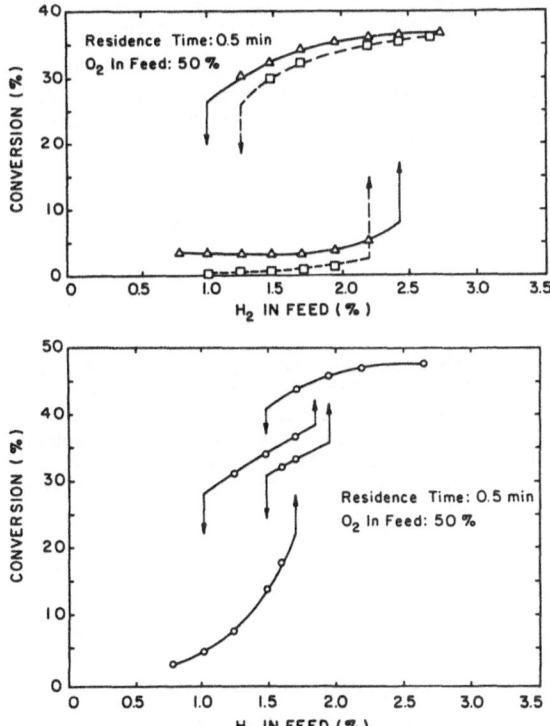

Fig. 9 Steady-state hysteresis loops for two isolated pellet experiments (Triangles represent data for one pellet, squares for the other.)

Fig. 10 Steady states of a two-pellet system showing four stable branches

solid and the dashed curves, each for a separate particle, both show bistability and simple hysteresis. They are similar to each other, but not identical, apparently attesting to the fact that no two particles are exactly alike.

In a follow-up experiment these same two particles were strung on the wire together, in direct contact, so that the behavior of the two-particle assembly could be observed. The resulting steady states of the bulk gas are shown in Fig. 10.

Since each particle can have two stable states, the system clearly could have four stable states. These correspond to both particles being in a state of high or low activity and to each being in separate activity states. Of course, if particle-particle communication were very good, the latter states would not be observed. As shown in Fig. 10, our experiments yielded a more complex picture of steady-state multiplicity and hysteresis, and hence of ignition and extinction, than is possible with either particle singly. Clearly, this picture would become still more complex if additional particles were added.

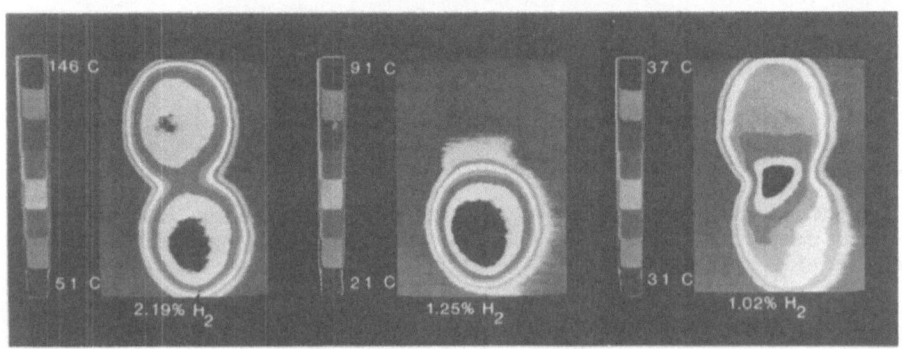

Fig. 11. Selected thermograms of steady states from three of the branches shown in Fig. 10 (From left to right, these are high, intermediate (second branch from top in Fig. 10) and low branches. The black region in these thermograms corresponds to zone 8 -- the zone of highest temperature.)

Thermograms for three of the states, one each on high, intermediate and low branches of Fig. 10, are shown in Fig. 11. The poor thermal communication between the particles is evident.

4.. Concluding Remarks

We have presented here selected samples of thermograms from our experiments to date with different types of catalyst specimens. Our results are not yet in definitive form, but nevertheless they serve to demonstrate certain spatial variations in activity on catalysts that are greater than most researchers would have expected. They also serve to demonstrate the utility of, and the potential for, the technique of thermal imaging for studying catalytic processes and for gaining new information and insights about them.

5. References

1. Wicke, E. , Kumman, P. , Keil, W., and Scheifler, J. 1980, Ber. Bunsenges. Phys. Chem., 84, pp. 315-323.

2. Turing, A. 1952, Phil. Trans. of the Royal Society, B237, pp. 37-72.

3. Prigogine, I. and Nicolis, G. 1967, J. Chem. Phys., 46, pp. 3542-3550.

4. Prigogine, I. and Lefever, R. 1968, J. Chem. Phys., $\underline{48}$, pp. 1695-1700.

5. Othmer, H. and Scriven, L. 1974, J. Theor. Biol, $\underline{43}$, pp. 83-112.

6. Schmitz, R.A. and Tsotsis, T. T. 1983, Chem. Engin. Sci., $\underline{38}$, pp. 1431-1437.

7. Tsotsis. T. T., 1983, Chem. Engin. Sci., $\underline{38}$, pp. 701-717.

8. Sheintuch, M. and Pismen, L. 1981, Chem. Engin. Sci., $\underline{36}$, pp. 489-497.

9. Saratasky, B. J. and Bell, A. T. 1982, ACS Symposium Series 178, No. 5, pp. 105-141.

10. Elhaderi, A. E., and Tsotsis, T. T., 1982, ACS Symposium Series 196, No. 7, pp. 77-88.

11. Belyaev, V. D., Slin'ko M.M., and Slin'ko, Proceedings of the International Conference on Catalysts, London, 1976, pp. 13-15.

12. Zuniga, J. R. and Luss, D., 1978, J. Cat., $\underline{53}$, pp. 312-320.

13. Rajagopolan, K. and Luss, D., 1980, J. Cat., $\underline{61}$, p. 289.

14. Kurtanjek, Z., Sheintuch, M., and Luss, D. 1980, J. Catalysis, $\underline{66}$, pp. 11-17.

15. Schmitz, R. A., Renola, G. and Garrigan, P. 1979, Ann. N. Y. Acad. Sci., $\underline{316}$, pp. 638-651.

16. Othmer, H. and Scriven, L. 1971, J. Theor. Biol., $\underline{32}$, pp. 507-537.

17. Winfree, A. T., 1974, Sci. Am., $\underline{230}$, pp. 82-95.

18. Walker, J., 1978, Sci. Am., $\underline{239}$, pp. 152-160.

19. Schmitz, R. A., D'Netto, G. A., Razon L. F., and Brown, J. R., Chemical Instabilities -- Applications in Chemistry, Engineering, Geology and Material Science, Ed. G. Nicolis and F. Baras, D. Reidel Publishing Co., Dudrecht, 1984.

20. D'Netto, G. A., Brown, J. R., and Schmitz, R. A., Proceedings of the Eighth International Symposium on Chemical Reaction Engineering, Institution of Chemical Engineers Symposium Series No. 87, pp. 247-254, 1984.

Dynamics of the Heterogeneous Catalytic Oxidation of Carbon Monoxide on Zeolite Supported Palladium

N.I. Jaeger, K. Möller, and P.J. Plath

Institut für Angewandte und Physikalische Chemie, Forschungsgruppe Angewandte Katalyse, Universität Bremen, Bibliothekstraße, NW 2, D-2800 Bremen 33, Fed. Rep.of Germany

Introduction

The complex temporal behavior of the heterogeneous catalytic oxidation of carbon monoxide on palladium and platinum catalysts has been studied intensively in recent years [1-12].

Phase-transitions regarding the structure and reactivity of adsorbate layers were discussed as a model for the interpretation of the experimental results [2,9,10]. They were observed under vacuum conditions on single crystal surfaces in the case of platinum [13]. The properties of a highly dispersed palladium phase and the necessary cooperation of a huge number of palladium crystallites in the formation of temporal structures in the CO-oxidation reaction was studied using a palladium-loaded faujasite type zeolite. Within the network of channels and cages of this a-lumosilicate, a well-defined highly dispersed palladium phase with a narrow size distribution of the Pd-crystallites can be prepared [14,15] (Fig. 1). The Pd-phase is formed by autoreduction of $Pd(NH_3)_4^{2+}$-ions which were ion-exchanged into the zeolite.

Depending on the experimental condition for the autoreduction process and on the degree of ion exchange, the size of the Pd-crystallites and their number per unit volume were controlled. The Pd-phase within the zeolite matrix consists of single crystals as was demonstrated by electron diffraction.

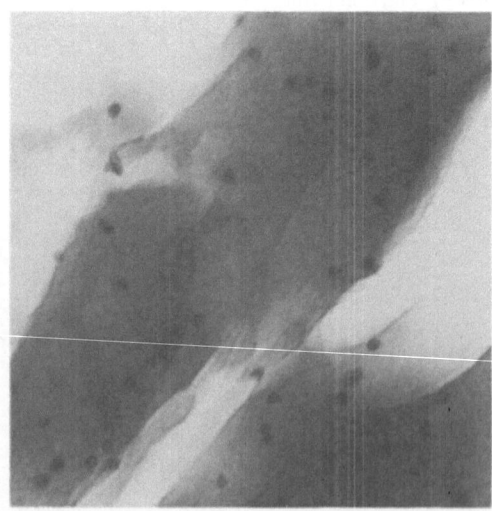

Fig. 1   Transmission electron micrograph of a palladium-loaded zeolite crystal (10 nm Pd-crystallites)

REACTOR

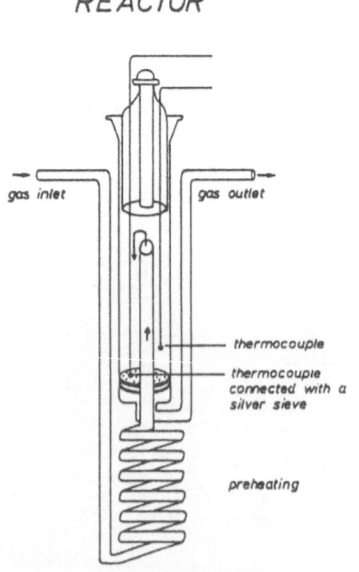

Fig. 2   Flow reactor

Experiments

The reaction was carried out in a differential flow reactor (Fig. 2) at temperatures between 130° and 250°C and CO-concentrations in the range of 0.2 - 5% by vol. in synthetic air. A linear flow rate of 2.5 cm/sec was used in most of the experiments (WHSV 230 h^{-1}). 20 - 50 mg of the catalyst were placed on the silver sieve under shallow bed conditions. The time-dependence of the average temperature of the catalyst support and the CO_2 conc. at the outlet of the reactor was continuously recorded by a thermocouple connected to the silver sieve and by infrared spectroscopy respectively.

Results

After a short period of forming the effective catalyst,the characteristic temporal pattern shown in Fig. 3 is obtained. The temperature and the CO_2-production are strongly coupled. The curves are shifted against each other in the graph for better visualization. The oscillation patterns are characterized by sharp peaks of decreased CO_2-production or temperature,culminating in an almost total absence of CO_2 in the outlet. Among the series of the continuously growing largest peaks, a similarly growing series of smaller peaks can be observed. The temperature curve shows the same behavior on a less sensitive scale.

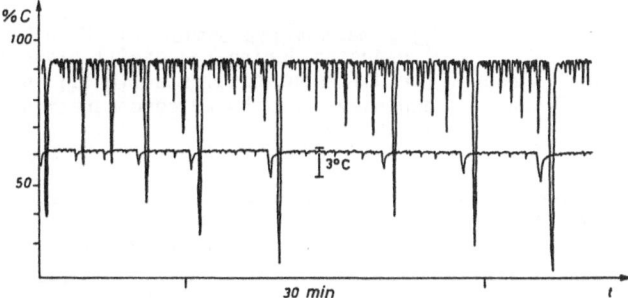

OSCILLATION PATTERN WITH SELFSIMILARITY

Fig. 3 Typical oscillation pattern with selfsimilarity of the CO_2-production (% conversion of CO) and the temperature of the silver sieve

The structure of the experimental curves seems to resemble patterns of selfsimilarity as discussed by Mandelbrot [16]. A discrete mathematical model creating a similar pattern is presented by Dress et al. [17].

The amplitudes of the largest peaks of reduced CO_2 production depend on the size and the density of the Pd-crystallites within the zeolite matrix,as well as on the CO concentration in the feed (Fig. 4,5; Table 1).

The reaction is found to oscillate within the hysteresis curves depicted in Fig.6 where ΔT, which corresponds to the rate of reaction, is plotted as a function of the CO concentration in the feed. For each catalyst there exists a threshold value of the CO concentration. Beyond this value,the heat production of the reaction drops to a low value. At this point,total coverage of the palladium surface by CO prevents further oxidation. With increasing density of the Pd-crystallites,the threshold value rises. However, looking at the number of CO-molecules per Pd-surface atom,the conclusion can be drawn that a smaller number of CO-molecules is needed to suppress the oxidation at higher densities of Pd-crystallites (Table 2).

A similar conclusion can be drawn from the peaks of maximal depression of CO_2-production for various catalysts (Fig. 5). The strongest depression is observed at the highest number of Pd-crystallites per unit volume and at the smallest particle size.

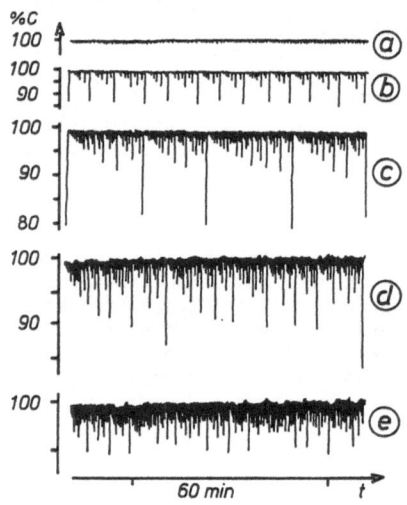

a) 0,32 VOL% b) 0,87 VOL% c) 1,08 VOL%
d) 2,82 VOL% e) 3,90 VOL%   T=200°C

Fig. 4 Time series of CO_2-production (% conversion of CO) in dependence on the CO concentration in the feed

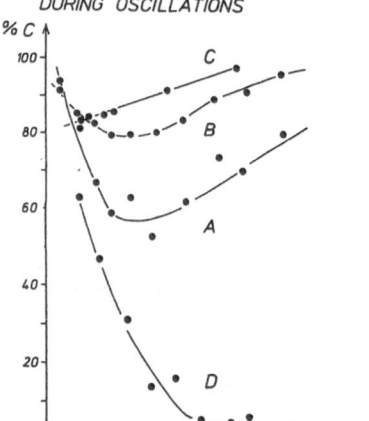

MAXIMUM DECREASE OF CONVERSION DURING OSCILLATIONS

Fig. 5 Maximum depression of CO to CO_2 conversion during the oscillations for different catalysts (Tab.1) in dependence on the CO concentration in the feed

Table 1

catalyst	Pd % b.wt.	Pd-crystallites size (nm)	Pd-crystallites number X10^{16}/ g of catalyst	Pd-surface m^2/g catalyst
A	14.7	10	2.34	7.38
B	4.4	10	0.71	2.22
C	2.4	10	0.38	1.18
D	14.7	4	36.50	18.34

Table 2

catalyst	CO-concentration % b.vol. extinction point	number CO-molecules/ Pd-surface atom·s	Pd-surface m^2/g catalyst
A	1.20	0.15	7.38
B	0.78	0.33	2.22
C	0.78	0.57	1.18

Following Turner et al. [18,19], we suggest a cyclic oxidation-reduction mechanism to account for the oscillation of the reaction on a single Pd-crystallite. At a high rate of conversion, the increasing temperature leads to a phase-transition from Pd to PdO. The oxide does not catalyse the oxidation of adsorbed CO-molecules. As a consequence of the suppressed oxidation, the local temperature decreases giving rise to the reduction of the PdO to the catalytically active Pd-phase by CO.

The different Pd-crystallites are coupled by the temporal and local excess of CO in the neighborhood of a Pd-crystallite blocked by its phase-transition to PdO.

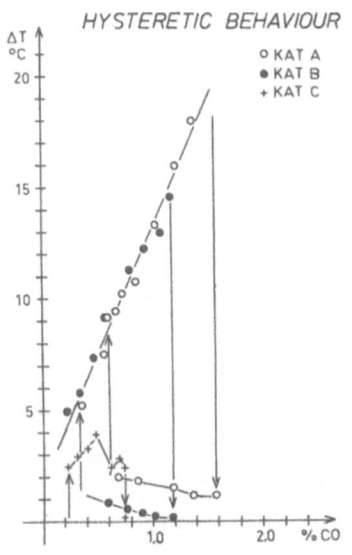

HYSTERETIC BEHAVIOUR

○ KAT A
● KAT B
+ KAT C

Fig. 6 Heat production in the catalytic oxidation in dependence on the CO concentration in the feed

▼ Fig. 7 Time series and corresponding phase portrait, Poincaré-section and Poincaré-map

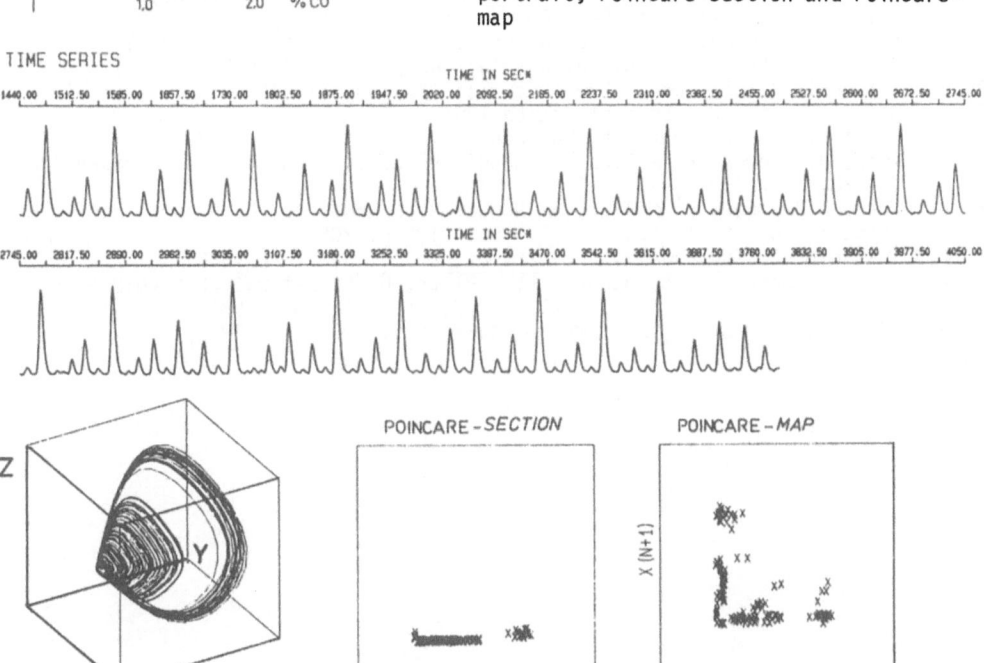

TIME SERIES

TIME IN SEC*

POINCARE-*SECTION*

POINCARE-*MAP*

To obtain a deeper insight into the temporal behavior of this reaction, the 3-dimensional phase-space was reconstructed using variables $x=c(t)$, $y=c(t-\tau)$ and $z=c(t-2\tau)$ where c is the concentration of CO_2 in the outlet and τ is a generic delay-time constant (Fig. 7). In Fig. 7 the attractor of a typical reaction-pattern of the system is shown. In order to decide whether or not a strange attractor of a chaotic movement is observed numerical treatments of the experimental data such as Fourier analysis, Poincaré-sections and Poincaré-mappings were carried out.

The Poincaré-sections show that the attractor is almost flat due to the strong dissipative character of the chemical system. Although we think that we can describe the basic dynamics of our system by a deterministic model, we did not succeed in obtaining a function in the X(N) versus X(N+1)-plane of the Poincaré-map. Nevertheless, the representation of the dynamics of our system can be understood on the basis of the discrete mathematical model mentioned above. This model will be discussed more extensively in further publications.

Acknowledgement

Financial support from the Stiftung Volkswagenwerk is gratefully acknowledged.

References

1 M. Sheintuch, R.A. Schmitz: Catal. Rev.-Sci. Eng. 15, 107 (1977)
2 W. Keil, E. Wicke: Ber. Bunsenges. Phys. Chem. 84, 377 (1980)
3 E. Wicke, P. Kummann, W. Keil, J. Schiefler: ibid., p. 315
4 J. Rathousky, J. Puszynski, V. Hlavacek: Z. Naturforsch. 35a, 1238 (1980)
5 N. Jaeger, K. Möller, P.J. Plath: Z. Naturforsch. 36a, 1012 (1981)
6 W. Adlhoch, H.G. Lintz, T. Weisker: Surf. Sci. 103, 576 (1981)
7 D. Barkowski, R. Haul, U. Kretschmer: Surf. Sci. 107, L 329 (1981)
8 P.C. Liao, E.E. Wolf: Chem. Eng. Commun. 13, 1315 (1982)
9 G. Ertl, P.R. Norton, J. Rüstig: J. Phys. Rev. Lett., 49, 177 (1982)
10 M.P. Cox, G. Ertl, R. Imbihl, J. Rüstig: Surf. Sci. 134, L 517 (1982)
11 J. Puszynski, V. Hlavacek: Chem. Eng. Sci. 39, 681 (1984)
12 K. Möller: Thesis, Universität Bremen 1984
13 R. Imbihl: Thesis, Universität München 1984
14 D. Exner, N. Jaeger, K. Möller, G. Schulz-Ekloff: J. Chem. Soc., Faraday Trans. I, 78, 3537 (1982)
15 D. Exner, N. Jaeger, K. Möller, R. Nowak, H. Schrübbers, G. Schulz-Ekloff, P. Ryder: Stud. Surf. Sci. Catal., vol. 12, p. 205
16 B.B. Mandelbrot: The Fractal Geometry of Nature, New York 1983
17 A. Dress, M. Gerhardt, N. Jaeger, P.J. Plath, H. Schuster: this volume
18 J.E. Turner, B.C. Sales, M.B. Maple: Surf. Sci. 103, 54 (1981)
19 J.E. Turner, B.C. Sales, M.B. Maple: Surf. Sci. 109, 591 (1981)

Toroidal Oscillations During the Oxidation of Methanol on Zeolite Supported Palladium

N.I. Jaeger, P.J. Plath, and P. Svensson

Institut für Angewandte und Physikalische Chemie, Forschungsgruppe Angewandte Katalyse, Universität Bremen, Bibliothekstraße, NW 2, D-2800 Bremen 33, Fed. Rep. of Germany

In the course of studying oscillating heterogeneous catalytic reactions, the oxidation of methanol was found to exhibit various types of dynamical behavior. Experiments with palladium on amorphous carrier materials (Al_2O_3/SiO_2) as catalyst showed oscillations of the reaction temperature with a period-doubling sequence or bistable reaction regimes. The observations were explained by a phase-transition of the catalytically active palladium particles [1-3].

In order to obtain a better-defined catalyst, the carrier material has been changed now to a Faujasite-type NaX zeolite. The system of channels and cavities of the zeolite was loaded with a distribution of palladium crystals all around 4 nm in diameter [4]. In contrast to the amorphous catalyst, the palladium particles are exclusively located within the zeolite matrix and not on the carrier material's surface [5].

Besides   period-one-, period-two- and chaotic oscillations, a new type of dynamical behavior could be observed experimentally [6]: For lower flow rates, the amplitude of the reaction-temperature oscillations undergoes a beating-like modulation (Fig. 1).

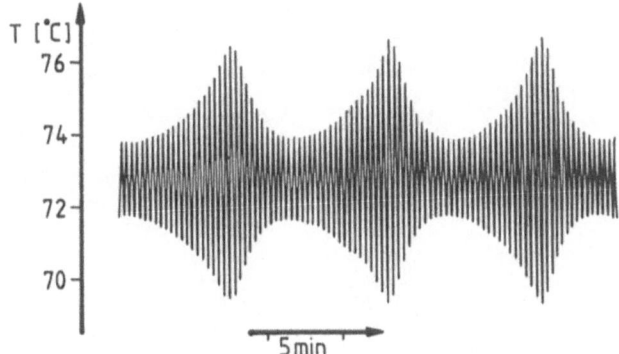

This kind of behavior has been predicted theoretically by Rössler and called "toroidal oscillation" with respect to the underlying attractor of the system [7,8].

A torus, being the cartesian product of two circular motions, gives rise to the assumption of two existing oscillators with independent but only slightly different frequencies, as they are observed in Fourier-spectra of the observed experimental oscillations [6]. If one identifies one of these oscillators with the chemical oscillation due to the palladium phase-transitions cited above, the question arises whether the second oscillator could be assigned to intracrystalline transport properties of the zeolite.

As it is known from other experiments [9], the diffusivity inside NaX-zeolite crystallites diminishes significantly with increasing water content.

Since the catalytic oxidation of methanol is combined with intense water production, and the temperature of the system in the oscillating state is always very low (below 100°C), the conclusion might be drawn that the transport of methanol to the active sites might depend on the rate of conversion within the zeolite cavity system.

Chemical and diffusional oscillation then could be coupled by concentration terms of the substances concerned.

References

1 N. Jaeger, P. Plath, E. v. Raaij
      Z. Naturforsch. 36a , 395 (1981)
2 P. Plath, A. Haberditzl, N. Jaeger, H. Prüfer, E. v. Raaij in:
      Dynamische Systeme, Proc. 5th 5-Tage-Kurs USP Mathematisierung, University
      of Bielefeld, 1982, p. 187
3 A. Haberditzl, N. Jaeger, P. Plath
      Z. phys. Chemie 265, 449 (1984)
4 D. Exner, N. Jaeger, K.Möller, R.Nowak, H.Schrübbers, G.Schulz-Ekloff, P.Ryder
      Stud. Surf. Sci. Catal., 1982, 12), (Metal Microstructures in Zeolites,
      P. Jacobs et al., Eds.) p. 205
5 G. Schulz-Ekloff, D. Wright, M. Grunze
      Zeolites 2 , 70 (1982)
6 P. Svensson
      Diplomarbeit, Universität Bremen 1984
7 O. Rössler
      Z. Naturforsch. 32a , 299 (1977)
8 O. Rössler
      Ann. of the N.Y. Acad. of Sci. 316 , 376 (1979)
9 A. Germanus, J. Kärger, H. Pfeifer
      Zeolites 4 , 188 (1984)

Oscillatory Behavior of Cobalt Electrodes During Transition from the Active to the Passive State

N.I. Jaeger, P.J. Plath, and Ngo Quoc Quyen[*]

Institut für Angewandte und Physikalische Chemie, Forschungsgruppe Angewandte Katalyse, Fachbereich 2, Universität Bremen, Bibliothekstraße D-2800 Bremen 33, Fed. Rep. of Germany

Oscillatory behavior in electrochemical systems has been known for a long time [1] and has been investigated intensively in the region of transition from the active to the passive state [2]. The model established for the potential or current oscillations of the iron electrode in sulfuric acid was among the first for heterogeneous systems [3]. The oscillating electrode switches autonomously between the active and the passive state.

Cobalt is among the metals which exhibit the transition from the state of active anodic dissolution to the state of passivity, where non-porous oxide layers on the metal surface cause the rate of dissolution of the metal to drop by orders of magnitude.

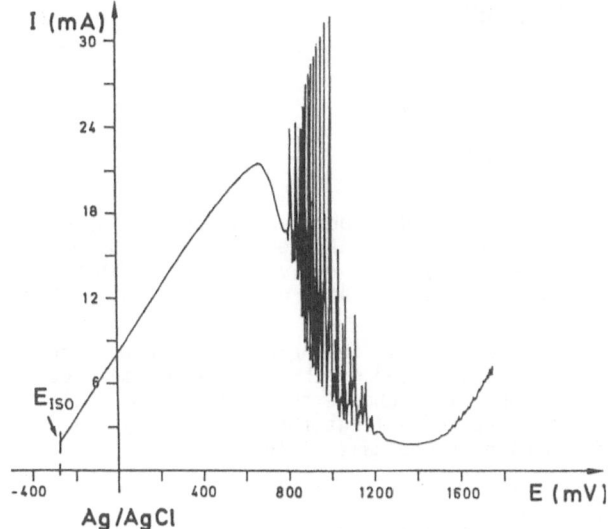

Fig. 1 Current voltage curve in the region of primary passivity of cobalt. Voltage scan: 5 mV/s. Potential vs. Ag/AgCl reference electrode. E(I=0): -270 mV

Oscillations of the potential of cobalt electrodes in solutions containing hydrochloric and chromic acid have been reported [4]. In the work presented here, the behavior of cobalt was studied in phosphate solutions (pH=2,55) at potentials where secondary passivity is observed. In this region, the metal surface is covered by a Co_3O_4 barrier layer on top of a less protecting CoO film under stationary conditions [5]. Transitions between different states of passivity can be observed.

The experiments were carried out using Co-wire electrodes (length 0.5 cm, diameter 0.05 cm/99.99% Goodfellow Metals). To assure comparable experimental conditions for the study of the oscillations, a freshly-prepared electrode was used in each experiment. Fig. 1 depicts the experimental current-voltage curve obtained by a vol-

[*]National Center for Scienctific Research, Institute of Chemistry, Hanoi, Vietnam

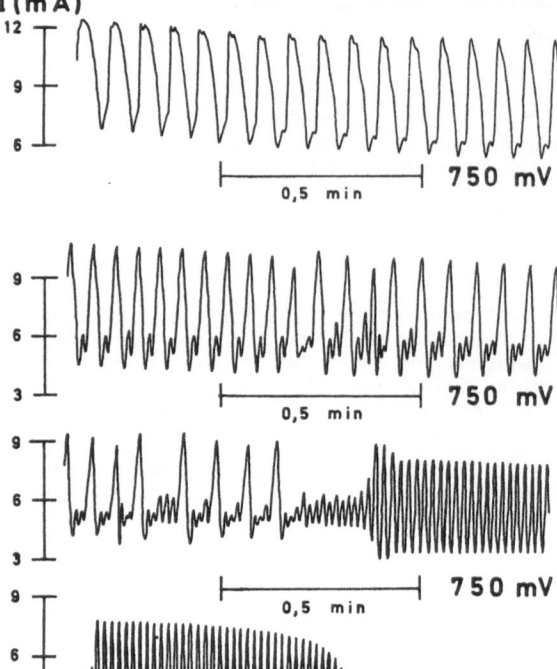

I(mA)

0,5 min   750 mV

0,5 min   750 mV

0,5 min   750 mV

0,5 min   750 mV

Fig. 2   Evolution of the current oscillation of the cobalt electrode at constant potential (750 mV vs. Ag/AgCl reference electrode)

tage scan. Oscillations at constant potential could be observed in the region of negative resistance,under the experimental condition that the electrode was swept by a potential scan,starting in the active region up to a given potential in the transition region. The current peaks reach the extrapolated dynamic current-voltage curve.

Fig. 2 depicts the temporal behavior at 750 mV vs. the Ag/AgCl reference electrode used in the experiments. Due to the dissolution of the anode,especially during the current peaks,the experimental conditions change continuously until the electrode has dissolved. Depending on the applied voltage,the electrode system is found to run through a number of states with period-multiplying.

1   A.T. Fechner: Schweiggers J. Chem. Phys. 53, 141 (1828)
2   U.F. Franck: Angew. Chem. 90, 1 (1978)
3   U.F. Franck, R. FitzHugh: Z. Electrochem. 65, 156 (1961)
4   U.F. Franck, L. Meunier: Z. Naturforsch. 8b, 396 (1953)
5   N. Sato in: Passivity of Metals (R.P. Frankentahl, J. Kruger, Eds.) Princeton 1978, p. 29

Periodic and Aperiodic Regimes in Forced Chemical Oscillations

M. Marek

Department of Chemical Engineering, Prague Institute of Chemical Technology
166 28 Prague 6, Czechoslovakia

1. Introduction

We shall discuss experiments 1) on phase-resetting of the limit cycle type chemical oscillations by a pulse of a reacting component and 2) on periodic forcing of oscillations in an isothermal CSTR by a periodically varying inlet reactant concentration. Oscillations in the Belousov-Zhabotinski reaction (BZR) in a CSTR and catalytic oxidation of carbon monoxide on a Pt-type catalyst in a recycle reactor with high recycle ratio were used as experimental systems. Periodic and aperiodic regimes observed ad 2) are interpreted on the basis of a simple mathematical model (phase advance map) evaluated from the phase-resetting experiments, considering recent theoretical results.

Periodic external excitation (forced oscillation) is an important tool in the investigation of dynamical systems with an oscillatory character. Spectral analysis of linear dynamical systems is based on the investigation of the response of the system to the input periodic signals with different frequencies. The results are often utilized to construct responses to arbitrary inputs, using the superposition principle. The superposition principle does not apply in nonlinear dynamical systems. However, forced oscillations are commonly used in analyses of electrical and mechanical systems [1],in the studies of biological systems (cf. reviews in [2-4]) and recently also in chemical systems [4-8]. Phase-resetting experiments are reported in biology since 1868 [9],but in chemical systems they have been used only recently for the study of the BZR [10,11] and the Briggs-Rauscher oscillator [7,12].

2. Periodic and Aperiodic Regimes in Forced BZR

A continuous flow stirred reactor was used in the experiments. The experimental arrangement is shown in Fig. 1. The reactor has a cooling jacket, electronically controlled stirrer, temperature control and is provided with measuring electrodes and a capillary serving for the introduction of the inlet concentration perturbations. Two

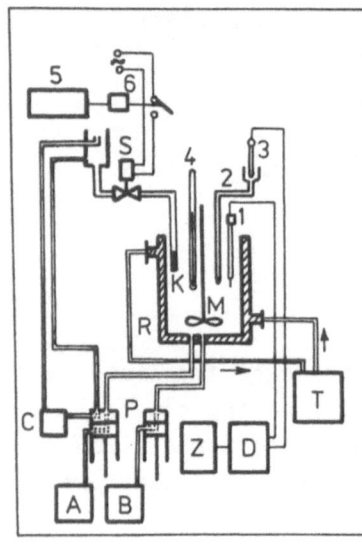

Fig. 1: Experimental arrangement
for the study of forced
BZR

R - continuous stirred tank reactor;
M - stirrer; T - thermostat; S - solenoid valve; D - signal conditioning,
Z - recording; P - pumps; A,B,C -
stock solutions, 1 - Pt electrode,
2 - salt bridge, 3 - calomel electrode, 4 - thermometer, 5,6 - electronics

separate streams of the inlet solution (solution of Ce^{4+} ions + sulphuric acid and solutions of BrO_3^- ions + malonic acid) were delivered by peristaltic pumps to the bottom of the reactor. The basic mechanism of the oscillations in the BZR is now well established [13].
In this mechanism, Br^- ions play the role of a control intermediate
[14].

The solution of Br^- ions used for stimulation was delivered via a
capillary located close to the liquid level in the reactor, and its
flow was controlled by a solenoid valve. The temperature in the reactor and the frequency of the turbine-type stirrer were kept constant. The inlet concentrations of the reaction components were (if
not specified otherwise): Ce^{4+} ions $(Ce(SO_4)_2)$: 0.001 [M], $KBrO_3$:
0.05 [M], malonic acid (MA): 0.05 [M], sulphuric acid: 1.5 [M]. The
stirring rate was 500 rpm, the residence time in the reactor was
1080 s , reactor volume 100 ml and the volume of the pulse 1 ml .
Pt electrode-calomel electrode couple was used to follow the course
of oscillations. The potential of the Pt electrode is proportional
to $\ln (Ce^{4+}/Ce^{3+})$ and the periods of oscillations of Br^- and of the
redox potential are the same. First, a steady oscillatory regime
with the period T_B has been set up in the reactor (the regime was
considered steady when fluctuations in T_B were less than one percent).
The magnitude of T_B was controlled by variation of the temperature
in the reactor. Concentration forcing was performed by the pulse
of solutions containing Br^- or Ce^{4+} ions (the residence time was
kept constant). The concentration pulses were either periodically

repeated with the period T_F (periodic forcing) or a single pulse applied at the specified phase of the oscillation was used (phase-shift experiments).

2.1 Periodic Forcing Experiments

The redox-potential oscillations are of a relaxational character under the experimental conditions used. They consist of very fast upward jumps of the potential followed by a slow potential decrease. This forms a basis for the measurement of the period and of the phase of the oscillations. The sequence of response "periods" δT_n is defined as a sequence of the time intervals separating two succesive jumps of the potential. The phase φ at the moment of the stimulation (considered modulo T_B) is the time interval between the pulse and the preceding jump of the potential. We shall conveniently measure all quantities in the units of the natural period T_B and denote

$$\tau_F = T_F/T_B, \quad \Delta\tau = \delta T/T_B, \quad \phi = \varphi/T_B.$$

Two types of concentration forcing were used: periodic repeating of pulse additions (reported here), and rectangular periodic variations of the inlet concentration of the chosen reaction component [4]. Both Br^- and Ce^{4+} ions were used in the forcing experiments. The forcing amplitude and dimensionless forcing period τ_F were varied. We now present several examples of experimental results.

The results of the experiment, where the forcing period of Ce^{4+} ions is twenty times higher than the natural period, $\tau_F = 20$ (slow forcing) are shown in Fig. 2. The sequence δT_n is periodic and the resulting period is equal to the forcing period. Hence, in this way we can form a signal with a modulated frequency. An opposite case of the fast forcing, when $\tau_F = 0.187$ is shown in Fig. 3.

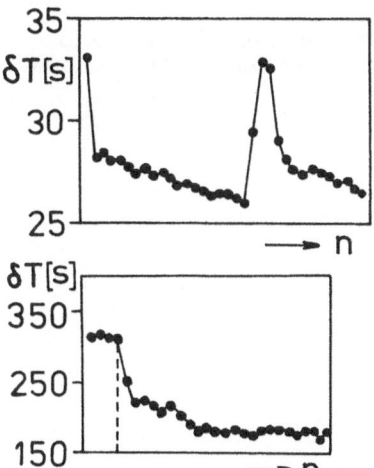

Fig. 2: Sequence of response "periods" of oscillations, slow forcing (periodic forcing by Ce^{4+} pulses)
$\tau_F = 20$; BrO_3^- - 0.1 [M]; MA - 0.4 [M]; Ce^{4+} - 0.0005 [M]; H_2SO_4 - 2 N .

Fig. 3: Sequence of response "periods" of oscillations: fast forcing, (periodic forcing by Ce^{4+} pulses)

We can again observe that the resulting behaviour is periodic, and
the resulting period is either equal to the forcing period or to its
small integer multiple. Both above examples are typical for a number of experiments performed under similar conditions.

A sequence of response "periods" of oscillations and the corresponding power spectra for the periodic forcing by the pulses of Br^-
ions with an amplitude equal to 1.43×10^{-4} [M] and the value of
$\mathcal{T}_F = 2.5$ is presented in Fig. 4. The resulting behaviour is periodic, as it is clearly confirmed by the power spectrum. Several examples of experimental results for the periodic forcing by Br^- ions
with an amplitude 5×10^{-5} [M] are shown in the Table 1 and in Figs.
5a-h. When in the course of q perturbations the number of responses
of the system is p, i.e.

$$q\, T_F = \sum_{n=1}^{p} \delta T_n \; ,$$

then a periodic regime with the phase-locking ratio $\rho = p/q$ is
established in the system.

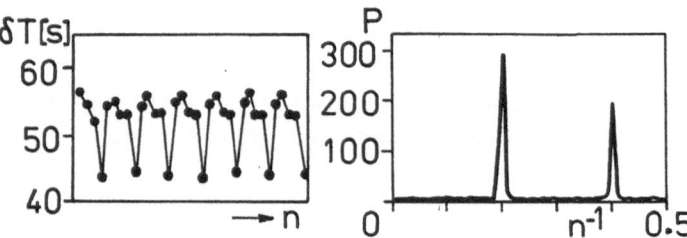

Fig. 4: Sequence of response "periods" of oscillations and corres-
        ponding power spectra, periodic forcing by pulses of Br^-
        ions; amplitude of Br^- pulses 1.43×10^{-4} [M], $\mathcal{T}_F = 2.5$

Table 1

Types of dynamical behaviour of BZR in dependence on $\mathcal{T}_F$. Periodic
(phase-locked with the ratio p/q) or aperiodic regimes (with some
p/q cycles interspersed) are observed

$\mathcal{T}_F$	character of behaviour
2.51	aperiodic (+5/2)
2.50	periodic, 5/2
2.39	aperiodic (+5/2, 12/5)
2.31	periodic, 7/3
2.24	aperiodic
2.17	aperiodic, intermittent
1.77	periodic, 2/1
1.66	aperiodic (+5/3)
1.46	periodic, 3/2
1.18	aperiodic, intermittent

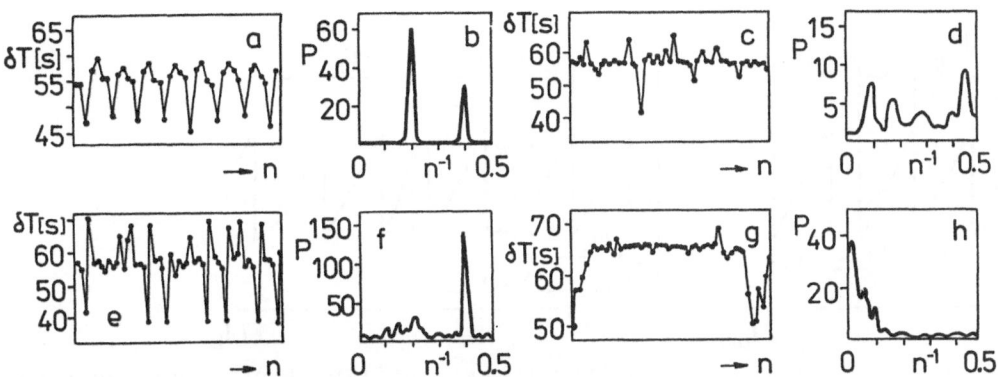

Fig. 5: Sequences of response "periods" (a, c, e, g) and corresponding power spectra (b, d, f, h); periodic forcing by pulses
of Br ions; amplitude of Br pulses [Br] = 5 x 10^{-5} M
a) τ_F = 2.5, periodic pattern with 5/2 phase-locking
c) τ_F = 2.24, chaotic oscillations
e) τ_F = 1.66, chaotic oscillations
g) τ_F = 1.18, intermittent chaos

There exists a sequence of increasing ratios p/q corresponding to
periodic trajectories observed in experiments with increasing values
of τ_F, cf. Table 1. An example of a sequence of response "periods"
for 5/2 resonance is shown, together with the corresponding power
spectrum in Fig. 5a,b.

However, for many values of τ_F are the response sequences $\{\delta T_n\}$
aperiodic, cf. Figs. 5c, e, g. The corresponding power spectra show
broad-band noise, cf. Figs 5d, f, h. Two qualitatively different
types of aperiodic behaviour can be distinguished - developed chaos
(Figs 5c-f) and intermittency (Figs 5g, h) [15 - 17].

2.2 Phase-Resetting Experiments

Single pulse stimulation does not change the natural period T_B but
shifts the phase. The periods before the pulse are $\delta T = T_B$ and after
the pulse δT converges quickly to T_B. In our experimental situation,
it was found that within experimental errors only the first response
period δT_1 (i.e., the period where the stimulation occured) was
affected, (cf. Figs 9a, c). The phase-transition curve (PTC) [18]
$\Theta = \phi + \Delta\tau(\phi)$, where $\Delta\tau = [T_B - \delta T]/T_B$ can be constructed on the
basis of experimental results, cf. Figs 6a, b. The shape of the PTC
depends on the perturbation amplitude. Three examples of the PTC's
are shown in Figs 7a - c. The initially non-decreasing curve (Fig.7a)
becomes nonmonotonous with the increasing perturbation amplitude.
As the response "periods" following after the stimulated period are
almost unaffected by the pulse (cf. Fig. 4 and Fig. 9), the PTC's
can be used to model dynamical behaviour of the phase in the perio-

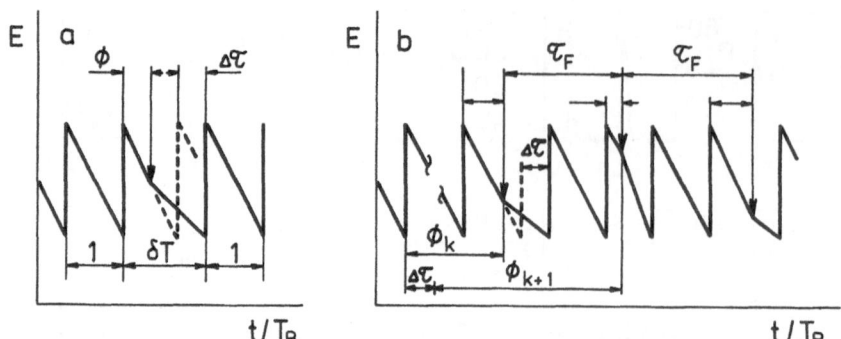

Fig. 6: a) Single pulse addition (phase-shifting experiment) schematically; δT response period; ϕ - old phase; $\Delta\tau$ - phase
shift; $E \sim \ln (Ce^{4+}/Ce^{3+})$
b) Periodic forcing, schematically; $\tau_F = T_F/T_B$, dimensionless forcing period

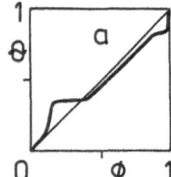 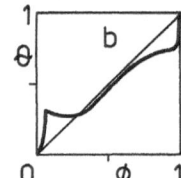 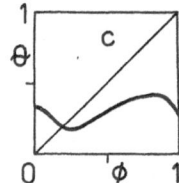

Fig. 7: Phase-transition curves evaluated from phase-shifting
experiments; pulse amplitudes
a) $[Br^-] = 1.43 \times 10^{-5}[M]$, b) $[Br^-] = 5 \times 10^{-5}$ [M],
c) $[Br^-] = 1.43 \times 10^{-3}[M]$

dically stimulated (forcing) experiments. The behaviour of the phase is then governed by a simple difference equation

$$\phi_{k+1} = \gamma(\phi_k) = \tau_F + \Theta(\phi_k) \quad \text{mod } 1 \tag{1}$$

known as an equation for the Poincaré map of the phase (phase-ad-
vance map) [19,20]. Equation (1) can be used to predict the experimental results given in Table 1 using the experimentally-measured
PTC's from Fig. 7b (cf. Fig. 8) or Fig. 7a (cf. Fig. 9). A continuous and piecewise differentiable approximation of the PTC's by
polynomials was used. A sequence $\{\phi_k\}$ was evaluated directly from
the experimental data and the values of ϕ_{k+1} were plotted against ϕ_k.
Figs 8a-d (corresponding to the data shown in Figs 5a,c,e,g) show
consistency between the data and the model function γ. In Fig. 8a
there are two clusters of points corresponding to 5/2 phase-locking;
in Figs 8b,c the experimental points are distributed along the curve γ which corresponds to chaotic oscillations. Fig. 8d shows an
example of the intermittent chaos. The experimental conditions are
close to 1/1 phase-locking, which forms a basis of the laminar pha-

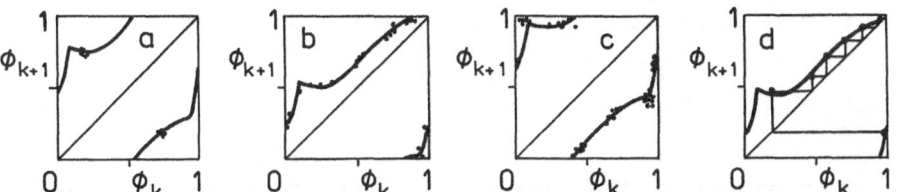

Fig. 8: Comparison of experimental data from Figs 5a,c,e,g with the
function γ constructed from PTC in Fig. 7b
a) τ_F = 2.5, phase-locking, b) τ_F = 2.24, chaos,
c) τ_F = 1.66, chaos   d) τ_F = 1.18, intermittency

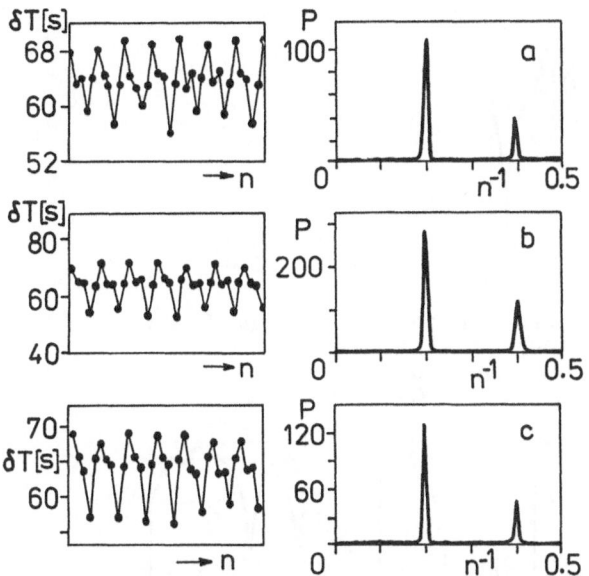

Fig. 9: Results of simulation of periodic forcing by the function γ
constructed from PTC in Fig. 7a, τ_F = 2.5
a) without noise, b) 2 % noise considered, c) effects of
the perturbation on the second "period" are considered

se (in Fig. 5g a laminar phase - nearly periodic behaviour - between two turbulent bursts is shown). The approach to laminar ($\sim$ periodic) behaviour is illustrated by the steps in Fig. 8d. Numerical
simulation has revealed that an alternation of turbulent and laminar
phases is almost regular (weak chaoticity) or quite regular (overall
periodic motion with the phase-locking ratio slightly higher than
one). Added noise, (cf. later) makes the alternation more irregular, which better corresponds to the experimental results.

The dynamical behaviour of the model (1) is complicated. According to theoretical results [21,16,17] we can expect quasiperiodic
or periodic (phase-locked) solutions for monotonous PTC's and chaotic or periodic solutions for PTC's with extrema. A sequence of

phases $\{\phi_k\}$ can be characterized by the rotation number ρ, defined as

$$\rho = \lim_{k \to \infty} 1/k \, [\gamma^k (\phi_0) - \phi_0] \, . \tag{2}$$

The value of ρ expresses the average increase of the phase on one iteration and is independent of ϕ_0. If ρ is rational (i.e., $\rho = p/q$) we have a periodic orbit with the period q and every point on this orbit is lifted by p after q iterations of γ. The dependence of the rotation number on τ_F is shown in Figs 10-12.

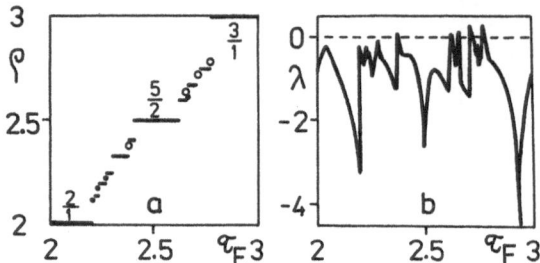

Fig. 10: Dependence of the rotation number ρ and the Liapunov number λ on τ_F. PTC curve in Fig. 7b
—··· periodic solutions ($\lambda < 0$)
oooo chaotic solutions ($\lambda > 0$)

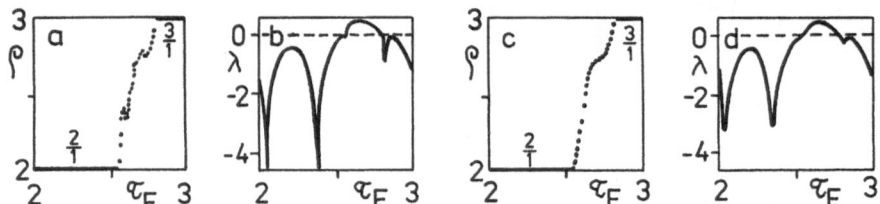

Fig. 11: Dependence of the rotation number ρ and of the Liapunov number λ on τ_F; PTC curve in Fig. 7c; a,b) without noise, c,d) 2 % rms noise level

The full lines denote the regions of constant rational values of ρ and the circles the irrational values of ρ (chaotic behaviour). We can infer from the Figs. 10,11 that wide regions of the parameter τ_F, where $\rho = 2$ and $\rho = 3$, exist. In addition, for lower forcing amplitude (fig. 10) we observe also other synchronized regimes with the rotation numbers $\rho = 5/2, \, 7/3, \, 9/4 \, \dots$

Another criterion which is used to differentiate between a periodic, quasiperiodic and chaotic behaviour is the value of the Liapunov exponent

$$\lambda = \lim_{N \to \infty} N^{-1} \sum_{k=1}^{N} \ln|d \gamma(\phi_k)/d\phi| \tag{3}$$

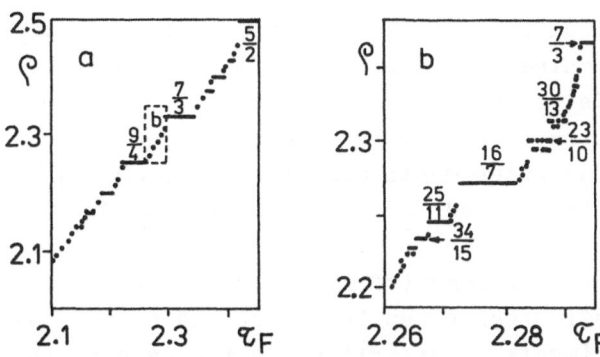

Fig. 12: Dependence of the rotation number ρ on τ_F; PTC curve in
Fig. 7a; b) area in the rectangle shown in a);
— ... periodic solutions

Negative values of λ correspond to periodic solutions and positive
values to chaotic ones. The dependence of the Liapunov exponent on
the parameter τ_F is again depicted in Figs 10,11.
We can observe that at the lower forcing amplitude (Fig. 10), the
periodic regimes prevail and the aperiodic behaviour occurs for narrow regions of the values of τ_F. At the higher forcing amplitude, a
(Fig. 11) relatively large range of τ_F values exists, where the
behaviour is aperiodic.
    If a stochastic term representing the experimental noise is added
to the deterministic model (1), we obtain a model in the form

$$\phi_{k+1} = \Theta(\phi_k) + \tau_F + \varepsilon X \qquad (4)$$

Here X is a random number uniformly distributed in the interval
(-1, 1) and ε is a chosen noise-amplitude. The dependences of ρ
and λ on τ_F for $\varepsilon = 0.02$ are also given in Figs 10,11. We can observe that even if the finer structure (e.g. higher resonances) has
disappeared, the overall character of the dependencies is preserved.
The detailed structure of the interspersed regions of resonances
(called "devil's staircase") is illustrated in Fig. 12. A more detailed discussion of the effects of noise and application to the
problem of synchronization of two coupled cells with mutual mass
transport was made recently [22].

3. Forced Oscillations at CO Oxidation
Oscillations in the outlet concentration of CO oxidized on the Pt-
catalyst are well known [23]. A differential reactor with recycle
(recycle ratio 1:150) working as a CSTR was used in our study of the
effects of periodic forcing by a varying inlet CO concentration on
the character of outlet oscillations [24]. Pt on alumina porous ca-

talyst pellets were placed in an isothermal bed. Oscillations occuring at constant CO inlet concentration (5 % CO at 428 K) were observed, with a period T_B = 1410 [s]. The change of the capacity of the
recycle and of the gas volume in the apparatus did not affect the
period of oscillations. Forcing experiments consisted of a 60 [s]
pulse of 7 % CO with the forcing period T_F = 600 s, hence $\mathcal{T}_F$=0.426.
Multiple periodic (cf. Fig. 13a) course of oscillations was observed
at T_F = 600 s and aperiodic oscillations (cf. Fig. 13b) were found
at T_F = 400 s. Multiple periodic regime can be understood on the basis of the single-pulse experiments. Similarly, as in the case of the
BZR, we have constructed a phase-transition function $\mathcal{Y}$. In the situation when $\mathcal{T}_F$ < 1 is $\mathcal{Y}$ defined by the relation

$$\phi_{k+1} = \mathcal{Y}(\phi_k) = (\phi_k - \Delta\mathcal{T}(\phi_k) + n\,(\phi_k)\mathcal{T}_F) \quad \text{mod 1} \tag{5}$$

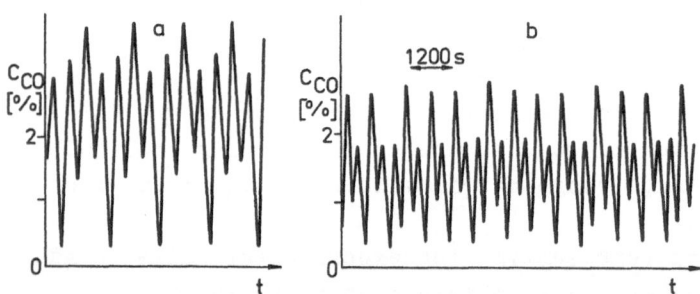

Fig. 13: Dependence of the outlet CO concentration on time
a) 3-periodic course of concentration, b) aperiodic
course of concentration

Here $n\,(\phi_k)$ gives the number of pulses at the given "period" δT_k.
It is evident that the function $\mathcal{Y}(\phi)$ will be generally disconti-
nuous, because of the varying number of pulses which fit into given
δT_n. The experimentally-determined function $n\,(\phi)$ is $n\,(\phi)$ = 2 for
$0 \leqslant \phi \lesssim 0.05$ U $0.18 \leqslant \phi \lesssim 0.426$ and $n\,(\phi)$ = 3 for
$0.05 \leqslant \phi \lesssim 0.18$. The function $\mathcal{Y}(\phi)$ is shown in Fig. 14. The
iteration of $\mathcal{Y}$ starting at any initial value of ϕ (cf. dashed lines
in Fig. 14) leads to one of two stable fixed points $\phi_2^* \simeq 0.03$ and
$\phi_3^* \simeq 0.16$. The approach to the first one, corresponding to the
phase-locking ratio 2/1 is very sensitive to the choice of the initial condition and to the effects of noise. Hence, the synchronization at the phase-locking ratio 3/1 (corresponding to ϕ_3^*) was observed in the experiments. A one-dimensional map constructed from a model of the BZR was used recently [25,26] to discuss periodic and
aperiodic regimes in the CSTR operated under constant inlet conditions.

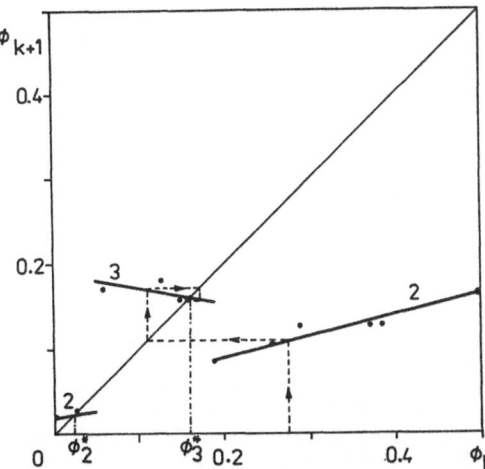

Fig. 14: Discontinuous phase-transition function $\Psi(\phi)$
stable fixed points $\phi_2^* = 0.03$, $\phi_3^* = 0.16$,---- example
of approach to ϕ_3^*

1 N. Minorsky: Nonlinear Oscillations (Van Nostrand, 1962).
2 A.T. Winfree: The Geometry of Biological Time (Springer Verlag, New York 1980).
3 L. Glass, A.T. Winfree: Am.J.Physiol. 246 (Regulatory Integrative Comp. Physiol. 15) R 251 (1984).
4 M. Dolník, D. Suchanová, M. Marek, P. Lichnovský: Sci. Pap. of the Prague Inst. Chem. Technol. K 18, 87 (1983).
5 A.M. Zhabotinski: Concentration Oscillations (Nauka, Moscow 1974), in Russian.
6 M. Dolník, I. Schreiber, M. Marek: Phys. Lett. 100A, 316 (1984).
7 E. Dulos, P. de Kepper: Biophys. Chem. 18, 211 (1983).
8 I. Gonda, B.F. Gray: Proc.R.Soc.Lond. A389, 133 (1983).
9 J. Breuer: Ber.Akad.Wiss. Wien 58, 909 (1868).
10 H. Ševčíková, D. Suchanová, M. Marek: Sci. Pap. Prague Inst. Chem.Technol. K 17, 137 (1982).
11 M. Ruoff: J. Phys.Chem. 88, 2851 (1984).
12 M. Ivanova, R. Christova: Acta Chim. Hung. 112, 59 (1983).
13 R.J. Field, E. Körös, R.M. Noyes: J.Am.Chem.Soc. 94, 8649 (1972).
14 R.M. Noyes: J.Am.Chem.Soc. 102, 4644 (1980).
15 P. Collet, J.-P. Eckman: Iterated Maps on the Interval as Dynamical Systems (Birkhäuser, Boston 1980).
16 J. Guckenheimer, P. Holmes: Nonlinear Oscillations, Dynamical Systems and Bifurcations of Vector Fields (Springer Verlag, New York 1983).
17 M. Marek, I. Schreiber: Stochastic Behaviour of Deterministic Systems (Academia, Praha 1984), in Czech.
18 A.T. Winfree: Arch. Biochem. Biophys. 149, 388 (1972).

19 M. Zaslavsky: Phys.Lett. 69A, 145 (1978).
20 L. Glass, R. Perez: Phys.Rev.Lett. 48, 1772 (1982)
21 M.T. Herman, in: Geometry and Topology Lect. Notes in Mathematics, Vol. 597 (Springer, Berlin 1977).
22 M. Dolník, I. Schreiber, M. Marek: preprint, CHISA Congress, Prague, Sept. 1984.
23 M. Sheintuch, R.A. Schmitz: Catal.Rev.Sci.Eng. 15, 107 (1977).
24 P. Knedlík, I. Schreiber, M. Marek: to be published.
25 J. Rinzel, I.B. Schwartz: J.Chem.Phys. 80, 5610 (1984).
26 I.B. Schwartz: Phys.Lett. 102A, 25 (1984).

Periodic Perturbation of the BZ-Reaction in a CSTR: Chemical Resonance, Entrainment and Quasi-Periodic Behavior

F. Buchholz, A. Freund, and F.W. Schneider

Institut für Physikalische Chemie, Universität Würzburg
D-8700 Würzburg, Marcusstraße 9/11, Fed. Rep. of Germany

Introduction

Chemical oscillations may occur in open non-linear kinetic mechanisms far from chemical equilibrium. They may show a variety of interesting phenomena when they are periodically perturbed by an external source. These phenomena include chemical resonance, entrainment, quasi-periodic behavior and even chaotic responses. While a number of theoretical discussions have been given on the subject of periodic perturbations of oscillating isothermal chemical reactions [1-10], there exist only few experimental verifications of some of these phenomena [11-15] in chemical systems. Here we report the results of initial experiments on the periodic perturbation of the Belousov-Zhabotinsky (BZ) reaction in an isothermal continuous flow stirred tank reactor (CSTR). We describe two sets of experiments. In the first set, the BZ reaction is altered to be a damped oscillator (in the presence of in-flowing KBr) and, in the second part, it behaves as a stable limit cycle. Both instances are investigated with the same experimental technique. A periodic perturbation is applied to the reacting system in the form of sinusoidal flow-rate changes as effected by a regulated high-precision syringe pump. We observe a single broad resonance curve for the damped BZ-reaction. For the undamped BZ-reaction, several sharp resonances in various entrainment bands, as well as quasiperiodic behavior are observed. In the transition region between fundamental entrainment and quasi-periodic response, the Fourier spectra slowly change with time.

I. The BZ-Reaction as a Damped Oscillator

The normally undamped BZ oscillation may be converted to a damped oscillator by adding KBr to one of the in-flowing solutions and by raising the in-flow rate to a critical value. The experimental set-up is given schematically in Fig.1.

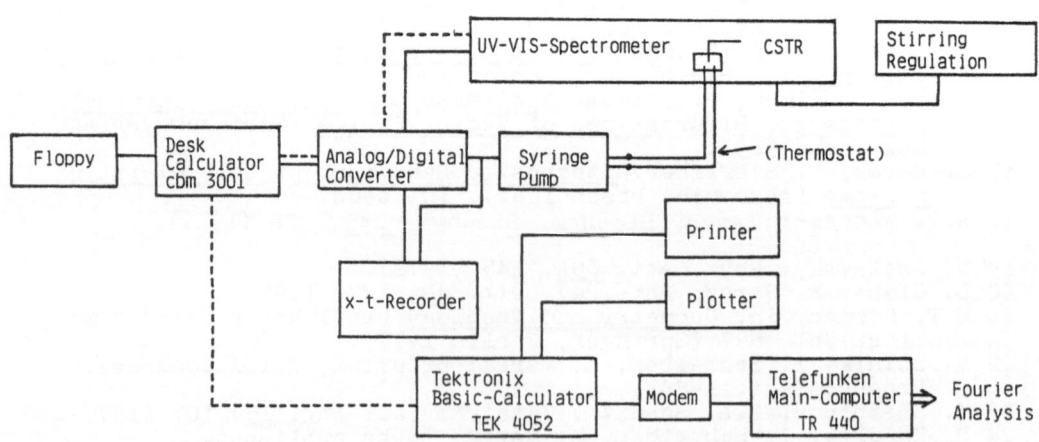

Fig. 1 For the "chemical" Fourier transformation the lower part is omitted

A thermostated syringe pump (Infors Precidor) simultaneously delivers two solutions from two 50 ml Hamilton syringes into a thermostated (25.0° C) CSTR at flow
rates that are regulated via a stepping motor by a CBM Commodore computer. The two
solutions contain 2 x 10^{-6} M KBr, 0.9 M malonic acid in 1.0 M H_2SO_4 and 0.30 M
$KBrO_3$, various Ce^{3+}, respectively. The resulting stable steady state which has
been reached by a damped oscillation is subsequently subjected to periodic perturbations in the flow-rate according to

$$k_f = k_f^o \, (1 + \alpha \cos \omega_p t) \tag{1}$$

where k_f^o is the flow-rate constant in the absence of a periodic perturbation, α is
the perturbation amplitude and ω is the perturbing frequency in rad/s. We use
$k_f^o = 1.1$ x 10^{-3} s^{-1} which corresponds to an average residence time of $\tau_r = 15.0$
min. The forced response frequency is identical with the forcing frequency, since
the reacting system behaves like a damped oscillator in this case. We use the
absorption of Ce^{4+} at 350 nm (Beckman Spectrophotometer)in order to monitor the
amplitude of the forced Ce^{4+} oscillations. After a steady state has been attained
the amplitude (ΔOD) of the Ce^{4+} oscillations is determined at each perturbing
frequency ω at constant α. Then the square of the Ce^{4+} response amplitude is
plotted point by point versus ω . Relatively broad chemical resonance curves are
obtained,which represent power spectra of the reaction (Fig.2).This procedure is
tantamount to a "chemical" Fourier transformation. For various Ce^{4+} concentrations, the measured resonance curves have been fitted by Lorentz functions, whose
resonance maxima increase with increasing Ce^{4+} where they are slightly shifted to
higher frequencies,in good agreement with the frequencies of the damped oscillations as measured in the time domain (not shown). Furthermore, the damping constants (as determined from the envelope of the damped oscillation) agree well
with the values obtained from the half width of the resonance curves. The experimental phase-shift between the perturbing function and the forced response may
be used to obtain the dispersion spectrum as well as the absorption spectrum of
the chemical reaction [15]. If the in-phase component between sinusoidal in-flow
and the measured response is plotted versus perturbing frequency, a typical
dispersion curve is obtained [15] whereas the 90° out-of-phase component of the
response oscillation gives the absorption spectrum of the damped BZ reaction. The
sum of the squares of dispersion and absorption curve is equal to the power
spectrum which does not contain any phase information. We have fitted Lorentz
functions to the experimental data and obtained good agreement,with the exception
of the data at the lowest perturbing frequencies,where a strong experimental
upturn of unknown origin is observed. In order to ascertain whether this deviation
is due to the organic part of the BZ reaction,or to the Oregonator model itself, we
performed a numerical Fourier transformation of the modified Oregonator model
using the rate constants of Showalter et al [16]. This model does not include

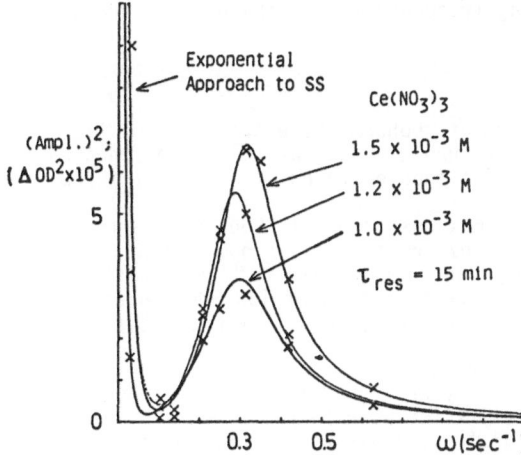

Fig.2 Power Spectra of the damped
BZ reaction. Concentrations are
given in the text; α= 0.5, 25.0°C

the complex organic chemistry of the BZ reaction in detail. The calculations did not show any upturn at low frequencies [15]. All 7 eigenvalues of the modified damped Oregonator have negative real parts, two eigenvalues being complex. Although the contribution of the negative real eigenvalues could lead to an upturn at $\omega \approx 0$ in principle, the eigenvectors of the model are too small to show such an effect. Therefore,we tentatively conclude that the experimental upturn is due to an exponential decay, which is known to transform into the positive part of a Lorentz curve,whose own frequency is equal to zero. This aspect of the BZ reaction will be further investigated. It is important to notice that the damped oscillation only occurs in a narrow experimental range of flow rates,which we found on the basis of our numerical model calculations. Below this range,the present system shows limit cycle behavior,and above this range exponential relaxations are observed.

II. The BZ Reaction as a Limit Cycle - A Fourier Analysis

In the following experiments the combined set-up of Fig. 1 (which is self explanatory) has been used. The BZ reaction behaves as a limit cycle at the present concentrations when KBr is absent from the in-flowing solution. The solutions in the two syringes contained 0.6 M malonic acid, 2×10^{-3} M Ce^{3+} and 0.175 M $KBrO_3$ in 0.75 M H_2SO_4, respectively. In order to perturb the oscillating reaction in the CSTR, the rate of in-flow was changed sinusoidally according to eq. 1 at various frequencies and amplitudes. We regard the Fourier analysis as a particularly useful method to investigate the response behavior of the BZ reaction. This method consists of the numerical calculation of a Fourier spectrum at a given experimental forcing frequency ω_p and amplitude α by recording the Ce^{4+} oscillations (Fig.3a, for example) and numerically Fourier transforming them into the frequency domain. The resulting Fourier spectrum (Fig.3b) gives information about the type of response the experimental system displays, whether it shows entrainment, quasiperiodic behavior or transition behavior, etc. A phase diagram (α vs. ω_p) may be constructed (Fig.4) which displays the various regions and their borders belonging to a given type of response. Each Fourier spectrum may also be decomposed into its individual Fourier components (Fig.5) according to [3]:

$$Ce^{4+}(t) = \sum_{1=1}^{1=\infty} Ce^{4+}_{1/k} \exp\left(i\, \frac{1}{k}\, \omega_p\, t\right) \tag{2}$$

where the $Ce^{4+}_{1/k}$ are the Fourier components of order $1/k$ registered at the frequencies $\frac{1}{k} \cdot \omega_p$, where $\omega_p = 2\pi/T$, T being the corresponding period. A given Fourier component of order $1/k$ is determined from a number of experimental Fourier spectra by reading off the amplitude at the precise value of $\frac{1}{k} \cdot \omega_p$ in an individual Fourier spectrum and then plotting the amplitudes as a function of ω_p. The general relation between the order $1/k$, the forcing frequency ω_p and the natural frequency ω_0 of the oscillator is:

$$1/k \times \omega_p \simeq \omega_0 \tag{3}$$

For example, to measure the Fourier spectrum of subharmonic entrainment of 1/2 order, the experimental perturbing frequency should be set at twice the fundamental frequency; the largest frequency band will then be found at about ω_0 (Fig.3b) in the Fourier spectrum.
  We performed a number of Fourier transformations to obtain Fourier spectra using ten oscillations for each numerical transformation. This number was found to be sufficient to obtain Fourier spectra of high quality. The Ce^{4+} oscillations are not sinusoidal; they ocasionally show some fine structure of low amplitudes at higher frequencies that are presently not of interest,since they are outside our range of perturbation frequencies by a factor of about 5.
So far a number of different responses were found by Fourier analysis:
Entrainment. The experimental forcing frequency was found to synchronize or entrain the natural frequency if eq.3 was fullfilled. One main frequency band appears in the Fourier spectrum,with integral overtones at high α. Initial experi-

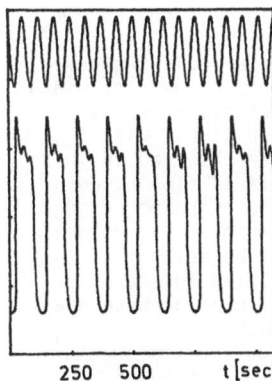

Fig.3a Upper trace: Sinusoidal perturbation of period 64.0 sec at $\alpha = 0.3$. Lower trace: Ce^{4+} oscillations at 350 nm as function of time, subharmonic entrainment of order 1/2.

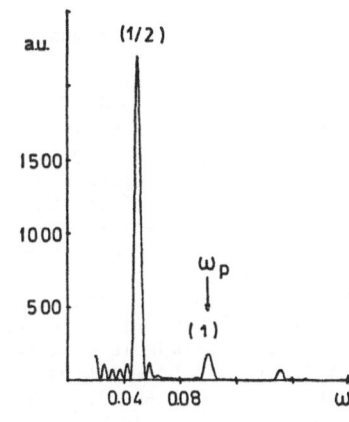

Fig.3b Fourier spectrum showing subharmonic entrainment of order 1/2; $\omega_p = 0.0982$ s^{-1}, $\alpha = 0.3$. Main component at $\sim \omega_0$.

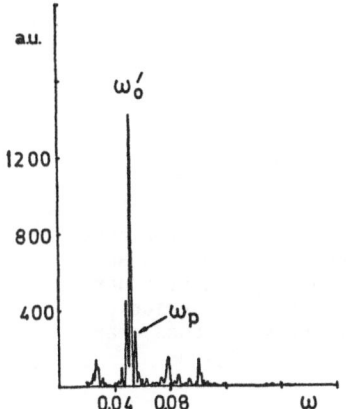

Fig.3c Fourier spectrum in the quasi-periodic region, $\omega_p = 0.0553$, $\alpha = 0.1$, ω_p and $\omega_0'(= 0.0520)$ are observed.

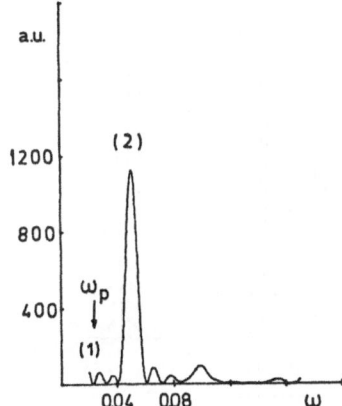

Fig.3d Fourier spectrum showing superharmonic entrainment (order2), $\omega_p = 0.0250$, $\alpha = 0.2$. Main component at $\sim \omega_0$.

ments revealed the following harmonic entrainments: first order at all values of α (= 0.1-0.5) (fundamental entrainment, $l = 1$, $k = 1$, $\omega_p \simeq \omega_0$); second order at low α (= 0.1-0.2, $l = 2$, $k = 1$, $\omega_p \simeq \omega_0/2$); 1/2 order (Fig.3a and 3b) at all α (subharmonic entrainment, $l = 1$, $k = 2$, $\omega_p \simeq 2\omega_0$); 1/3 order (subharmonic entrainment, $l = 1$, $k = 3$, $\omega_p \simeq 3\omega_0$) and 2/3 or 3/5 order (ultra-subharmonic entrainment, the two cases are difficult to distinguish). Entrainment was established after only few forcing oscillations. Phase-locking occurred, i.e. the phase-angle between perturbing function and forced response was constant at entrainment. We found the experimental entrainment bands to be very narrow. Their width increased with increasing α; at $\alpha = 0.5$ the entrainment width is approximately $\pm$ 2-3 % of ω_p. The period of fundamental entrainment was 128.0 sec at 25.0 + 0.05 º$\overline{C}$. Our experimental precision was $\pm$ 0.2 % with regard to the forcing period.

Quasi-Periodic Region. This region occurs at a distance of about $\pm$ 10 % in frequency from the entrainment center in the BZ reaction at small values of α. The corresponding Fourier spectra consist of at least two frequency bands, one corresponding to ω_p and the other to a somewhat modified fundamental frequency ω_0' (Fig.3c).

Transition Region between Fundamental Entrainment and Quasi-Periodic Region. This transition region was found to be wider than the entrainment bands. Its width increased with increasing α. It was difficult to determine precisely where it starts and ends, due to the fact that the Fourier spectra changed slowly with time over several hours, i.e. slowly fluctuating Fourier spectra were observed in this region. This situation is reminiscent of the phenomenon of critical slowing down which has been predicted to be significant and disturbing at the entrainment edges for other limit cycles [7]. Weaker harmonic combination bands were observed here. The frequency bands in the true transition region were found to be neither at ω_p nor ω_0 but at intermediate frequencies. This region will be studied further.

A summary of the experimental results so far obtained is given as a phase diagram in which the perturbing amplitude is plotted versus ω_p (Fig.4). Most of phase space remains yet to be explored. The measured Fourier components $Ce_{1/k}^{4+}$ (eq.2) of the orders 1, 2, 1/2 and 1/3 are plotted versus ω_p in Fig.5.

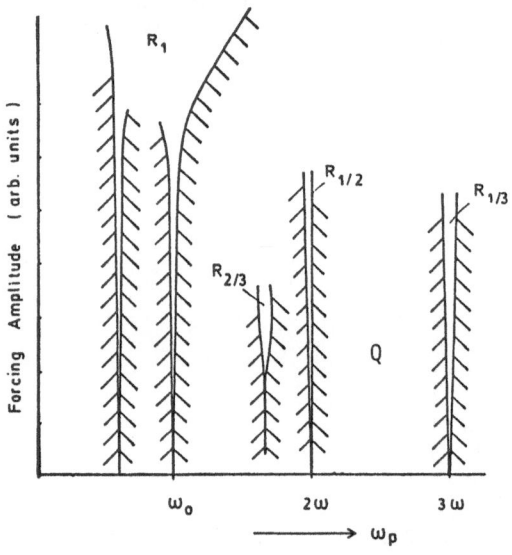

Fig. 4 Tentative phase diagram: α (arb.units) vs ω_p. R_i are entrainment regions of order i; Q stands for quasi-periodic behavior. Cross-hatched areas are transition regions.

Amplitudes of Fourier Components

Amplitude at:

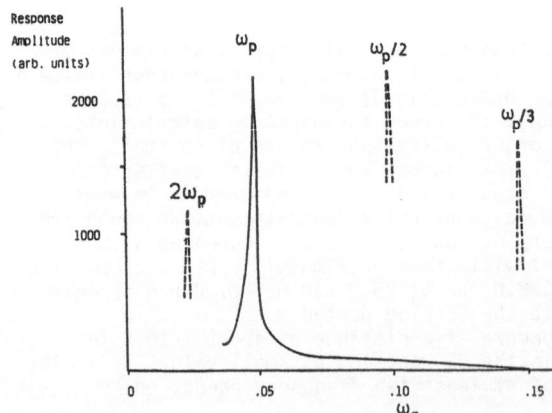

Fig.5 Various Fourier Components as shown by experiment so far; obtained from individual Fourier spectra by reading off their amplitudes at $\frac{1}{k} \cdot \omega_p$ and plotting them as a function of ω_p.

Entrainment is optimal at the resonance maximum of a given Fourier component. Generally speaking, the extreme sensitivity of the experimental system towards very small changes (≈ 0.5 %) of the perturbation period is noteworthy, as exemplified by the narrow resonances in the entrainment bands.

Conclusions

So far a number of phenomena have been shown to occur in the periodically perturbed BZ reaction. For the damped oscillating BZ reaction, a single broad chemical resonance curve was measured. For the limit cycle BZ oscillation, narrow chemical resonances in narrow entrainment bands of orders 1, 2, 1/2, 1/3 and 2/3 (or 3/5) were obtained. Quasi-periodic behavior was observed, and slowly fluctuating Fourier spectra occured in the transition range between fundamental entrainment and quasi-periodic behavior. More extensive work is planned to reveal further phenomena, possibly including chaotic frequency responses.

Acknowledgement

We thank the Fonds der Chemischen Industrie for support of this work.

References

1. K. Tomita, T. Kai and F. Hikami, Prog.Theor.Phys. 57, 1159 (1977)
2. A. Ito, Prog.Theor.Phys. 61, 45 (1979)
3. T. Kai and K. Tomita, Prog.Theor.Phys. 61, 54 (1979)
4. P.H. Richter and J. Ross, J.Chem.Phys. 69, 5521 (1978)
5. P.H. Richter, P. Rehmus and J. Ross, Prog.Theor.Phys. 66, 385 (1981)
6. Y. Termonia and J. Ross, P.N.A.S., USA, 79, 2878 (1982)
7. P. Rehmus and J. Ross, J.Chem.Phys. 78, 3747 (1983)
8. P. Rehmus, W. Vance and J. Ross, J.Chem.Phys. 80, 3373 (1984)
9. P.H. Richter, Physica 10D, 353 (1984)
10. T.W. Taylor and W. Geiseler, This Volume
11. A. Boiteux, A. Goldbeter and B. Hess, P.N.A.S. USA, 72, 3829 (1975)
12. M. Markus and B. Hess, P.N.A.S. USA, 81, 4394 (1984)
13. E. Dulos and P. DeKepper, Biophys.Chem. 18, 211 (1983)
14. M. Dolnik, I. Schreiber and M. Marek, Phys.Lett. 100A, 316 (1984)
15. F. Buchholtz and F.W. Schneider, J.Am.Chem.Soc., 105, 7450 (1983)
16. K. Showalter, R.M. Noyes, K. Bar-Eli, J.Chem.Phys. 69, 2514 (1978)

Periodic Perturbation of Limit Cycles in an Isothermal CSTR

T.W. Taylor[1] and W. Geiseler

Institut für Technische Chemie, Technische Universität Berlin
D-1000 Berlin 12, Fed. Rep. of Germany

1. Introduction

Oscillating chemical reactions, subjected to external periodic perturbations, can give rise to interesting non-equilibrium phenomena such as frequency entrainment, resonance effects, phase-locking, and chaotic motion [1,2]. Previously investigated chemical systems, besides simple artificial models, were mostly extremely complex and even the unperturbed ones were not completely understood. The investigation of a less complicated chemical oscillator, which is well understood by a detailed mechanism, can therefore contribute to a better understanding of such phenomena.

A chemical system belonging to this category of oscillators is the stirred flow oscillating reaction of bromate, bromide, and cerous ions in sulfuric acid solution [3]. This reaction, referred to as a minimal oscillator [4,5] and almost quantitatively described by the NFT mechanism [6], was investigated computationally under conditions where limit-cycle oscillations appeared. The oscillations occurring under certain constant operating conditions were subjected to sinusoidal variations of either the total input rate or one of the reactant inflow concentrations. Both the perturbation amplitude and frequency were varied over wide ranges. Emphasis was put on the determination of frequency bands where peculiar non-equilibrium phenomena were predicted.

2. Mechanism and Model

The kinetics of the minimal oscillator are adequately described by the NFT mechanism, which is made up of seven reversible reaction steps. A detailed description is given elsewhere [3]. The model for the flow reaction is an ideal one, a well mixed, isothermal continuously stirred tank reactor (CSTR). Only mass balances of the system, then needed to be considered. Assuming mass action kinetics for the nine chemical species included in the mechanism, the dynamics of the reactor can be described by the overall mass balances of species i (for details see Ref. [3])

$$dC_i/dt = R_i(C) + k_0(C_{i0} - C_i). \tag{1}$$

In order to model the response of the limit cycle oscillations to periodic perturbations of the operating conditions (input rate k_0 and inflow concentrations C_{i0}), either the sinusoidal function

$$k_0 = \bar{k}_0 (1 + \alpha \sin \omega t) \quad \text{or} \quad C_{i0} = \bar{C}_{i0} (1 + \alpha \sin \omega t) \tag{2a,b}$$

was incorporated to Eq. (1). $\bar{k}_0$ and $\bar{C}_{i0}$ are the averages of the input rate and the inflow concentrations, respectively, α is the fractional amplitude of the perturbation oscillation, and ω is the perturbation frequency. As the dynamics of the experimental system can easily be followed by measuring the ceric and bromide ion concentration [3], the computational results are given in terms of these two species.

---

[1]Present address: 3M Company, CRPTL, St. Paul, Minnesota 55144, U.S.A.

3. Computational Results

The computational results reported here are representative ones, and were obtained when the limit cycle oscillations occurring under the following constant operating conditions

$$[BrO_3^-]_0 = 0.150 \text{ M}, \qquad [Br^-]_0 = 4.1 \times 10^{-4} \text{ M}, \qquad [Ce^{3+}]_0 = 3 \times 10^{-4} \text{ M}$$

$$[H^+]_0 = 0.75 \text{ M}, \qquad k_0 = 5 \times 10^{-3} \text{ s}^{-1}, \qquad T_c = 25 \text{ }^\circ C$$

were sinusoidally perturbed. The frequency and period of the natural (unperturbed) oscillations were $\omega_0 = 5.48 \times 10^{-3}$ s^{-1} and $T_0 = 1.146 \times 10^3$ s, respectively.

When either the input-rate or one of the reactant inflow concentrations was sinusoidally varied according to Eq. (2a) or (2b), several interesting non-equilibrium phenomena were predicted. Depending on the perturbation amplitude and frequency applied, the CSTR gave rise to either (1) Harmonic entrainment in a range of frequency close to the natural frequency of the system, (2) Subharmonic entrainment of order 1/n (n being an integer) at increased frequencies, (3) Regular transient behavior, (4) Chaotic oscillations, and (5) Low amplitude harmonics. Typical results are shown in Fig. 1 where the CSTR was subjected to periodic input rates. Both the forcing oscillation and the response of the reactor in terms of the instantaneous concentration of ceric ions are depicted. Generally, the new dynamic pattern was established after a short transient time following the switch from constant to periodic operation of the reactor; however, in certain special cases the transient time increased considerably. All the predicted phenomena are not restricted to certain fixed forcing frequencies but exist within more or less wide bands of frequencies. In Fig. 2 typical frequency bands of distinct dynamical behavior are shown for the CSTR where the inflow concentration of bromide was sinusoidally varied. In the frequency range investigated ($0.002 < \omega < 0.2$ s^{-1}) a large number of entrainment bands was identified. Moreover, inside certain gaps between neighbouring entrainment bands, rather narrow

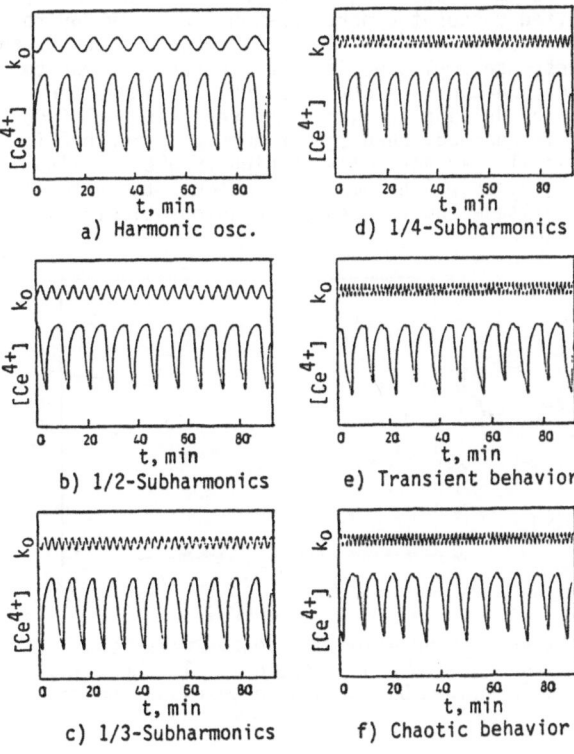

a) Harmonic osc.

d) 1/4-Subharmonics

b) 1/2-Subharmonics

e) Transient behavior

c) 1/3-Subharmonics

f) Chaotic behavior

Fig. 1 Non-equilibrium phenomena arising in an oscillating CSTR due to sinusoidal input rate: $k_0 = \bar{k}_0 (1 + \alpha \sin \omega t)$ with $\bar{k}_0 = 0.005$ s^{-1}, $\alpha = 0.05$, and distinct values of ω (a − 0.012, b − 0.025, c − 0.04, d − 0.055, e − 0.07, and f − 0.069 s^{-1})

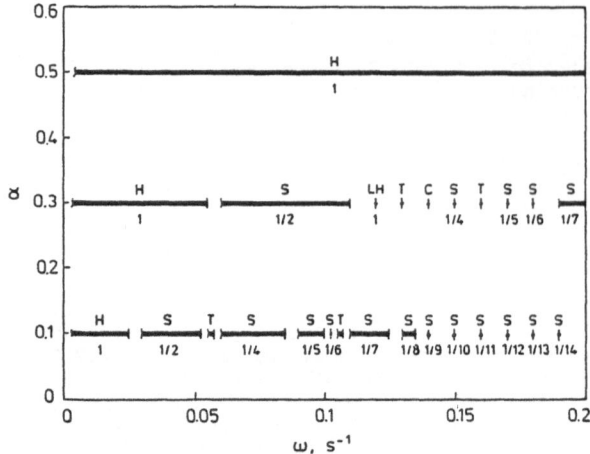

Fig. 2 Frequency bands of distinct non-equilibrium phenomena: H harmonic and S subharmonic entrainment, T regular transient behavior, C chaotic oscillation, and LH low amplitude harmonic entrainment. Periodic perturbation according to $C_{Br^-,0} = \overline{C}_{Br^-,0} (1+\alpha \sin \omega t)$ with $\overline{C}_{Br^-,0} = 0.00041$ M and $\alpha = 0.1, 0.3,$ and 0.5

bands of regular transient behavior, chaotic oscillations, and low amplitude harmonics are predicted. We remark here that the extent of the frequency bands shown are costly to determine accurately, because of the computing time required before transient effects decay and the system settles into a new dynamic pattern.

Finally, the amplitude of system oscillations is considered as a function of the driving frequency. In Fig. 3 the reduced amplitude A/A_0 of ceric ions (ratio of perturbed over natural amplitude) is plotted versus the perturbation frequency for the CSTR where the bromide inflow concentration was varied at two different values of the perturbation amplitude. The largest response amplitude is always observed close to or slightly below the natural frequency, and can exceed the natural amplitude more than 50 percent. With the increase of the driving frequency the response amplitude generally drops until a minimum value is reached. When the frequency is further increased, the amplitude rises again. Finally, at sufficiently high frequencies ($\omega \sim 10$ s^{-1}) a response oscillation is obtained, which completely matches the natural oscillations, i.e. both the natural amplitude and frequency are asymptotically ap-

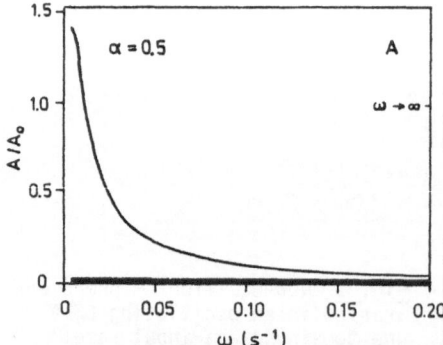
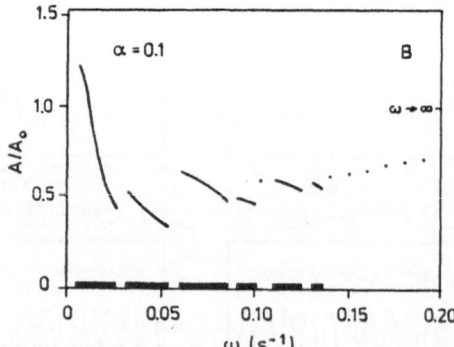

Fig. 3 Reduced response amplitude A/A_0 vs. perturbation frequency ω for two different values of the perturbation amplitude. (A) $\alpha = 0.5$, (B) $\alpha = 0.1$. Operating conditions and periodic perturbation as in Fig. 2. Dark lines on abscissa indicate the extent of frequency bands

proached. Obviously, the reaction system cannot follow the perturbation any longer and "feels" only the average value of the forcing oscillation.

The results of Fig. 3 imply that when a system exhibits limit cycle oscillations, application of a sinusoidally varying input of the proper amplitude and frequency can significantly increase or decrease the system oscillations. This could, for instance, be useful in stabilizing systems at regions where undesired limit cycle oscillations occur.

The excellent agreement between experimental observations and model predictions found in previous studies [3,4] provides confidence that the dynamic phenomena reported here can be verified experimentally in a future work. The authors feel encouraged to such work by the preliminary results of Lachmann [7] who conducted experiments under conditions similar to the ones used here. A good qualitative and semi-quantitative agreement was reported.

Acknowledgement

The authors are indepted to the National Science Foundation (USA) for a NATO Post-doctoral Fellowship (to T.W.T.) to support this research.

References

1. A. Boiteux, A. Goldbeter, B. Hess, Proc. Natl. Acad. Sci.(USA) 72, 3829 (1975)
2. P. Rehmus, J. Ross, J. Chem. Phys. 78, 3747 (1983)
3. K. Bar-Eli, W. Geiseler, J. Phys. Chem. 87, 3769 (1983)
4. W. Geiseler, Ber. Bunsenges. Phys. Chem. 86, 721 (1982)
5. M. Orban, P. DeKepper, I.R. Epstein, J. Am. Chem. Soc. 104, 2657 (1982)
6. R.M. Noyes, R.J. Field, R.C. Thompson, J. Am. Chem. Soc. 93, 7315 (1971)
7. H. Lachmann, University of Würzburg, personal communication

Oscillations in a CSTR Affected by the Pump Periodicity

K. Bar-Eli

Department of Chemistry, Tel-Aviv University, 69978 Tel-Aviv, Israel

Most experiments on chemical oscillations are executed in a CSTR, i.e. continuously stirred tank reactor. The reactants are usually fed into the reactor by means of a peristaltic pump. The pump does not work in a continuous manner, but, as its name implies, pushes the material in a wave-like manner. Thus one must assume that the flow into the reactor is not continuous but involves some time periodicity, which depends on the pump. The usual differential equations used in such problems are:

$$\dot{c} = R(c) + k_o(c_o - c)$$

where c is the vector of concentrations, $R(c)$ is the vector of chemical rate terms, c_o is the vector of feed concentrations and k_o is the flow rate - taken usually to be a constant value.

In this work we assumed that k_o depends on time as:

$$k_o(t) = \overline{k}_o(\sin \frac{2\pi t}{T} + 1)$$

$\overline{k}_o$ is the average flow and T is the pump period. The flow thus oscillates between $2\overline{k}_o$ and zero and thus approximates the real operation of the pump. In principle, other periodic functions could have been taken, but the results obtained here suffice to demonstrate the involved problems.

We have calculated the oscillations produced by the Field Körös Noyes [1] mechanism of the Belousov-Zhabotinskii reaction [2] for the constraints $[BrO_3^-]_o = [MA]_o$ = 0.03 M, $[Ce^{3+}]_o = 3 \times 10^{-3}$ M, $[H^+]_o = 1.5$ M and $\overline{k}_o = 1.5 \times 10^{-3}$ sec^{-1} for a series of pump periods T as shown in Fig. 1.

At $T = \infty$, Fig. 1a, i.e. the pump works continuously, the oscillations have a period $\tau = 708$ sec with two little bumps in the middle. This pattern recurs only at T shorter than 1 sec - Fig. 1l.

In between these values of T one observes dramatic changes both in period and in the general pattern of the oscillations. Thus at large T's a burst of oscillations is followed by a quiescent period. At $T \sim 40$ sec, a short period of about 380 sec is obtained for the oscillations. A long period composed of two shorter cycles is also seen in a few cases.

The results shown here have important practical implications:
a) Results of experiments conducted under the same constraints may be different if different pumps are used.
b) If the same pump is used and the same flow is achieved by using different tubes with different pump speeds, one may obtain different oscillations.
c) Small changes in pump speeds due to mechanical irregularities or to tubes fatigue may result in large changes in period and pattern of oscillations, and may even be recorded as chaos.
d) Different oscillation patterns may be obtained whether the materials are introduced directly into the liquid, or are dropped from above; the period in the first case depends on the construction of the pump, while in the second case it depends on the drop size, tube diameter, surface tension etc. These periods may be different, thus producing different oscillations.

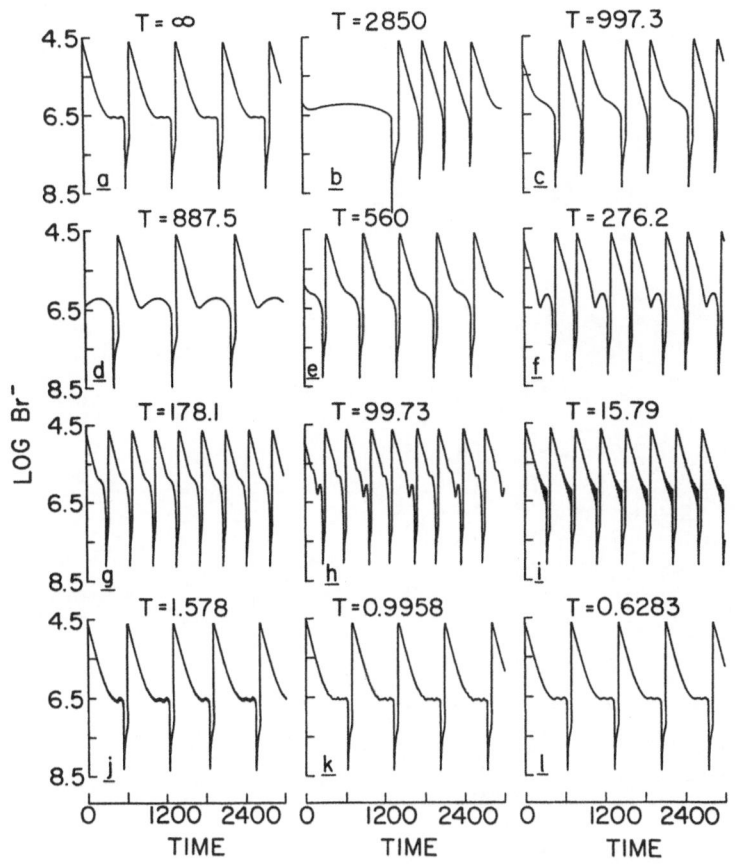

Fig. 1 Calculated oscillations for a series of pump periods T

e) The described effects are most pronounced near the limits of the oscillation region.

References

1 R.J. Field, E. Körös and R.M. Noyes: J. Am. Chem. Soc. 24, 8649 (1972)
2 a) B.P. Belousov Sb. Ref. Radiats. Med. 1958 Medgiz, Moscow 145 (1959)
  b) A.M. Zhabotinskii Dokld. Akad. Nauk. SSSR 157, 392 (1969)
  c) A.N. Zaikin and A.M. Zhabotinskii, Nature (London) 225, 535 (1970)

Part II

Dynamics of Biological Systems

Synchronization, Phase-Locking and Other Phenomena in Coupled Cells

Hans G. Othmer

Department of Mathematics, University of Utah, Salt Lake City, UT 84112, USA

1. Introduction

Knowledge of how coupling affects the collective behavior of aggregates of more or less identical cells or dynamical systems is important for understanding both normal and aberrant processes in numerous physical and biological systems. These include pattern-formation in developmental biology, the spatial and temporal coordination of cellular activity in the heart and the nervous system, spatial and temporal pattern-formation in chemically-reacting systems, and many more. In some instances, the aberrant behavior of the system is directly attributable to alterations in intercellular communication. For example, epileptic seizures consist of uncontrolled synchronized discharges in the central nervous system, and epileptogenic drugs that cause the rapid depolarization of neurons characteristic of the onset of a seizure act on chemical [1,2] or electrical synapses [3], thereby altering the type or intensity of communication between neurons. At a more accessible level, coupled electrical circuits can be used as model systems to demonstrate the dramatic effect that changes in coupling can have [4].

At the present state of knowledge of dynamical systems, one cannot predict a priori how different modes of coupling affect the dynamical behavior of a coupled network, except at the extremes in the strength of coupling. As we shall see in a later section, analytical results can be obtained at sufficiently strong coupling or sufficiently weak coupling, but some of the most interesting behavior lies between these extremes. Thus it is necessary to analyze specific systems, using a combination of analytical and numerical techniques. In [5] the dynamics of a pair of oscillators that are coupled by a linear resistive or diffusive coupling are studied over the entire range of coupling strength, and a number of new phenomena that arise at intermediate coupling strengths are described. In the continuum limit, diffusively-coupled systems of this type can be described by reaction-diffusion equations, and it is well known that such equations can lead to a variety of spatio-temporal patterns.

In [5] and in most other studies of coupled cells [see, for example, [6] and references therein), the cells are directly-coupled, in the sense that there is no storage capacity in the coupling mechanism. However, there are a number of biologically-important examples in which cells communicate indirectly via a passive medium that introduces a capacitance, and hence a distributed time-delay, into the coupling. Examples include suspensions of the cellular slime mold Dictyostelium discoideum [7], the aggregation phase in the same system [8], electrotonic coupling of neurons in the hippocampus [9], and suspensions of yeast cells [10]. The last example provided the motivation for some previous work on the dynamical significance of indirect coupling [11,12]. It will be seen in the fourth section that such indirect or capacitive coupling can be used to model patches of inexcitable medium in an otherwise excitable medium, and this viewpoint may prove helpful in understanding the reflection mechanism of cardiac arrhythmias [3].

In the following section we present a general formulation for describing coupled systems of cells or other autonomous subunits, and derive a criterion that guarantees synchronization of the network. In Section 3 we briefly review the results for weak coupling of directly-coupled cells, and describe some new results for the weak-coupling limit of indirectly-coupled systems. In the last section we present some numerical results on several model systems.

2. Synchronization in Cellular Networks

Consider a system of N autonomous subunits or cells, and suppose that in the absence of coupling the state $x^{\alpha}(t) = (x_1^{\alpha}(t),\ldots, x_n^{\alpha}(t))$, $\alpha = 1, \ldots, N$, evolves according to the equation

$$\frac{dx^{\alpha}}{dt} = R^{\alpha}(x^{\alpha}). \tag{1}$$

When coupling between the cells is introduced (1) must be modified to account for the effect of other cells. One can associate a directed graph G with the entire aggregate by associating a vertex of G with each cell, and introducing a directed edge (α,β) from vertex α to vertex β if there is a non-vanishing flux from cell α to cell β for some non-zero x_i^{α}. In general, the relation between cells defined by the coupling between them is not symmetric, because we do not require that there be a flux between α and β in both directions. We suppose that there are r edges in G and denote the flux of x_i on the αth edge by J_i^{α}, $\alpha = 1$, $\ldots$, r, where J_i^{α} is a non-negative smooth function. Without loss of generality, we may assume that there is only one directed edge from α to β, for if there are more the fluxes on them can be combined, and all but one edge can be deleted. The vector $(J_1^{\alpha}, \ldots, J_n^{\alpha})$ defines a flow on the graph G in the sense used in [14].

The evolution of the entire network is governed by the equations

$$\frac{dx^{\alpha}}{dt} = \sum_{\beta=1}^{r} E_{\alpha\beta} J^{\beta} + R^{\alpha}(x^{\alpha}) \qquad \alpha = 1, \ldots, N. \tag{2}$$

The quantities $E_{\alpha\beta}$, which are the entries of the incidence matrix of G, are defined as follows [14].

$$E_{\alpha\beta} = \begin{cases} +1 & \text{if edge } \beta \text{ is incident at vertex } \alpha \text{ and directed toward it} \\ -1 & \text{if edge } \beta \text{ is incident at vertex } \alpha \text{ and directed away from it} \\ 0 & \text{otherwise} \end{cases}$$

The state of the entire network is given by $x(t) = (x^1(t),\ldots, x^N(t))$, and the network is said to be __synchronized__ or in a synchronous state if $x(t) = (\phi(t),\ldots,\phi(t))$, where $\phi = (\phi_1(t),, \ldots\phi_n(t))$ and $\phi_i(t) = $ constant is allowed. We admit the possibility that cell α is a pacemaker, which means that (1) has an asymptotically stable periodic solution for label α. Since we can also stipulate that it receives no inputs from the remainder of the network, the foregoing formulation applies to periodically-forced networks, and in that context a synchronous state is one in which all cells track the pacemaker with no phase lag. In the remainder of this section we only consider networks of identical cells. To allow for the possibility of phase-differences between cells we say that a network is __entrained__ or __phase-locked__ in a periodic state if $x(t + T) = x(t)$ for some positive T and all $t \geqslant 0$. This definition covers cases in which $x(t) = (\phi(t-T_1), \ldots, \phi(t-T_n))$ as well as cases in which each $x_i(t)$ is different. Conditions under which entrainment occurs for a certain class of forced networks are given in [14].

Equation (2) can be written more compactly as

$$\frac{dx}{dt} = (E \otimes I_n)J(x) + R(x) \tag{3}$$

where ' $\otimes$ ' denotes the tensor product [15], I_n is the n x n identity matrix, and $R(x) = (R(x^1), \ldots, R(x^N))$. The flow J is said to be __balanced__ at x if $(E \otimes I_n)J(x) = 0$, __i.e.__, if $\Sigma\, E_{\alpha\beta}J^\beta(x) = 0$ for $\alpha = 1, \ldots, N$, which simply means that the net flux into any cell vanishes at x. Balanced flows are the network analogues of solenoidal flows in fluid mechanics, and like solenoidal flows, they have simple representations. It is shown in [14] that any nonnegative balanced flow can be represented as a linear combination of the directed cycles in G.

For any state x, define the average cellular state as $\overline{x} = \Sigma\, x^\alpha/N$; then the network is synchronized if and only if $x^\alpha = \overline{x}$ for all α. In a synchronized network, the state is $\xi = 1_N \otimes \overline{x}$, where 1_N is an N-component vector of ones. Therefore, if the difference $z = x - \xi \to 0$ as $t \to \infty$ for some $x(0)$, the network synchronizes for those initial conditions. If the flow is balanced at ξ then

$$\frac{dz}{dt} = (E \otimes I_n)(J(z+\xi) - J(\xi)) + R(z+\xi) - 1_N \otimes \frac{1}{N} \sum_\alpha R(z^\alpha + \overline{x}),$$

and therefore

$$\frac{1}{2}\frac{d}{dt}\|z\|_2^2 = -\langle z, -(E \otimes I_n)(J(z+\xi) - J(\xi))\rangle + \langle z, R(z+\xi) - 1_N \otimes \frac{1}{N} \sum_\alpha R(z^\alpha + \overline{x})\rangle$$

The first term can be bounded above by $-\delta \|z\|_2^2$, where $\delta \equiv \inf \|-\partial(E \otimes I_n)(J(z+\xi) - J(\xi))/\partial z\|_2$, and the infimum is taken over all z with zero mean and all synchronized states ξ . Similarly, the second term can be bounded above by $\kappa \|z\|_2^2$, where $\kappa \equiv 2 \sup\|\partial R/\partial x\|$. Therefore, if the flux relations are <u>uniformly monotone</u> or <u>passive</u>, in the sense that $\delta > 0$, then $\|z\|_2^2 \to 0$ exponentially in t whenever $\delta > \kappa$, i.e., the network synchronizes. Since δ and κ are effectively first-order rate constants, their reciprocals define relaxation times. Thus the foregoing can be summarized as follows: If the flow on G is balanced whenever the network is synchronized, and if the flux relations are passive, then the network relaxes to a synchronized state provided that the relaxation time δ^{-1} for transport is shorter than the kinetic relaxation time κ^{-1} . This result generalizes a criterion known for reaction-diffusion equations [16,17] to networks with more general flux relations, and a comparison of the present approach with that in [16] shows that the present result can be generalized to nonlinear transport relations in distributed systems as well.

As an example, suppose that G is symmetric, in the sense that whenever the edge $(\alpha,\beta) \in G$, then $(\beta,\alpha) \in G$. Further, suppose that the flux relations are linear with constant coefficients, and that the flux on (α,β) depends only on x^α . Then $J = Dx^\alpha$ where D is a matrix of 'diffusion' coefficients, and one finds that (3) can be written

$$\frac{dx}{dt} = (\Delta \otimes D)x + R(x).$$

Here Δ is the discrete Laplacian operator for G . Since we have assumed that there are no sources or sinks in the network, $\Delta 1_{2r} = 0$, and the flow is balanced at any synchronized state.

3. Weakly-Coupled Systems

The case in which (1) has an unstable rest point and a unique globally attracting periodic solution for every α is important in a number of applications. In this case, we say that all cells are oscillatory. When they are also identical and the flux relations satisfy the conditions given previously, the network will synchronize in a stable periodic state for almost all initial conditions. Furthermore, when the cells are not too different and the coupling is sufficiently strong, the network will relax to a phase-locked state. At the opposite extreme in which the coupling is weak, one can also predict the asymptotic states of the coupled system, but in this case it is possible that dynamic behavior that is not present in the individual cells will arise in the coupled system.

To provide a measure of the strength of coupling, we introduce a parameter δ and write (3) as follows.

$$\frac{dx}{dt} = \delta(E \otimes I_n) J(x) + R(x) \tag{4}$$

If each subsystem governed by (1) has a stable, globally attracting periodic solution ϕ^α, then for $\delta = 0$ (4) has an asymptotically stable invariant torus T_0^N that is generated by the product of the ϕ^α's. One can show that this torus persists for sufficiently small δ and remains attracting [5], and therefore the asymptotic behavior of the weakly-coupled system is determined by the flow in a neighborhood of the perturbed torus T_δ^N.

In a neighborhood of T_δ^N we can write (4) as

$$\frac{dr}{dt} = A(\theta)r + R(r,\theta) + \delta F(r,\theta,\delta)$$

$$\frac{d\theta}{dt} = \Omega(\delta) + \delta G(r,\theta,\delta) \tag{5}$$

where $r \in R^{(n-1)N}$ and $\theta = (\theta_1, \ldots, \theta_N)$. In these coordinates T_0^N corresponds to $r = 0$, and by hypothesis the Floquet multipliers of $A(\theta)$ are all less than one in magnitude. The vector Ω is called the frequency or rotation vector, and Ω is said to be rational when $\Omega = \lambda(n_1, \ldots, n_N)$ for some $\lambda \epsilon R$ and integers n_1. Ω is non-resonant if there are positive constants c and p such that

$$\sum_{i=1}^{N} k_i \Omega_i > \frac{c}{|k|^p}$$

for all $k = (k_1, \cdots, k_N) \epsilon Z^N$ such that $|k| \neq 0$. When $\Omega(0)$ is rational the uncoupled flow

$$\frac{dr}{dt} = A(\theta)r + R(r,\theta)$$

$$\frac{d\theta}{dt} = \Omega(0)$$

has a periodic solution on T_0^N, and weak coupling will lead to phase-locking in a region of $\Omega-\delta$ space. On the other hand, when $\Omega(0)$ is non-resonant the flow on T_δ^N is equivalent to an irrational flow for sufficiently small δ [15], and all solutions of the coupled system will be quasi-periodic or chaotic.

The typical structure of the phase-locking regions for a pair of weakly-coupled oscillatory cells is shown in Fig. 1. In the hatched regions labeled m:n the system is phase-locked in an m:n periodic solution, which means that in one period the phase of one cell changes by $2m\pi$ while in the other it changes by $2n\pi$. A region of m:n phase-locking emanates from each rational number m/n at $\delta=0$, but generally only those for m and n small integers are important in applications. It should be noted from the figure that an m:n periodic solution may not persist as δ increases at fixed Ω_1/Ω_2; it may be necessary to vary both δ and Ω_1/Ω_2 to remain within an m:n region. Furthermore, the regions may overlap, and other more complicated phenomena can occur for large δ. Examples of such phenomena for identical coupled oscillators are given in [5].

To determine the structure of the phase-locking regions more precisely for small δ one has to know the vectorfield in (1) and the coupling mechanism. The two-dimensional axially symmetric vectorfield

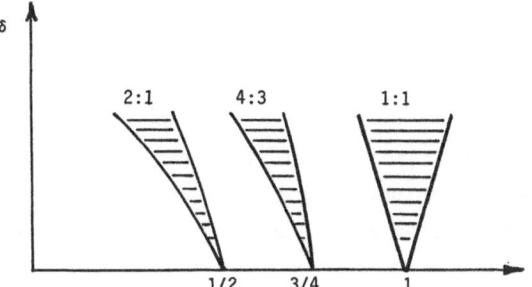

Fig. 1 The phase-locking regions as a function of the frequency ratio and the coupling strength.

$$R_1(x,y) = \alpha x + \Omega y - x(x^2 + y^2)$$

$$R_2(x,y) = -\Omega x + \alpha y - y(x^2 + y^2) \tag{6}$$

is the easiest to treat analytically. Under linear diffusive coupling one obtains the following system for the polar form of the governing equations [5]:

$$\frac{dr_1}{dt} = \alpha r_1 - r_1^3 + \text{Re}\{\delta_1(r_2 e^{i(\theta_2-\theta_1)}-r_1) + \delta_2(r_2 e^{-i(\theta_2+\theta_1)} - r_1 e^{-2i\theta_1})\}$$

$$\frac{d\theta_1}{dt} = -\Omega_1 + \frac{1}{r_1}\text{Im}\{\delta_1(r_2 e^{i(\theta_2-\theta_1)}-r_1) + \delta_2(r_2 e^{-i(\theta_2+\theta_1)} - r_1 e^{-2i\theta_1})\} \tag{7}$$

plus a similar pair for (r_2,θ_2). Here δ_1 and δ_2 are complex constants related to the diffusion coefficients and Re and Im denote the real and imaginary part of a complex number. These equations can be cast into a form that is suitable for averaging [19], and one finds that the averaged equation for nearly identical oscillators is

$$\frac{d\,\Theta}{d\theta_2} \equiv \frac{d(\theta_1-\theta_2)}{d\theta_2} = \frac{\Omega_1-\Omega_2}{\Omega_2} + \frac{\delta(D_{11}+D_{22})}{\Omega_2}\sin\Theta + \mathcal{O}(\delta^2)\,. \tag{8}$$

The constants D_{11} and D_{22} are the diagonal elements of the diffusion matrix, scaled so that the largest is $\mathcal{O}(1)$. If $\Omega_1-\Omega_2 \sim \mathcal{O}(\delta)$ and Ω_2 is regarded as fixed, then there is 1:1 phaselocking in the region bounded by the lines

$$\frac{\Omega_1}{\Omega_2} = 1 \pm \delta\,\frac{(D_{11}+D_{22})}{\Omega_2}\,.$$

If $\Omega_1/\Omega_2 = m/n$ at $\delta=0$ there is no phase-locking at $\mathcal{O}(\delta)$ and higher-order averaging is required. One finds that these higher resonances lead to phase-locking only when $n\Omega_1-m\Omega_2 \sim \mathcal{O}(\epsilon^2)$.

New problems arise when oscillatory cells communicate indirectly via a passive medium. These cannot be treated in detail here, and to simplify the discussion we shall only consider the network shown in Fig. 2, wherein 1 and 2 are oscillatory cells and 0 is a passive medium. The governing equations are

$$\frac{dx^1}{dt} = \delta P(x^0-x^1) + R(x^1) \qquad \frac{dx^2}{dt} = \delta P(x^0-x^2) + R(x^2) \qquad \frac{dx^0}{dt} = \epsilon\delta P(\bar{x}-x^0), \tag{9}$$

135

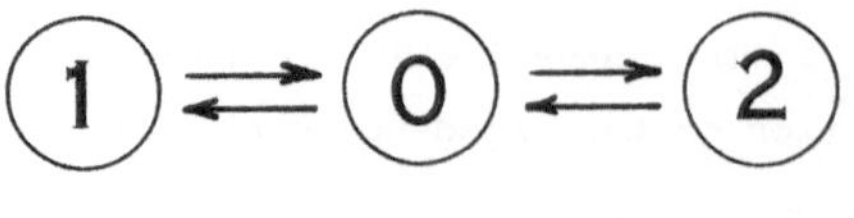

Fig. 2

where δ is a scalar parameter, P is an nxn matrix of permeability
coefficients, $\bar{x} \equiv (x^1 + x^2)/2$, and $\varepsilon = 2V_c/V_0$ is the ratio of the total volume of
active cells to the volume of the passive medium. As $\varepsilon \to \infty$, $x^0 \to \bar{x}$, except
perhaps for an initial transient. In this case, the governing equations reduce to
those for a directly-coupled system with linear diffusion coupling and a
diffusion matrix given by $D = P/2$. At the opposite extreme of $\varepsilon \to 0$ there is no
communication between the oscillatory cells, because the capacitance of the
passive medium is infinite. In this case, there is an additional constant input
to each cell for $x^0 \neq 0$, and an additional first-order decay term from the
transport into the passive medium.

When $\delta = 0$ and $\varepsilon < \infty$ the cells are uncoupled, but there is no invariant torus
as in the directly-coupled case. However this doesn't preclude the existence of
an invariant torus for $\delta > 0$; it only implies that it cannot be found by a
perturbation argument. Suppose that n=2 and that R is given by (6),
wherein Ω may be different for the two cells. If we write the cell variables in
polar coordinates and define ξ_j via $r_j = \sqrt{\alpha} + \sqrt{\delta}\,\xi_j$, $j = 1,2$, we obtain the
system

$$\frac{d\xi}{dt} = -2\xi + \sqrt{\delta}X(\xi,\theta,x^0,\delta)$$

$$\frac{d\theta}{dt} = -\Omega + \delta Y(\xi,\theta,x^0,\delta) \tag{10}$$

$$\frac{dx^0}{dt} = \varepsilon\delta Z(\xi,\theta,x^0,\delta).$$

Here $\xi = (\xi_1,\xi_2)$, $\theta = (\theta_1,\theta_2)$, $\Omega = (\Omega_1,\Omega_2)$ and the functions X, Y and Z are
polynomials in x^0, ξ, $\sin\theta_j$ and $\cos\theta_j$.

When $\varepsilon \sim 0\,(1/\sqrt{\delta})$ one can prove that for sufficiently small δ (10) has a
unique invariant two-dimensional torus T_δ^2 that has the representation

$$\xi = F(\theta,\delta)$$

$$x^0 = G(\theta,\delta). \tag{11}$$

Furthermore, this torus is attracting and therefore one can determine the phase-
locking characteristics of (10) by studying the flow on T_δ^2. This is done as in
the directly-coupled case, but one finds that the phase-locking behavior is
different. Namely, whereas 1:1 phase-locking occurs when $\Omega_1 - \Omega_2 \sim (\varepsilon)$ in the
directly-coupled case, this in not true in the indirectly-coupled case. The flow
is quasi-periodic when $\Omega_1 - \Omega_2 \sim 0(\varepsilon)$, but 1:1 phase-locking does occur in
regions of width $0\,(\delta^{3/2})$. The phase-locking characteristics at higher-order

136

resonances have not been determined at present, and this and other aspects of the problem are currently being studied. In the following section we discuss the dynamical behavior of indirectly-coupled systems at smaller values of ε and stronger coupling.

4. Further Results on Indirectly-Coupled Systems

A recurrent question in trying to understand real networks of cells is whether the observed dynamical behavior is inherent in the individual cells of the network, or whether it stems from the interactions between cells. This question arises in the context of the oscillations observed in cellular suspensions [10], in the behavior of neural networks [20], in certain cardiac arrythmias [21] and in many other applications. Often it is difficult to resolve this question in vivo if single-cell dynamics are not observable. Thus it is useful to know what types of phenomena arise from various modes of interaction in mathematical models of networks. Phase-locking is clearly one example of a network phenomenon and others have been identified in the context of pattern formation in directly-coupled cells [16], [22], [23]. In this section we discuss nonhomogeneous steady states and periodic solutions that arise in networks of two cells like that shown in Fig. 2, but now we assume that each of the cells relaxes to a unique steady state in the absence of coupling. For simplicity, we restrict the discussion to two-species systems. We no longer assume that the coupling is weak and therefore we set $\delta=1$ in (9).

Static solutions of (9) solve the nonlinear system

$$P(x^0-x^1) + R(x^1) = 0$$

$$P(x^0-x^2) + R(x^2) = 0 \tag{12}$$

$$\varepsilon P(x^0-\overline{x}) = 0 .$$

If P is invertible then $x^0 = \overline{x}$, and (x^1,x^2) satisfies the system

$$\frac{1}{2} P(x^2-x^1) + R(x^1) = 0 \qquad \frac{1}{2} P(x^1-x^2) + R(x^2) = 0 \tag{13}$$

Since these are just the equations for a directly-coupled system with $D = P/2$, the number of static solutions of (9) can be deduced from the directly-coupled case, but the stability of solutions may be different. In particular, there is always a static solution of the network problem for every zero of R. Let K denote the Jacobian of $R(x)$ at one of its zeroes, which we denote x^s. Then it is easily shown [12] that the stability of the solution $(x^1,x^2,x^0) = (x^s,x^s,x^s)$ of (9) is governed by the eigenvalues of K-P and the matrix L defined as follows.

$$L = \left[\begin{array}{c|c} K - P & P \\ \hline \varepsilon P & -\varepsilon P \end{array} \right]$$

It is also shown in [12] that if eigenvalues of L cross the imaginary axis as ε is varied, they can only do so pairwise. Therefore changes in ε generically lead to oscillatory instabilities, and under some mild nondegeneracy conditions, to Hopf bifurcation of periodic solutions. Since it is the eigenvalues of L that become complex, the existence of this instability depends only on the interaction of a cell with the passive medium, and thus exists when there is only one cell. When there are more than one, all <u>cells</u> are synchronized, and we call these periodic solutions cell-synchronous solutions.

The locus in parameter space on which this instability occurs can be found analytically when P is diagonal and only one of the diagonal elements is nonzero. Suppose that at the homogeneous steady state $k_{11} < 0$, $k_{22} k_{21} < 0$ and $k_{11} + k_{22} < 0$. Then the steady state x^s of (1) is asymptotically stable. However, if P_2 is zero then it can be shown that L has a pair of complex conjugate eigenvalues with zero real part on the curve whose equation is

$$-\varepsilon^2 P_1^2 T_1^K - \varepsilon P_1 [P T_1^K + P_1 k_{22} - (T^K)^2] - (T^K - P) \det(K-P) = 0 \ ,$$

where $T^K \equiv k_{11} + k_{22}$. This curve is shown in Fig. 3 for the case $k_{11} + 2k_{22} > 0$; other cases are shown in [12]. In a two-cell network static instabilities are also possible by virtue of eigenvalues of K-P crossing the imaginary axis. In fact, the foregoing conditions on K and P are sufficient for the existence of a Turing instability of the homogeneous steady state at $P_1 = \overline{P}_1$, where $\overline{P}_1$ is such that $\det(K-P) = 0$ when $P = \mathrm{diag} (\overline{P}_1, 0)$. Thus the line $P_1 = \overline{P}_1$ is the locus of static instability in Fig. 3, and this line intersects the curve on which oscillatory instabilities occur at $(\overline{P}_1, 0)$ and $(\overline{P}_1, \overline{\varepsilon}_1)$. The latter point corresponds to the standard static/Hopf interaction [24], but the former is a codimension five singular point that has not been unfolded at present.

An explanation of the fact that the oscillatory instability only exists for intermediate values of ε can best be given in the context of a specific reaction scheme and a network comprising one cell and the passive medium.

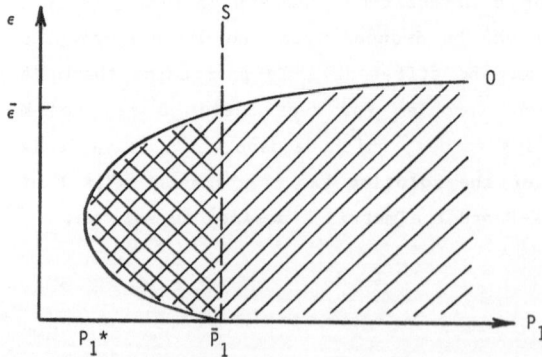

Fig. 3 The boundaries of oscillatory (O) and static (S) instability in the P-ε plane. xxx: oscillatory instability; ////: oscillatory and static instabilities.

Consider the two-dimensional system

$$x = \delta - \kappa x - xy^2 \qquad y = \kappa x + xy^2 - y \qquad (14)$$

which models one step in the glycolytic sequence [22]. The curves $x = 0$ and $y = 0$ for (12) are shown in Fig. 4(a). The steady state is $(x^s, y^s) = (\delta/\kappa + \delta^2, \delta)$, and this is stable when $0 < \delta < \delta_-$ or $\delta > \delta_+$, where

$$\delta_\pm = \frac{1-2\kappa \pm \sqrt{1-8\kappa}}{2}.$$

When $\delta \epsilon (\delta_-, \delta_+)$ (14) has a unique periodic solution. Let $\delta > \delta_+$ be such that $k_{11} + k_{22} < 0 < k_{11} + 2k_{22}$, in which case Fig. 3 applies, and suppose that $P_y = 0$. Then $\overline{P}_x = \det K/k_{22} = (\kappa + \delta^2)^2/(\delta^2 - \kappa) \sim \mathcal{O}(1)$ if $\kappa \ll 1$ and $\delta \sim \mathcal{O}(1)$.

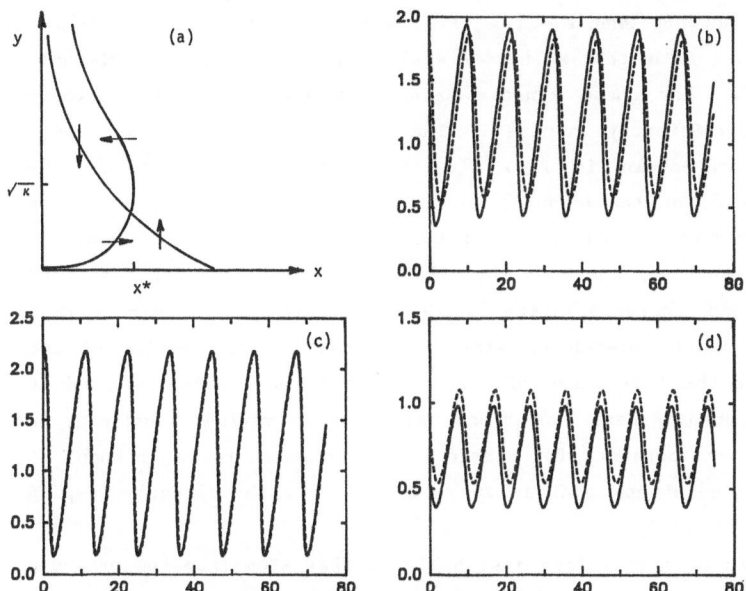

Fig. 4 (a) The zero isoclines for (14) (b)-(d): The periodic solutions of the two-cell network for $\kappa = 0.1$, $\delta = 0.8$, $P_y = 0$, and (b) $P_x = 1.0$; (c) and (d), $P_x = 6.0$. Solid lines denote x^1 in (b)-(d); dashed lines denote x^0 in (b) and (c) and x^2 in (d).

The curve $y = 0$ is unchanged when only x diffuses into a passive medium, but $x = 0$ now depends on the concentration x^0 of x in the medium. If ϵ is small x and y vary rapidly compared to x^0, and the initial evolution is to the solution $((\delta + P_x x^0)/(\kappa + P_x + y), y)$, where x^0 takes its initial value and y is the solution of a cubic equation. On a longer time-scale this quasi-static solution evolves to the steady state (x^s, y^s, x^s). Enroute it may pass through the unstable region, and slowly-varying oscillations may appear, but the final state is stable.

When ε is large the initial effect is to average x over the cell and medium, but this has the effect of isolating the cell since the flux then vanishes. However, the isolated cell is stable by hypothesis, and therefore the network is also stable at sufficiently large ε.

The oscillatory instability that can arise at intermediate ε is evidently similar to the static Turing instability that occurs at $\overline{P}_x$ for a two- (or more) cell system in that the diffusivity of the self-activating species must be smaller than that of the self-stabilizing species, but the oscillatory instability can exist for $P_x \ll \overline{P}_x$. However, as the foregoing argument has confirmed, such instabilities cannot exist in directly-coupled cells when the kinetic mechanism has only two species.

Figs. 4(b), (c) and (d) show some of the periodic solutions that can exist in a one- or two-cell network when ε = 1 . Fig. 4(b) shows x in a cell and in the medium at P_x = 1. Note that x^0 has a smaller amplitude and is slightly phase-shifted from x^1 . For these kinetic parameters several stable periodic solutions coexist at sufficiently large P_x. The x^1 and x^0 components of a cell synchronous solution are shown in Fig. 4(c) for P_x = 6. These are the continuations, to P_x = 6, of the solution shown in Fig. 4(b). These solutions can exist in either a one- or multi-cell network, but the solution shown in Fig. 4(d) is genuinely nonhomogeneous in that the two cells are out of phase and have different amplitude characteristics. This solution arises via a Hopf bifurcation from a secondary branch of static solutions. The details of the bifurcation diagram for the full range of ε have not yet been worked out, and it is not known how the nonhomogeneous solutions vary as ε is varied. However, it seems likely that they persist for large ε and connect to a branch of nonhomogeneous periodic solutions that is known to exist in the directly-coupled problem ε = ∞ [22].

Since the existence of oscillatory instabilities that stem from the presence of the passive medium depends only on the local properties of the kinetic vectorfield, they can exist in excitable systems under the proper conditions. To illustrate the character of the solutions that can result far from the bifurcation point, we shall present some numerical results for the following kinetic vectorfield.

$$\theta \frac{dx}{dt} = x(1-x) - fy \left(\frac{x-\mu}{x+\mu}\right) \qquad \frac{dy}{dt} = x-y \qquad (15)$$

This was derived as a two-dimensional approximation to a model of the Zhabotinskii - Belousov reaction in [25], and is similar to a chemical switching mechanism proposed in [26]. Eq. (15) has a unique steady state for all choices of the parameters, and we choose them so that it is stable. A sketch of the curve x = 0 shows that it has the typical 'cubic-like' shape characteristic of excitable systems.

Since x is self-activating, the instability exists when P_y is sufficiently large relative to P_x. For $P_x = 0$ one finds that $\overline{P}_y = 50.1$, $P_y* = 1.86$, and $\overline{\varepsilon} = 2\varepsilon* = 105$ (cf. Fig. 3). Numerical computations done along the curve of oscillatory instability show that the Hopf bifurcation is always subcritical and unstable. Continuation of the bifurcating solutions in either ε or P_y, using the software package AUTO [27], shows that the unstable bifurcating branches connect to stable branches far from the curve on which instability occurs. For example, at $\varepsilon = 50$ the unstable bifurcating solutions exist for $P_y \sim 10^{-3}$, where they connect to a stable branch of solutions. An example of solutions on the stable cell-synchronous branch for $P_x = 0$, $P_y = 2 \times 10^{-3}$ is shown in Figs. 5(a) and 5(b). Note that the time course of x, which is the 'active' or 'voltage-like' species, has a strong relaxation character, unlike the recovery variable y.

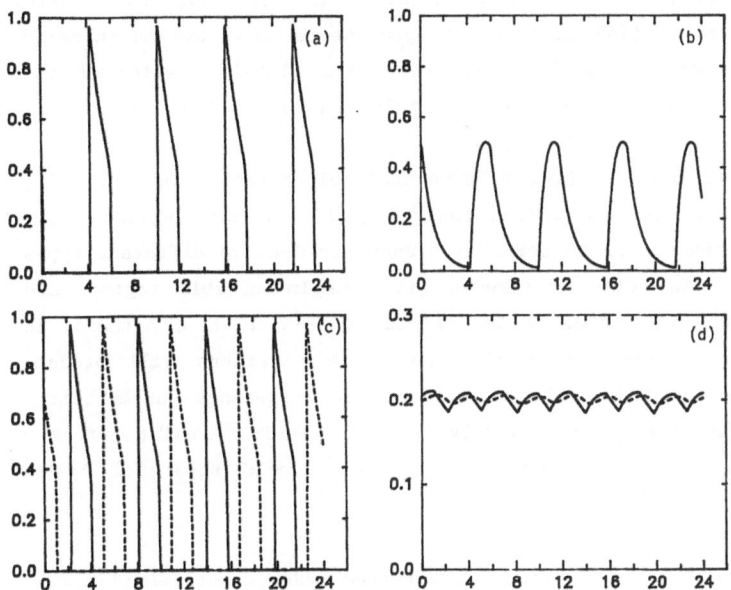

Fig. 5 The periodic solutions for the kinetics in (15), with $f = 0.505$, $\mu = 10^{-3}$, $\theta = 10^{-2}$, $\gamma = 1.006$, and $\varepsilon = 50$. The x^1 [(a)] and y^1 [(b)] components of a cell synchronous solution for $P_x = 0$, $P_y = 2 \times 10^{-3}$. (c) The x^1 (solid line) and x^2 (dashed line) components of an asynchronous solution for $P_x = P_y = 2 \times 10^{-3}$. (d) The x^0 (solid line) and y^0 (dashed line) components of the solution in (c).

These results demonstrate that stable cell-synchronous solutions can exist for very small values of P_y, even though the oscillatory instability occurs at a much larger value of P_y. Furthermore, we have found that these solutions exist even when $P_x = P_y = 2 \times 10^{-3}$, and have x vs time plots for the cells that are indistinguishable from Fig. 5(a). (Of course the x^0 and y^0 vs time plots are different.)

In addition to the stable cell-synchronous solutions that exist at $P_x = P_y = 2.10^{-3}$, there are stable asynchronous solutions in which the <u>cells</u> are a half period out of phase. The x-components of these solutions are shown in Fig. 5(c), and the x^0- and y^0-components are shown in Fig. 5(d). All of these solutions exist in a region of parameter space in which the static synchronized state of the network is stable.

The foregoing shows that coupling can have a dramatic effect on the dynamics of a network, and that indirect coupling can lead to new types of solutions. For instance, there is no analogue of the cell-synchronous solutions in a directly-coupled network using the foregoing kinetic parameters. Thus these solutions must necessarily disappear at some finite ε. The asynchronous solutions seem to be related to the echo waves sought unsuccessfully for directly-coupled cells with Fitzhugh-Nagumo kinetics in [28], and which were described for the kinetics at (15) in [25]. At present, we only have an informed guess, based on some numerical results, concerning the global behavior of the cell-synchronous and asynchronous branches. This aspect is currently under study.

A network such as is shown in Fig. 2, with excitable kinetics in each cell, provides a simple model of cardiac tissue containing inexcitable regions. Our results show that the presence of inexcitable tissue can lead to different types of sustained rhythmicity in cells contiguous with the inexcitable region, and thus they may be relevant to studies of the origin of re-entrant arrhythmias in cardiac tissue [13]. One can expect that inexcitable regions will produce similar effects in large networks and thus may lead to long-range automaticity. The simplest model in which to investigate this would consist of two one-dimensional, directly-coupled networks of excitable cells separated by an inexcitable cell.

Acknowledgments   This research was partially supported under Grant #GM29123 from the National Institutes of Health.   The author is indebted to J. Keener for an introduction to the problems associated with cardiac arrhythmias.

References

1. G. F. Ayala, M. Dicther, R. J. Gumnit, H. Matsuomoto, and W. A. Spencer: Brain Res. <u>52</u>, 1 (1973)

2. R. L. MacDonald and J. L. Barker: Neurology <u>28</u>, 325 (1978)

3. S. G. Rayport and E. R. Kandel: Science <u>213</u>, 462 (1981)

4. J. P. Gollub, T. O. Brunner, and B. G. Danley: Science <u>200</u>, 48 (1978)

5. D. G. Aronson, E. J. Doedel, and H. G. Othmer: in preparation.

6. G. B. Ermentrout and N. Kopell: SIAM J. Math. Anal. <u>15</u>, 215 (1984)

7. G. Gerisch, D. Malchow, W. Roos, and U. Wick: J. Exp. Biol. <u>81</u>, 33 (1979)

8. G. Gerisch: Ann. Rev. Physiol. <u>44</u>, 535 (1982)

9. J. G. R. Jefferys and H. L. Haas: Nature <u>300</u>, 448 (1982)

10. J. Aldridge and E. K. Pye: Nature <u>259</u>, 670 (1976)

11. H. G. Othmer and J. A. Aldridge: J. Math. Biol. <u>5</u>, 169 (1978)

12. H. G. Othmer: <u>Dynamics of Synergetic Systems</u> (Springer-Verlag, Heidelberg 1980)

13. C. Antzelevitch, J. Jalife, and G. K. Moe: Circ. <u>61</u>, 182 (1980)

14. H. G. Othmer: <u>Chemical Applications of Topology and Graph Theory</u> (Elsevier, Amsterdam 1983)

15. H. G. Othmer and L. E. Scriven: J. Theor. Biol. <u>32</u>, 507 (1971)

16. H. G. Othmer: <u>Lectures on Mathematics in the Life Sciences</u> (American Mathematical Society, Providence 1977)

17. E. Conway, D. Hoff, and J. Smoller: SIAM J. Appld. Math. <u>35</u>, 1 (1978)

18. N. N. Bogoljubov, J. A. Mitropoliskii, and A. M. Samoilenko: <u>Methods of Accelerated Convergence in Nonlinear Mechanics</u> (Springer-Verlag Berlin 1976)

19. J. Hale: <u>Ordinary Differential Equations</u> (Wiley-Interscience New York 1969)

20. W. O. Friesen and G. S. Stent: Ann. Rev. Biophys. Bioeng. <u>7</u>, 37 (1978)

21. F. J. L. van Capelle and D. Durrer: Circ. Res. <u>47</u>, 454 (1980)

22. M. Ashkenazi and H. G. Othmer: J. Math. Biol. <u>5</u>, 305 (1978)

23. J. C. Alexander, preprint (1984)

24. W. F. Langford: <u>Nonlinear Dynamics and Turbulence</u> (Pitman Boston 1983)

25. J. J. Tyson: <u>Bifurcation Theory and Applications in Scientific Disciplines</u> (New York Academy of Sciences New York 1979)

26. O. E. Rössler: J. Theor. Biol. <u>36</u>, 413 (1972)

27. E. J. Doedel: Cong. Num. <u>30</u>, 265 (1981)

28. V. I. Krinskii, A. M. Pertsov and A. N. Reshetilov: Biofizika <u>17</u>, 271 (1972)

The Influence of Fluctuations on Sustained Oscillations

H. Engel-Herbert, W. Ebeling, and H. Herzel
Department of Physics, Humboldt-University Berlin
DDR-1086 Berlin, German Democratic Republic

1. Introduction

It is well known that under appropriate conditions nonlinear physical, chemical or biological systems show various instabilities such as, for example, transitions leading to multistability, limit cycle oscillations or chaotic motion [1-3]. These instabilities correspond to bifurcations of the underlying dynamic equations. A realistic treatment requires one to take into account fluctuations, which appear as thermal fluctuations or are due to the discrete change of particle numbers in nonlinear chemical reactions. Another origin of noise may be the influence of the environment —which in general varies more or less randomly —on the thermodynamically open systems. Usually the description of complex systems, e.g. biochemical oscillations, is restricted to simplifying models which represent the observed behaviour only qualitatively. Then fluctuations may appear as a result of "hidden" reactions neglected in the model.

Taking into account fluctuations smears out the predictions of the deterministic theory. For this reason it is sometimes believed that the effect of noise on nonlinear oscillations may be neglected because the stochastic analysis is much more complicated and does not contain any additional information about the behaviour of the oscillator not contained in the deterministic theory. In general this is not true. Even small fluctuations may drastically influence the dynamic behaviour of the oscillators. Before going into the details, let us mention some typical situations:

- If fluctuations are present in the system the question arises whether the deterministic predictions for the bifurcational behaviour remain valid in the stochastic case or not. For example, the threshold for the transition from nonoscillatory to oscillatory motion may be shifted by the fluctuations or new transitions may even appear. These effects, called noise induced kinetic transitions [4,5], are observed experimentally [6]. Another interesting effect is the stochastic stabilization of stationary states which has no analog in the deterministic picture [4]. The main origin of the observed behaviour is the coupling of the fluctuations to the nonlinear dynamics (multiplicative noise). Nevertheless, even additive noise generates similar effects in nonpotential systems (systems without detailed balance) [7].

- Stochastic realizations may overcome barriers between different domains of attraction. Examples are the tunneling through separatrices or unstable limit cycles — processes not possible in the deterministic picture.

- In systems with weak damping (slightly below a Hopf-bifurcation) oscillatory behaviour may be generated by fluctuations. That means, the amplitudes exceed the noise level by orders of magnitude and the correlation function is different from zero for many periods.

In the following we consider the effect of noise on mechanical oscillations and on a simple model for glycolytic oscillations.

2. Mechanical Oscillation. Noise Induced Modifications of the Threshold Behaviour

A whole class of nonlinear oscillations may be described by the equation of motion

$$m\ddot{q} + \varepsilon\gamma(q,\dot{q},u)\dot{q} + \frac{dV(q;u,\varepsilon)}{dq} = g_k(q,\dot{q};u,\varepsilon)\xi_k(t) \tag{2.1}$$

or in the equivalent form

$$\dot{q} = \frac{\partial H}{\partial p} \quad , \quad \dot{p} = -\frac{\partial H}{\partial q} - \varepsilon\frac{\gamma}{m}p + g_k\xi_k \quad , \quad H = \frac{p^2}{2m} + V \quad . \tag{2.2}$$

In (2.1,2) p, q are the generalized momenta and coordinates, $u = \{u_1,...,u_k\}$ is the set of parameters, m denotes the mass of the particle, γ the friction function and V the external potential. The g_k are coupling functions between the dynamics and the stochastic sources ξ_k; they define the dependence of the noise intensity on the state of the system. We assume that the correlation time τ of the fluctuations is much smaller than all other characteristic times of the system. Then ξ_k may be approximated by Gaussian white noise

$$<\xi_k(t)> = 0 \quad , \quad <\xi_k(t)\xi_\ell(t)> = \delta_{k\ell}\delta(\tau) \quad . \tag{2.3}$$

In (2.1) and in the following, summation over repeated indices is implied. The dot means differentiation with respect to time.

In general the Fokker-Planck equation for the probability density P(q,p,t;u) corresponding to (2.1-3) cannot be solved exactly even in the stationary case. Several approximate methods have been developed which are based on the existence of a hierarchy of well divided time scales in the system. Indeed, if ε is small

$$\tau \ll T \ll \frac{1}{\varepsilon} \tag{2.4}$$

then the relaxation time ε^{-1} of amplitude A and phase φ exceeds the period T of the oscillations performed in the limit $\varepsilon = 0$ by orders of magnitude. In this case of weak dissipation it is useful to transform to the amplitude phase representation [8] or to other slowly varying quantities, for example to H [9], to simplify the Fokker-Planck equation. The rapid oscillations of P (on the time scale T) can be eliminated in a first approximation by averaging the equation over T, supposing that A and φ do not change.

An explicit example treated in an earlier paper [10] is

$$\gamma(q,p;u) = u_1 + u_2p + u_3p^2 + u_4p^3 + u_5p^4 \quad , \quad V = \frac{\omega_0^2}{2}q^2 \quad . \tag{2.5}$$

Let us suppose that u_1 and u_2 are stochastic quantities which fluctuate with variances of order $\varepsilon^{-\frac{1}{2}}$ around some mean values $\bar{u}_1$, $\bar{u}_2$ respectively

$$u_1 \to u_1(t) = \bar{u}_1 + \varepsilon^{-\frac{1}{2}}\sigma_1\xi_1(t) \quad , \quad u_2 \to u_2(t) = \bar{u}_2 + \varepsilon^{-\frac{1}{2}}\sigma_2\xi_2(t) \quad . \tag{2.6}$$

We also take into account weak internal fluctuations modelled by the additional term

$$\varepsilon^{\frac{1}{2}}\sigma_0\xi_0(t) \quad . \tag{2.7}$$

Therefore

$$g_0 = \varepsilon^{\frac{1}{2}}\sigma_0 \quad , \quad g_1 = \varepsilon^{\frac{1}{2}}\sigma_1 p \quad , \quad g_2 = \varepsilon^{\frac{1}{2}}\sigma_2 p^2 \quad . \tag{2.8}$$

The probability density

$$P_A(A,t;u) = \int_0^{2\pi} P(A,\varphi,t;u)d\varphi \tag{2.9}$$

satisfies the Fokker-Planck equation [10]

$$\frac{1}{\varepsilon}\frac{\partial P_A}{\partial t} = -\frac{\partial}{\partial A}\left\{\left[\frac{1}{4}\sigma_0^2\frac{1}{A} - \frac{1}{2}\left(\bar{u}_1 + \frac{1}{8}\sigma_1^2\right)A - \frac{3}{8}\left(u_3 + \frac{7}{12}\sigma_2^2\right)A^3\right.\right.$$
$$\left.\left. - \frac{5}{16}u_5 A^5\right]P_A - \left(\frac{1}{4}\sigma_0^2 + \frac{3}{16}\sigma_1^2 A^2 + \frac{5}{32}\sigma_2^2 A^4\right)\frac{\partial P_A}{\partial A}\right\} \quad . \tag{2.10}$$

The extrema $\hat{A}$ of the stationary probability density of the amplitudes P_A^0 are defined as the roots of the algebraic equation

$$\frac{5}{16}u_5\hat{A}^6 + \frac{3}{8}\left(u_3 + \frac{7}{12}\sigma_2^2\right)\hat{A}^4 + \frac{1}{2}\left(\bar{u}_1 + \frac{1}{8}\sigma_1^2\right)\hat{A}^2 - \frac{1}{4}\sigma_0^2 = 0 \quad . \tag{2.11}$$

For vanishing noise ($\sigma_0 = \sigma_1 = \sigma_2 = 0$) (2.11) reduces to the equation of the stationary amplitudes A_d in the deterministic picture

$$\frac{5}{8}u_5 A_d^4 + \frac{3}{4}u_3 A_d^2 + u_1 = 0 \quad . \tag{2.12}$$

From (2.12) one easily determines the deterministic bifurcation set presented on Fig. 1. The noise induced modifications follow from (2.11) (compare Fig. 1 again). We realize that the thresholds for hard as well as for soft excitation of sustained oscillations are shifted due to the action of the fluctuations. The shift is proportional to the noise intensity σ^2. For soft excitation this behaviour was verified experimentally in an electrical system [6].

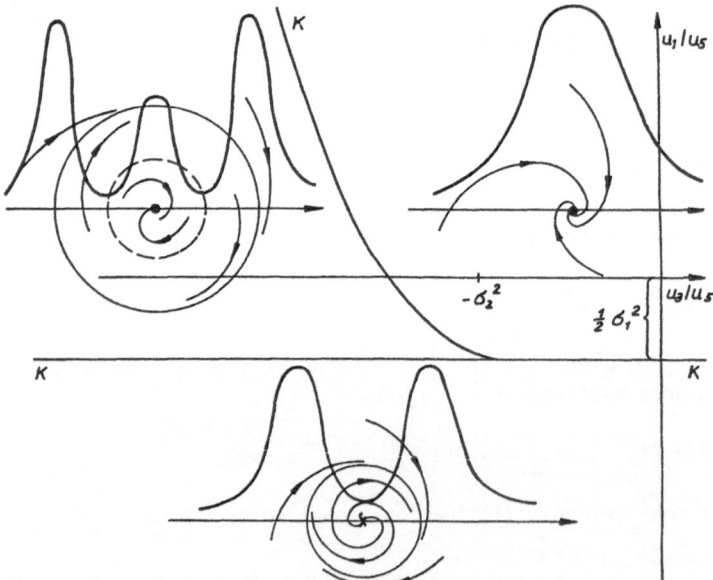

Fig. 1. Cross sections of the stationary probability density P^0 of the pumped nonlinear oscillator (2.1,5,6). Three regions of qualitative different behaviour must be distinguished in the u_1, u_3 -plane: damped oscillations, P^0 is single peaked; soft excitation of sustained oscillations, P^0 shows a crater; hard excitation of sustained oscillations, P^0 shows a crater with central peak

Some comments about the definition of a bifurcation in a dynamical system affected by noise are necessary.

In the deterministic theory the transition to sustained oscillations is connected with the appearance of a limit cycle in the phase plane. The deterministic threshold value is mathematically well defined as the bifurcation value of the parameter governing the transition. The bifurcations leading to limit cycles have been extensively studied by Andronov et al. [11].

If, however, the oscillator is affected by noise, the above definitions lose their sense. Now we use the stationary probability density P^0 as the basic quantity for the definition of the bifurcation. We connect instabilities and transition phenomena in stochastic systems with topological changes of the stationary probability density and of the stationary probability flow [10]. In fact we have to study how extremal probable states (points, or bits, ...) are created or annihilated if the parameters cross some critical values.

The transition to sustained oscillations becomes apparent in the formation of a probability crater on P^0. The rotating stationary probability flow indicates the oscillatory regime. Of course there are many possibilities for the change of a one-peak stationary probability density to a crater-shaped one, for example

($\oplus$- local maximum, $\ominus$- local minimum, o- saddle, ●- degenerate singular point of P^0).

3. Biochemical Oscillations

We study a simple model for biochemical oscillations which is a generalization of the first Lotka scheme. The raw material X reacts autocatalytically to form the intermediate Y. Y is transformed to a final product which is removed from the reactor with constant rate A_1

$$\xrightarrow{A_0} X \xrightarrow{\ulcorner F(Y) \urcorner} Y \xrightarrow{A_1} \quad . \tag{3.1}$$

The equations of chemical kinetics are given by

$$\dot{X} = A_0 - XF(Y) \quad , \quad \dot{Y} = XF(Y) - A_1 Y \quad . \tag{3.2}$$

$F(Y)$ denotes the feedback function and A_0 the raw material input per unit time. We would like to investigate the effect of fluctuations of the raw material input rate on the transition to oscillatory behaviour. We suppose that A_0 fluctuates with variance $\tilde{\sigma}$ around some mean value $\bar{A}_0$ according to

$$A_0(t') = \bar{A}_0 + \tilde{\sigma}\xi(t') \quad . \tag{3.3}$$

For $\xi(t')$ the idealization of Gaussian white noise is again made.

Introducing dimensionless time and concentrations by

$$t = A_1 t' \quad , \quad y = \frac{A_1}{\bar{A}_0} Y \quad , \quad x = \frac{A_1}{\bar{A}_0} X \quad , \quad \sigma = \frac{A_1^2}{\bar{A}_0^2} \tilde{\sigma} \tag{3.4}$$

and the function

$$f(y) = \frac{1}{A_1} F\left(\frac{\bar{A}_0}{A_1} y\right) ,$$
(3.5)

we get the following system of equations:

$$\dot{x} = 1 - xf(y) + \sigma\xi(t) , \quad \dot{y} = xf(y) - y .$$
(3.6)

We choose the feedback function f in the form

$$f(y) = u_0 + u_1 y + u_2 y^2 .$$
(3.7)

For $u_0 = u_2 = 0$, (3.6,7) reduces to the well known Lotka scheme; for $u_0 = u_1 = 0$, one gets a simple model of glycolytic oscillations proposed by Selkov [12]. The cases $u_0 = 0$ and $u_1 = 0$ are studied in [13] and [14] respectively using the discrete Master-equation approach.

Under fixed conditions on u_0, u_1 and u_2 [14] the system (3.6,7) undergoes a Hopf bifurcation. The stable focus $(x_0, y_0) = (1, f(1)^{-1})$ changes to an unstable focus and in the vicinity a stable limit cycle appears. It is useful to transform (3.6) to a representation similar to (2.1). Differentiating the second equation (3.6) with respect to time and using the first equation we obtain

$$\ddot{y} + \left[1 + f - (y + \dot{y}) \frac{1}{f} \frac{df}{dy}\right] \dot{y} + (y - 1)f = f\sigma\xi .$$
(3.8)

In general, for the biochemical oscillator well divided time scales don't exist. However, in an ε-vicinity of the Hopf bifurcation

$$\varepsilon = u - u_{cr} , \quad 0 < \varepsilon \ll 1$$
(3.9)

the amplitude of the oscillation is of order $\varepsilon^{\frac{1}{2}}$ and amplitude and phase are slowly varying quantities. In this parameter region we make for the deviations Δx, Δy from the stationary point

$$x = f(1)^{-1} + \Delta x , \quad y = 1 + \Delta y$$
(3.10)

the ansatz

$$\Delta y(t) = \frac{A(t)}{\omega} \cos\psi(t) , \quad \Delta\dot{y}(t) = -A(t) \sin\psi(t) , \quad \psi(t) = \omega \cdot t + \varphi(t) , \quad \omega = f(1) .$$
(3.11)

Now we again consider the Fokker-Planck equation for the probability density $P(A,\varphi,t;u)$. Explicit calculations for the choice $u_1 = 0$ in (3.7) yield after averaging over one period $T = 2\pi/(u_0 + u_2)$ of the rapid oscillations and integration over φ

$$\frac{\partial P_A}{\partial t} = \frac{\partial}{\partial A} \left\{ \left[-\frac{\sigma^2}{4} (u_0 + u_2)^2 \frac{1}{A} + \left(\alpha - \frac{\sigma^2}{8} \frac{u_2(3u_2 + u_0)}{u_0 + u_2}\right)A \right. \right.$$

$$\left. + \left(\beta - \frac{\sigma^2}{32} \frac{u_2^2}{(u_0 + u_2)^2}\right)A^3 \right] P_A + \frac{\sigma^2}{2} \left[\frac{(u_0 + u_2)^2}{2} + \frac{u_2(3u_2 + u_0)}{4(u_0 + u_2)} A^2 \right.$$

$$\left. \left. + \frac{u_2^2 A^4}{16(u_0 + u_2)^2} \right] \frac{\partial P_A}{\partial A} \right\}$$
(3.12)

where the abbreviations

$$\alpha = \frac{u_0 + u_2 - 1}{2} + \frac{u_0}{u_0 + u_2} \quad , \qquad \beta = \frac{u_2}{8(u_0 + u_2)} + \frac{u_0 u_2 (3u_2 - u_0)}{4(u_0 + u_2)^4} \tag{3.13}$$

are used.

Equating to zero the expression in the first curly brackets on the right-hand side of (3.12) we obtain a condition for the most probable stationary amplitudes $\hat{A}$. For vanishing noise $\sigma = 0$ it reduces to the right-hand side of the equation for the evolution of the amplitude in the deterministic description

$$\dot{A}_d = -\alpha A_d - \beta A_d^3 + o(A_d^5) \quad . \tag{3.14}$$

The deterministic threshold is $\alpha = 0$ and the amplitude of the sustained oscillations is, in a first approximation

$$A_d = \left(-\frac{\alpha}{\beta}\right)^{\frac{1}{2}} \quad . \tag{3.15}$$

The existence of the oscillations is restricted to negative values for α; this leads to the condition (compare 3.13)

$$u_2 - u_0 > (u_2 + u_0)^2 \quad . \tag{3.16}$$

Thus the results (3.14-16) of the deterministic theory [14] are recovered from (3.12).

If the strength of fluctuations of the raw material input rate is small

$$\sigma \sim o(\varepsilon) \tag{3.17}$$

then to lowest order in ε the extrema of P_A^0 are located at the roots of the equation

$$\beta \hat{A}^4 + \alpha \hat{A}^2 - \frac{\sigma^2}{4} (u_0 + u_2)^2 = 0 \tag{3.18}$$

(compare Fig. 2).

Summarizing, near the threshold biochemical oscillations may be appropriately studied in the framework of slowly varying quantities.

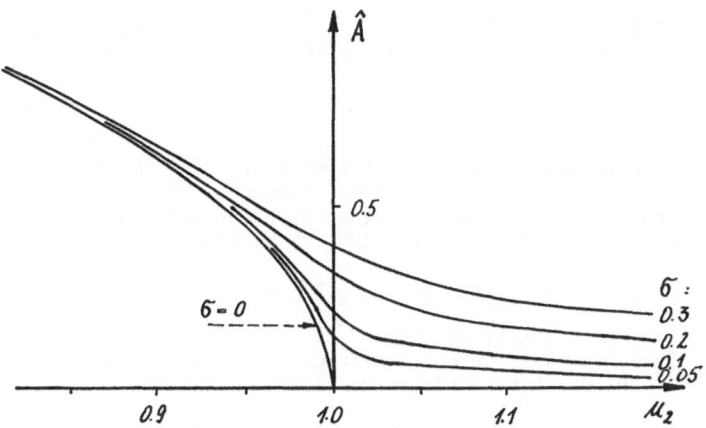

Fig. 2. Most probable amplitude of the glycolytic oscillations ($f(y) = u_2 y^2$) in the vicinity of the deterministic threshold $u_2 = 1$ for different strength σ of the fluctuations of the raw material input rate A_0

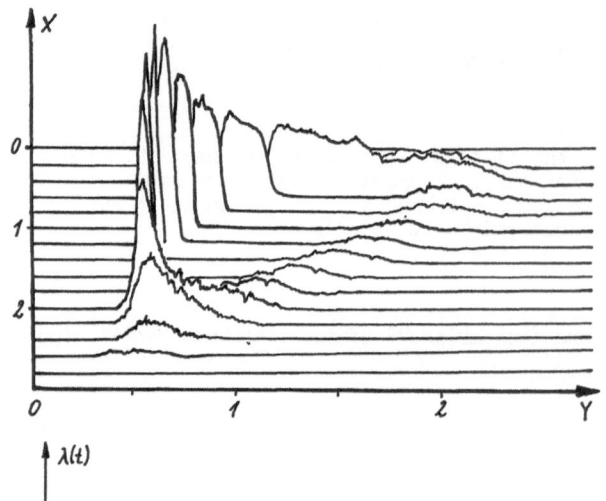

Fig. 3. Perspective of the stationary probability density for the Selkov oscillator in the region of developed oscillations $(f(y) = u_2 y^2,\ u_2 = 0.9)$

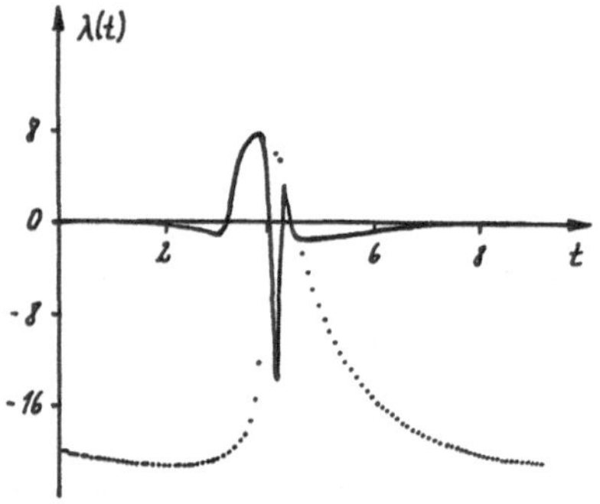

Fig. 4. The "local" maximal Lyapunov exponent along the limit cycle of the Selkov oscillator (3.20); the dotted line indicates $y(t)$; $u_3 = 0$

In the parameter region of fully developed oscillations these methods break down. Here numerical simulations of the Langevin equations may be applied [15]. Due to the different velocity of the dominant reaction along the orbit and due to the different strength of the amplitude fluctuations the probability crater becomes asymmetric (Fig. 3).

Different parts of the deterministic trajectory show different sensitivity to fluctuations. This is clearly expressed by the "local" Lyapunov exponent [16,17]

$$\lambda(t) = \frac{1}{\tau} \ln \frac{\| \delta x(t + \tau) \|}{\| \delta x(t) \|} \tag{3.19}$$

($\delta x(t)$ - deviation from the deterministic trajectory at time t, τ sufficiently small). The time average of $\lambda(t)$ gives the usual Lyapunov exponent. Figure 4 presents the time dependence of the maximal "local" Lyapunov exponent along the limit cycle of the Selkov oscillator affected by additive Gaussian white noise

$$\dot{x} = 1 - u_3 x - u_2 xy^2 + \sigma_1 \xi_1(t) \quad , \quad \dot{y} = u_2 xy^2 - y + \sigma_2 \xi_2(t) \quad . \tag{3.20}$$

Parts of the orbit with positive values $\lambda(t)$ correspond to parts of the stochastic realizations where the amplitude fluctuations are amplified. The fluctuations increase if the nonlinearity becomes dominant (around the minimum of $y(t)$).

Another interesting effect is the tunneling of the stochastic realization through the boundary of the deterministic domains of attraction. As an example Fig. 5 shows a stochastic realization for the Selkov oscillator (3.20). The corresponding de-

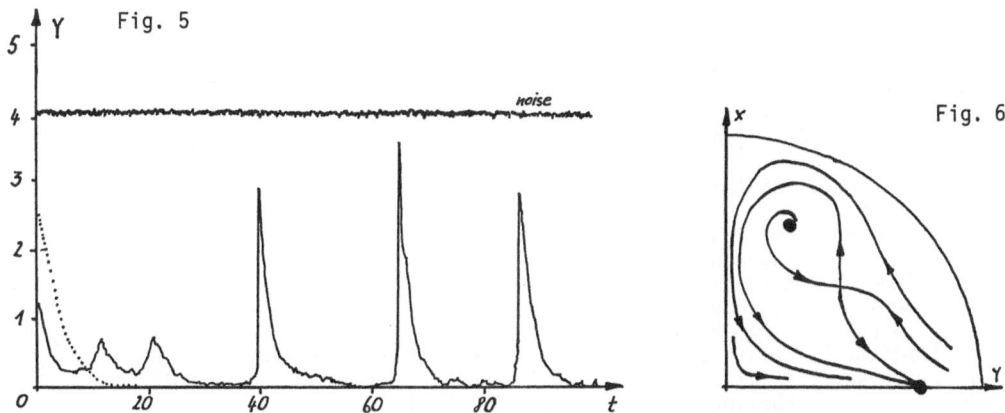

Fig. 5. Stochastic realization of (3.20), $u_2 = 1$, $u_3 = 0.125$, $\sigma_1 = \sigma_2 = 0.025$

Fig. 6. Schematic representation of the phase portrait of the deterministic oscillator (3.20) for $u_2 = 1$, $u_3 = 0.125$, $\sigma_1 = \sigma_2 = 0$ on the Poincaré sphere

terministic phase portrait is sketched in Fig. 6. The deterministic trajectory always approaches the stable node on the x axes. In contrast the stochastic realization resembles the burst oscillations known from radio-techniques. Due to the fluctuations the stochastic realization may overcome the separatrices of the saddle at irregular times.

Finally, we mention that the stochastic realizations slightly below the threshold ($|u - u_{cr}|^{-1} > T$) exhibit oscillatory behaviour generated by the fluctuations. This is demonstrated in Fig. 7: the amplitude of the oscillations exceeds the noise level by orders of magnitude. Similar results have been reported elsewhere [18].

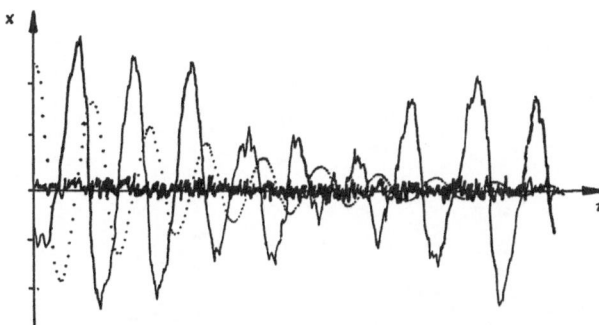

Fig. 7. Typical stochastic realizations x(t) of (3.20) below the deterministic threshold ($u_2 = 1.1$, $u_3 = 0$). The dotted line is a trajectory without noise, the noise level is indicated too ($\sigma_1^2 = 0.01$, $\sigma_2 = 0$)

References

1. H. Haken: *Synergetics. An Introduction. Nonequilibrium Phase Transitions and Self-Organization in Physics, Chemistry and Biology*, Springer Ser. Synergetics, Vol. 1, 3rd. ed. (Springer, Berlin, Heidelberg 1983)
2. G. Nicolis, I. Prigogine: *Self-Organization in Non-Equilibrium Systems* (Wiley, New York 1977)
3. W. Ebeling, R. Feistel: *Physik der Selbstorganisation und Evolution* (Akademie-Verlag, Berlin 1982)

4. W. Horsthemke: Ph. D. Thesis, Brüssel 1978
5. W. Horsthemke: "Nonequilibrium transitions induced by external white and coloured noise", in *Dynamics of Synergetic Systems*, ed. by H. Haken, Springer Ser. Synergetics, Vol.6 (Springer, Berlin, Heidelberg 1980) p.67
6. T. Kawakubo, S. Kabashima, Y. Tsuchiya: Suppl. Prog. Theor. Phys. **64**, 150 (1978)
7. L. Schimansky-Geier, A.V. Tolstopjatenko, W. Ebeling: Phys. Lett. (1984), to appear
8. R.R. Stratonovich: *Topics in the Theory of Random Noise* (Gordon and Breach, New York 1963)
9. H. Haken: Rev. Mod. Phys. **47**, 67 (1975)
10. W. Ebeling, H. Engel-Herbert: Physica **104**A, 378 (1980)
11. A.A. Andronov, A.A. Witt, S.E. Chaikin: *Theorie der Schwingungen* (Akademie-Verlag, Berlin 1965, 1969)
12. E.E. Selkov: Europ. J. Biochem. **4**, 79 (1968)
13. W. Ebeling, R. Feistel: "Limit Cycles in Reaction Systems with Second Order Autocatalysis" in VII Int. Conf. Nonl. Oscillations Proc. (Berlin 1975)
14. R. Feistel, W. Ebeling: Physica **93**A, 114 (1978)
15. W. Ebeling, H. Herzel, E.E. Selkov: stud. biophysica **98**, 147 (1983)
16. H. Herzel, W. Ebeling, L. Schimansky-Geier, E.E. Selkov: "The influence of noise on a biochemical oscillator", in *Methods and Applications in Connection with Lotka-Volterra Equations in Systems Analysis*, UNESCO Conf. Proc. (GDR Eisenach 1984)
17. I. Schimada, T. Nagashima: Prog. Theor. Phys. **61**, 1605 (1979)
18. K. Iwamoto, M. Seno: J. Chem. Phys. **70**, 5851 (1979)

Movement in Space

Chemomechanical Interfacial Instabilities and Waves:
Their Possible Role for the Cell Locomotion on Substrates

E. Sackmann

Physik Department E 22 (Biophysics Group) Technical University München
D-8046 Garching, Fed. Rep. of Germany

The chemically-driven Dupeyrat instability at the interface between an organic and an aqueous solution exhibits a number of dynamic regimes such as random or regular lateral motions, rotating interfacial waves or amoeban-like motions. The latter two are studied in a system composed of iodine in salicylaldehyde as organic phase and an aqueous solution of an ammonium soap. Striking analogies between the hydrodynamic interfacial motion and the locomotion of cells on solid substrates suggest that the latter may at least be triggered by an interfacial instability at the membrane substrate interface.

I Introduction

The chemically-driven instability at the interface between an aqueous phase containing a positive (ammonium) soap and a solution of an organic acid (e.g. picric acid) in a polar organic solvent (e.g. nitrobenzene) which was discovered by Dupeyrat et al.[1] is of interest for two reasons: First, it provides a relatively simple model in order to test theories of interfacial instabilities[2, 3]. Secondly, there is a striking analogy between the chemically-driven interfacial waves and the migration of cells such as macrophages on solid substrates,which suggests that a chemomechanical energy conversion similar to that of the Dupeyrat instability may provide an initial driving force for cell motility.

The most spectacular type of instability of the Dupeyrat-system is the rotational interfacial wave observed if the aqueous solution is superimposed on the organic one in a round beaker. Under suitable conditions, a wave forms at the interface which rotates with constant angular velocity. An example is shown in Fig. 1. Depending on the concentration and on the type of organic solution, amplitudes of up to 2 cm can be created.

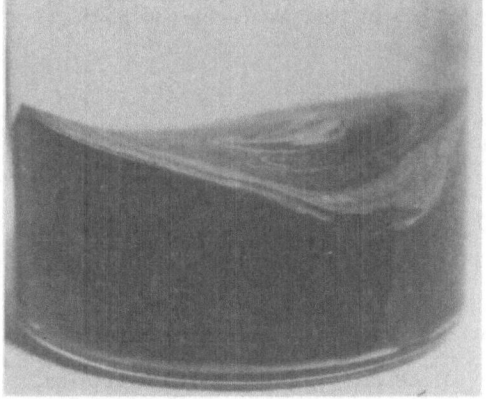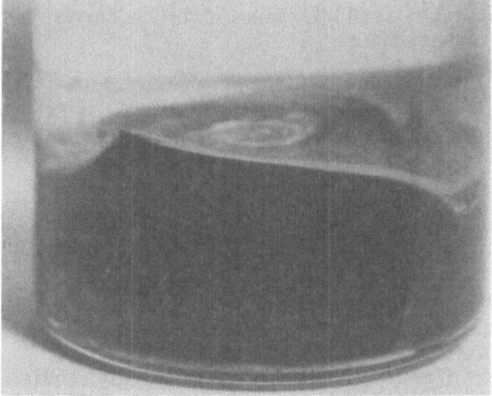

Fig. 1: Rotating wave formed at the interface between a solution of I_2 in salicylaldehyde (10^{-2} M; dark liquid at bottom) and a 10^{-2} M solution of ammonium soap (C16 Cl) in water (colorless liquid at top). Diameter of beaker: 2 cm; angular frequency $\nu \approx 1$ Hz.

In the following I will report about experiments with a system consisting of a solution of iodine (or salicylic acid) in salicylaldehyde (SAD) as organic phase and the same aqueous ammonium soap solution as used by Dupeyrat et al. [1] as water phase. In the second part of my contribution I will address the question whether this type of instability could play a role for the cell movement on solid substrates.

II SYSTEMS EXHIBITING HYDRODYNAMIC INTERFACIAL INSTABILITIES AND CHEMICAL REACTIONS INVOLVED

Several systems are known which are listed in Fig. 2a. In all cases the aqueous phase contains a surface active ammonium soap with a strong tendency to form a salt with the aromatic carboxylic acid which is much less soluble in water but much more soluble in the organic phase than the soap itself. Thus Dupeyrat et al. showed that the distribution coefficient of the soap between water and nitrobenzene decreases by a factor of 10^8 if it combines with picric acid. For that reason, the system is far from thermodynamic equilibrium.

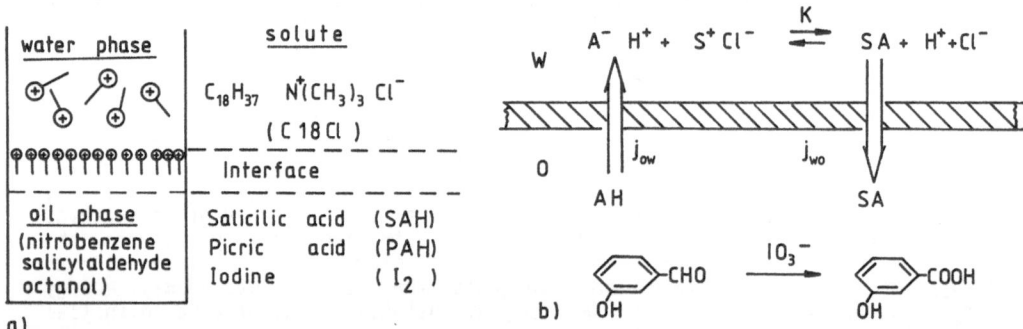

a)

Fig. 2:
a) A summary of systems exhibiting hydrodynamic interfacial instabilities. Possible organic solvents are given on the left and suitable solutes on the right side. In the case of octanol the aqueous phase is at the bottom. The critical micelle concentration of C16 Cl is 8×10^{-4} M.
b) Possible chemical events at interface. The organic acid (AH) dissociates and the anion forms a salt (AS) with the ammonium soap S^+. In the case of I_2 the organic acid is assumed to be formed by oxidation in the pre-reaction shown at the bottom.

The driving force for the instability is attributed to the chemical reaction shown in Fig. 2b. The organic acid is transfered into the aqueous phase where it dissociates and the aromatic anion A^- combines with the soap molecule S^+ to form SA. Due to its low water solubility, the SA molecule is expected to desorb from the interface and to go into the organic phase. As a result of this reaction one has two opposite mass-transfer processes: (1) a net flow j_{ow}, of H^+-ions into the aqueous phase and a transport, j_{wo}, of the neutralized soap into the organic phase. Experimental evidence for such a flow comes from the finding that the pH of the aqueous phase decreases very rapidly at the interface if the instability sets in.
The chemical events in the case of iodine as solute are more complicated. The organic acid is probably created in a pre-reaction by oxidation of the aromatic aldehyde. The solubility, C_{max}, of I_2 in water is very low ($C_{max} \approx 2 \cdot 10^{-3}$M) as compared to that in salicylaldehyde where it is $C_{max} \approx 0.5$ M in the absence and $C_{max} = 3$ M in the presence of 0.5 M KJ. However, in basic aqueous solutions I_2 is rapidly transformed into IO_3 according to

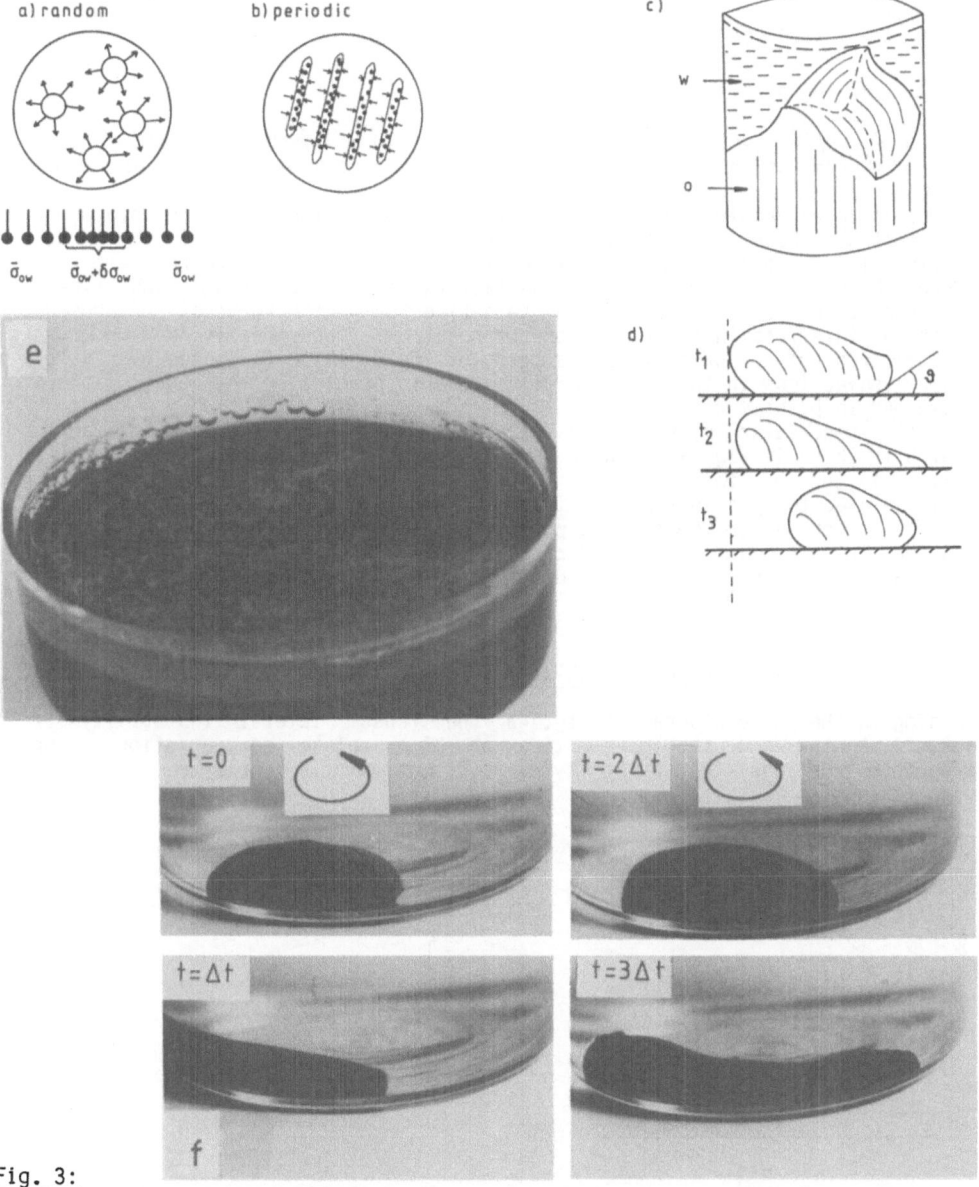

Fig. 3:
Types of interfacial instabilities: a) Localized, pulse-like lateral expansions occurring at random sites of interface. b) Regular streaming pattern. c) Rotating interfacial waves. One observes waves with one (solitary waves), two or three crests. d) Amoeba-like crawling motion of drop of organic phase on solid substrate. e) Photograph of random lateral expansions. Note the dark patches which are formed when the local expansions of the soap monolayer at the interface push aside the fine foam covering the liquid/liquid interface. f) Four subsequent pictures of a drop in the state of an amoeban motion. The time delay between two pictures is about o.2 sec. Note that the drop moves in a counterclockwise direction and that it performs just about one circulation between two pictures.

$$3 \; I_2 + 6 \; OH^- \rightleftarrows IO_3^- + 5 \; I^- + 3 \; H_2O \qquad\qquad (1)$$

Since the iodate ion is a strong oxidant, it could then oxidize salicylaldehyde into salicylic acid (cf Fig. 2b bottom).

III TYPES OF INTERFACIAL INSTABILITIES

Depending on the concentrations of the two solutions, or on the material of the solid substrate, several types of instabilities are observed which are shown schematically in Fig. 3.
At low soap and I_2 contents, one observes lateral convections in the plane of the interface which do not lead to substantial deformations in the normal direction. In most cases pulse-like lateral expansions of the interface are observed, which start at random sites and last for a fraction of a second. These local expansions are visible owing to a fine foam which forms at the aqueous side of the interface in the course of the reaction. Occasionally, one observes the formation of a wave-like pattern of foam which is caused by regular circular convections in the interfacial region. In particular, this wave pattern is observed in plastic beakers.
The most spectacular hydrodynamic instability is the rotating wave, an example of which is shown in Fig. 1. Regular rotating waves consisting of one, two (as in Fig. 1) or of three crests (separated by 120°) form after some time. Closely related to this instability is the amoeba-like motion of droplets of the organic phase in the aqueous phase. The drops move with astonishing speed along the inner edge at the bottom of a vessel as shown in Fig. 3f. On flat horizontal glass-surfaces, the drops crawl along the surface as shown schematically in Fig. 3d.

IV Properties of Interfacial Instabilities

According to the phase diagram of Fig. 4a the random lateral instability is observed at low and the wave-motion at high solute concentrations. For hydrophobic surfaces the first type of motion is observed at all concentrations.

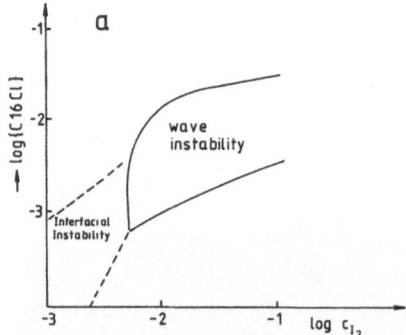

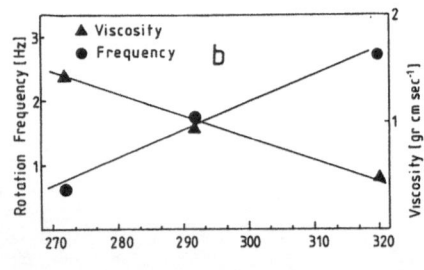

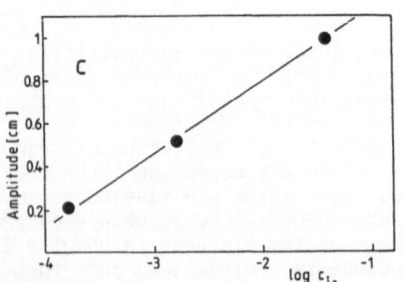

Fig. 4: a) Range of coexistence of random lateral and wave-like instability for system salicylaldehyde $+I_2/H_2O+C16Cl$ in a glass vessel.
b) Temperature dependencies of angular velocity, ω, of rotating wave and of viscosity, η, of aqueous phase.
c) Dependence of amplitude of rotating wave one iodine concentration for system salicylaldehyde $+I_2/H_2O+C16Cl$.

156

All types of instabilities are accompanied by a rapid increase in the H^+-concentration at the aqueous side of the interface. This leads to a sharp pH gradient into the direction normal to the interface, which is rapidly equalized in the case of the wave motion.

In the course of the reaction a fine foam is formed in the aqueous phase of the interface, which is helpful for the observation of the flow pattern (cf Fig. 3) but is not related to the instability.

The instabilities stop if the H^+-concentration decreases below a certain threshold value which is pH_{th} = 3.5 for the system salicylaldehyde + I_2 / H_2O + C16 Cl. The instabilities are reactivated by the addition of OH^--ions.

The type of instability is related to the contact angle formed between the interface and the solid substrate. The lateral motions occur if ϑ deviates substantially from 90° in either direction, while the wave motion is only observed if the contact angle is about $\vartheta \approx 90^o$. The amoeban motion is also observed if ϑ deviates substantially from 90°.

The vertical flow in the case of the wave motion is always directed from the organic to the aqueous phase. Thus if salicylaldehyde is replaced by octanol which swims on top of the water phase,the interface is deflected from the top to the bottom of the vessel.

The rotational wave motion exhibits the following properties.
1) The angular frequency, ω, increases linearly with the diameter of the glass cylinder. Therefore the group velocity of the rotating waves is constant:

$$v = \omega \ r$$

For the system salicylaldehyde + I_2/H_2O + C16Cl it is $v \approx$ 9 cm/sec at 20^oC. ω does not depend appreciably on the solute concentrations.
2) The temperature-dependence of ω is determined by that of the viscosity of the liquids. According to Fig. 4b it is

$$\eta \cdot \omega = const. \hspace{3cm} (2)$$

3) The amplitude of the wave increases about exponentially with the concentration of the iodine as shown in Fig. 4c.

V DRIVING FORCE AND ENERGY-DISSIPATION OF INSTABILITIES

The hydrodynamic flows at the liquid/liquid and liquid/solid interfaces are driven by negative fluctuations of the tensions (1) between the aqueous phase and the organic phase $\mathfrak{S}_{ow}$, and (2) between the first and the substrate $\mathfrak{S}_{ws}$. These depend on the state of the soap monolayer, and are therefore coupled to the chemical reaction. The destabilization of interfaces by negative fluctuations of the interfacial tension has indeed been predicted by linear stability analysis [2,3].
In the case of the lateral instability,the negative fluctuation of the tension $\mathfrak{S}_{ow}$ leads to a local expansion of the adsorbed interfacial layer. The situation for the wave motion is more complicated. Experimentally, this is observed when the contact angle between the liquid/liquid and solid/liquid interface (cf Fig. 5 for definition) is about ϑ = 90°, that is, when the two (average) tensions $\mathfrak{S}_{os}$ and $\mathfrak{S}_{ws}$ nearly compensate each other. As can be seen from Fig. 5 this occurs at a soap concentration of 10^{-2}M. At this point the total momentaneous interfacial tension at the line s/o/w is determined by the fluctuations in $\mathfrak{S}_{ow}$ and $\mathfrak{S}_{ws}$. It is clear that the fluid interface can yield to the expansive pressure caused by a negative fluctuation $\delta\mathfrak{S}_{ow}$ by deforming in the vertical direction. The contact angle will change according to

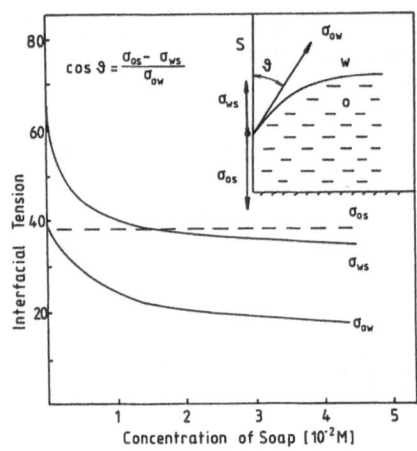

Fig.5:
Variation of average interfacial tension at solid/liquid and liquid/liquid interface as a function of C16 Cl-soap concentration for salicylaldehyde $+I_2/H_2O +$

C16Cl-system at $10^{-2}MI_2$.
The contact angle changes from $<\widehat{\Pi}/2$ to $>\widehat{\Pi}/2$ at about 10^{-2} M C16 Cl.
This threshold determines the onset of the wave motion.

$$\delta \cos \vartheta = \frac{1}{\sigma_{ow}} (\delta \mathcal{B}_{ad} - \cos \vartheta_o \, \delta \mathcal{B}_{ow}) \qquad (3)$$

where $\mathcal{B}_{ad} = \mathcal{B}_{os} - \mathcal{B}_{ws}$ is the adhesion tension and where ϑ_o is the average contact angle. The direction of the deformation depends on the sign of $\cos \vartheta_o$; $\delta \mathcal{B}_{ad}$ and $\delta \mathcal{B}_{ow}$. The experimentally observed upward deflection of the interface from the organic towards the aqueous phase observed for the system salicylaldehyde $+ I_2 / H_2O + C16Cl$ would result if the initial contact angle ϑ is slightly larger than $\vartheta = 90°$.

Let me now consider the shearing force for the observed vertical deflection by assuming that the fluid interface has a parabolic profile. The shearing force required for a deflection of amplitude h of the organic phase in the vertical direction can be estimated according to

$$\mathcal{B} = \frac{\Delta \varsigma \cdot g}{3} h R_o \qquad (4)$$

where R_o is the radius of the vessel and $\Delta \varsigma$ the density difference of the two liquids. For $h = 1/2$ cm, $\Delta \varsigma = 0.1$ gr cm^{-3} and $R_o = 1$ cm one obtains $\mathcal{B} = 16$ dynes/cm.
The energy dissipated per second by the rotating wave is given by the surface integral

$$\frac{dW}{dt} = - \eta \int \nabla \dot{v}^2 \, d\vec{a} \qquad (5)$$

∇v^2 is of the order of $\nabla v^2 \approx v^2/\lambda$ where λ is the wavelength of the rotating wave. For a cylindrical vessel of radius R_o the rate of energy dissipation is of the order of

$$\frac{dW}{dt} \approx - 4 \eta \frac{v^2 R_o^2}{\lambda} \approx 5.10^{-7} \, J \, sec^{-1} \qquad (6)$$

for $\lambda = 1$ cm, $\eta = 0.01$ gr cm^{-1} sec^{-1}, $v = 9$ cm sec^{-1}, $R_o = 1$cm.
The rate of chemical turnover can be described in terms of the rate of H$^+$-production in the aqueous phase. For the system salicylaldehyde $+ 10^{-2}$ M I_2 and $H_2O + 10^{-2}$ M C16Cl the pH changes from pH 5.5 to pH 3.5 in about 1000 seconds. Thus the rate of H$^+$-production is

$$d \ [\ H^+]/dt \ = \ 10^{-9} \ M \ sec^{-1}$$

The energy ΔE required per mole of H^+ turnover in order to account for the energy dissipation would be

$$\Delta E \ = \ \frac{dW/dt}{d[H^+]/dt} \ = \ 0.5 \ kJ \cdot M^{-1} \tag{7}$$

This is well below the value of the dissociation energy of the salicylic acid molecule.

VI POSSIBLE COUPLING BETWEEN CHEMICAL PROCESS AND INTERFACIAL MOTION

The soap monolayer adsorbed to the liquid/liquid interface is essential for the instability because the latter is abolished if the soap is replaced by a non-amphiphatic ammonium salt. The lipid-monolayer may play a twofold role: First it acts as a barrier for the fluxes of the reactants j_{ow} and j_{wo} (cf. Fig.2). Secondly, it determines the liquid/liquid interfacial tension. Both the permeability and the tension are closely related to the molecular structure of the monolayer which may be modulated by the chemical reaction. Thus the monolayer plays an important role for the coupling between the chemical reaction and the interfacial instability.
The molecular mechanisms relating the fluctuations of $\mathfrak{S}_{ow}$ and $\mathfrak{S}_{ws}$ to the chemical reaction are still an open question. One possibility is that the soap may exist in two possibly non-miscible configurations: for instance, as dissociated cation and as non-dissociated salt of the salicylic acid. The total lateral pressure Π_{ow} of the <u>charged</u> interface may be expressed as

$$\Pi \ = \ \alpha_1 \Gamma_1 \ + \ \alpha_2 \Gamma_2 \ + \ \beta_{12} \ \Gamma_1 \Gamma_2 - \frac{4 \Pi \ q_s^2}{\mathcal{E}_o \chi_o + \mathcal{E}_w \chi_w} \tag{8}$$

where Γ_1 and Γ_2 are the area concentration of the two species and where β_{12} is the coupling term of the regular solution theory. χ_o, $\mathcal{E}_o$ and χ_w, $\mathcal{E}_w$ are the Debye lengths and the dielectric constants of the organic and the aqueous phase and q_s is the surface charge density. The third term accounts for the lateral pressure of the Coulomb repulsion between the head group charge of the soap and the diffuse Gouy-Chapman layers of the organic and the aqueous phases.
As is well known, the diffuse layer may destabilize interfaces[2,3]. Thus,both a change in the composition or an increase in the surface charge density could account for the negative fluctuation in the tensions. Since the monolayer is already highly charged before the reaction starts,it is very likely that it is the change in composition which triggers the instability.
There appear to be two counteracting mechanisms: The first is a positive feedback between the change in the state of the monolayer and the flow of reactants from the organic into the aqueous phase. Thus,the mass transport could be facilitated by fluctuations in the lateral packing density of the monolayer arising when the S^+ soap ion is transformed into SA (s. Fig. 2b). The heterogeneity of mixed monolayers composed of two coexisting phases has recently been established [4]. Such lateral density fluctuations are well known to facilitate the passive transport through membranes. A negative feedback mechanism could be the formation of H^+ at the interface,which inhibits the transformation of I_2 into IO_3^-.

In this picture the decrease of $\mathfrak{S}_{ow}$ (corresponding to a positive fluctuation of Π_{ow}) could be attributed to the formation of AS since it is well known that a neutralization of charged head groups causes a dramatic increase in the packing density of the chains. If AS is produced at a certain site of the monolayer, it would cause a local lateral expansion which could then lead to a facilitated transport j_{on} and accelerate the chemical turnover. This would just correspond to a positive feedback. Since the production of IO_3^- is prevented by H^+ the reaction would soon cease, and the monolayer could go back into the resting state.

VII ANALOGIES TO CELL LOCOMOTION

Cells like amoeba, macrophages or fibroblasts move on surfaces by continuously extending and rectracting processes. The latter can be fingerlike (called pseudopodia), as in the case of ameoba or sheetlike (called lamellipodia) as in the case of macrophages [6].

There is one important difference between the motion of these two types of cells. The pseudopodia of amoeba can extend into the surrounding fluid and their formation is coupled to a flow of cytoplasm in the direction of the extension. The flowing cytoplasm transforms into a gel-like state at the tip of the newly forming pseudopod, while it changes into a more fluid state at its sources (situated at a retracting extension). This streaming is most probably driven by components of the cytoskeleton and the chemomechanical process would then correspond to that of muscle contraction [7] In the case of the motion of the vertebrate cells on solid substrates, however, the lamellipodia extend by spreading on the substrate. The advancement of the pseudopodia and the retraction of the trailing edge occurs in steps as shown in Fig. 6a. The cell changes its shape from a more spread to a more blown-up form much in the same way as the moving droplets of organic phase in the soap solution (Fig. 3f). Part of the plasma membrane remains attached to the surface, while the sheet-like extension is carried back towards the reacting end on the upper surface of the cell in a wavelike motion as shown in Fig. 6c.

The motion of the cell is a very complicated process which involves the concerted action of elements of the plasma membrane and of the cytoskeleton [5, 6]. The latter certainly plays an essential role in the stabilisation of the sheets on the lamellipodia. The spreading of the leading edge on the substrate is, however, initiated by a change of the contact angle formed with the substrate. It is well conceivable that the (necessary) change in the interfacial tension is coupled to one or several biochemical reactions at the membrane surface, which eventually lead to an interfacial instability.

The local change in the interfacial tension may be caused by a change in the local concentration of counterions at the cell surface at the site of membrane/substrate contact, or by a lateral reorganization of the membrane components. Evidence for the first process is provided by the finding of Nuccitelli et al. [8] that the formation of pseudopodia is accompanied by a steady current entering at the trailing edge and leaving at the tip of the pseudopod. The current has a density of 0.1 - 0.2 A cm^{-2} and is carried partially by Ca^{++}-ions.

The molecular mechanism of the change in interfacial tension is completely unknown. The outer surface of the plasma membrane is made up of an entangled network of charged polysaccharide and polypeptide formed by the head groups of glycoproteins (e.g. cell surface receptors) and glycolipids. The interfacial tension of this polyelectrolytic network could be easily modulated by local changes in the ion composition.

Evidence for the close relationship between cell migration on substrates and the change in contact angle comes from experiments by Dunn and Heath [9]. These authors observed that the cells cannot move over the edge of a prism-shaped substration if the angle ϑ (cf Fig.7) is greater than a threshold angle. This has been attributed to the stiffness of the bundles of microfilament of the cytoskeleton. However, an equally probable explanation is that the negative fluctuation of the membrane/water interfacial tension is not sufficient to produce a resultant interfacial force in the direction of motion.

Another interesting point is that the cell shows a preference for more adhesive materials like metals, which could be explained in terms of a facilitated cellular motion due to a reduction in the adhesion force.

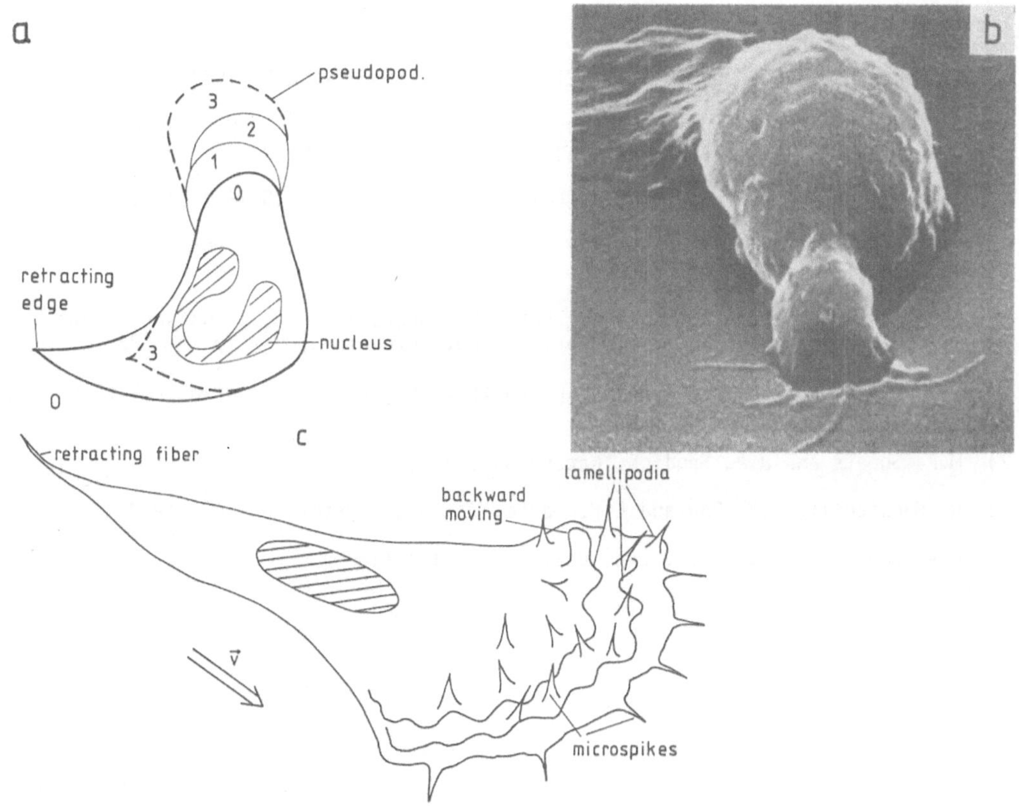

a

pseudopod.

3
2
1
0

retracting
edge

nucleus

3

0

retracting fiber

c

b

lamellipodia

backward
moving

$\vec{v}$

microspikes

Fig. 6:
a) Schematic view of granulocyte locomotion on solid substrate. The cell transforms from a round to a spread shape if it comes in contact with the substrate. The numbers indicate successive positions of the leading and the trailing edge. Note that the retracting edge follows by sudden movements (After Bessis[6]).
b) Scanning electron micrograph of macrophage in the dynamic state. A pseudopod is formed at the left upper corner. At the retracting end (uropod) filaments are formed which try to fix the cell.
c) Schematic view of cell (fibroblast) crawling on a surface. At the leading edge sheet-like exclusion -the lamellipodia- are formed which first spread on the surface and move then back towards the retracting end in a wavelike manner.

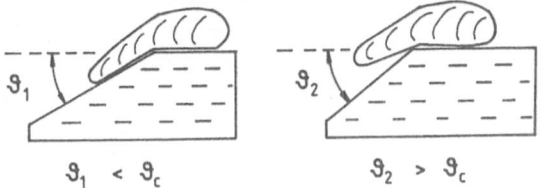

ϑ_1

ϑ_2

$\vartheta_1 < \vartheta_c$

$\vartheta_2 > \vartheta_c$

Fig. 7:
Schematic view of cell moving across the edge of a prism-shaped glass surface.

ACKNOWLEDGEMENT:

This work was performed during a stay at the Stauffer Laboratories at the Stanford University. The generous hospitality of Prof. H.M. McConnell is most gratefully acknowledged. Moreover, I am most grateful for the help of H. Riegler who prepared the experimental presentation.

LITERATURE:

1) E. Nackache and M. Dupeyrat, J. Electroanal. Chem. 141 283 (1982)

2) P.M. Bisch and A. Sanfeld, in Lecture Notes in Physics; J.C. Legros and J.K. Platten eds., Springer Verlag Berlin-New York 1978, p 112-119

3) C.A. Miller and L.E. Scriven, J. Colloid Interf. Sci. 3 360-371 (1970)

4) A. Fischer, M. Lösche, H. Möhwald and E. Sackmann, J. de Physique Lett. 45 L.785 (1984)

5) B. Alberts, D. Bray, J. Lewis, M. Raff, K. Roberts and J.D. Watson, "The Molecular Biology of the Cell" Garland Publishing Inc., New York, 1983

6) M. Bessis, "Living Blood Cells and their Ultrastructure", Springer Verlag, 1973

7) M.P. Sheets and J.A. Spudich, Nature 305, 31 (1983)

8) R. Nuccitelli, M.M. Poo and L.F. Jaffe, J. Gen. Physiol. 69 743 (1977)

9) G.A. Dunn and J.P. Heath, Exp. Cell Res. 101 1 (1976)

Contraction and Oscillations in a Simple Model for Cell Plasma Motion

Wolfgang Alt

Universität Heidelberg, Sonderforschungsbereich 123, Stochastische mathematische Modelle, Im Neuenheimer Feld 293, D-6900 Heidelberg, Fed. Rep. of Germany

1. Oscillatory activities of various cell types and their "motor"

In contrast to many experimental physico-chemical systems, as for instance the fluid interface experiment presented by Erich Sackmann [this symposium] with its similarities to amoeboid motion, biological systems like the intracellular plasma of moving cells (slime molds, fibroblasts, leukocytes etc.) cannot be so easily controlled by well defined external conditions.

The observed motions of cells [4,10,11,20] or cell plasma fragments [3,13] rather suggest that the cytoplasma contains an autonomous motile apparatus, a "motor", which is relatively insensitive to external perturbations. In several different experiments it has been documented that oscillatory moving patterns can be interrupted or suppressed by some mechanical or biochemical manipulation, but that afterwards, the original activity is completely regenerated or partially rearranged.

Take, for example, the polynuclear slime mold *physarum polycephalum* which grows and spreads on agar surfaces as a network of plasmodial veins (or strands). The leading fronts expand in an oscillatory manner, supported by a well observable and highly coordinated "shuttle

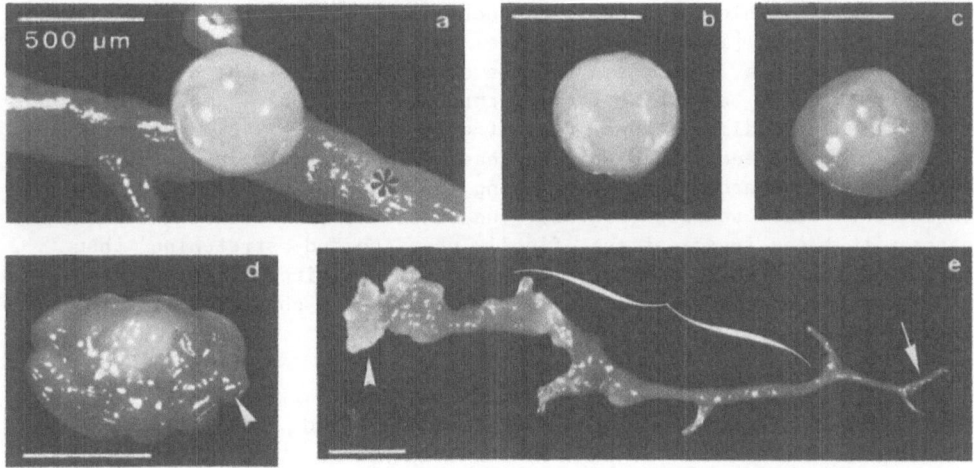

Figure 1: *Physarum polycephalum*: Transformation of a protoplasmic drop into a small plasmodium. From [2]:Fig.1 , with kind permission of the authors. See text below.

streaming" within the strands [11] . When the cell membrane is
punctured by a glass needle, it releases a drop of cytoplasma (Fig.1a)
which then continues its own life in the following stages:

After about 10 - 20 minutes it contracts (Fig.1b,c) and initiates
regular oscillations (Fig.2), until after 1 - 2 hours the drop flattens
onto the surface, extends pseudopods (Fig.1d) and finally spreads into
a locomoting plasmodium which (after about 3 hours) has only one vein
and one leading front, but already shows the characteristic features
of larger plasmodia (Fig.1e).

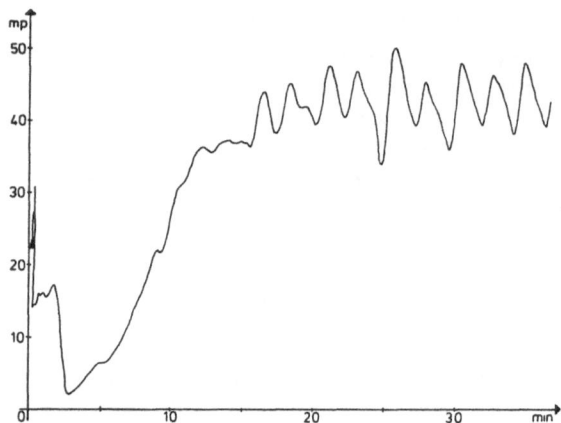

Figure 2: De novo generation of oscillating contractions in a protoplasmic drop.
Contractive force plotted over drop age. From [2]:Fig.4 .

Thus, *physarum* plasmodia represent a fully regenerative system
of "biological motion". Moreover, since it is one of the most studied
moving microorganisms, its "motor" has been found : filaments of
polymeric actin which probably attract each other by binding to
oligomeric myosin [12] . During the above - described regeneration
process, for example, actomyosin fibrils can be observed which, in the
early drop stages, are distributed through the whole protoplasma
(Fig.3a) then gradually become concentrated near the "plasmalemma".
This is the plasma membrane of an increasing system of invaginations
originating at the drop boundary, growing inside and partially fusing
with intracellular vacuoles (Fig.3b). The actin fibrils make intimate
contact with these invaginations, finally pressing and stretching them
like a sandwich (Fig.5c in [2]), thereby creating circular formations,
which later become the tubular "walls" of endoplasmic channels (see the
asterisks in Fig.3c).

Note that the rhythmic contractions already start (Fig.2 and
Fig.3b) when the separation betweeen "endoplasma" and the surrounding
"ectoplasma" is not yet performed. This means that the protoplasmic
actomyosin "motor" is able to create oscillations of the whole
plasmodial drop as soon as it provides sufficient interconnections
between different parts of the membrane system.

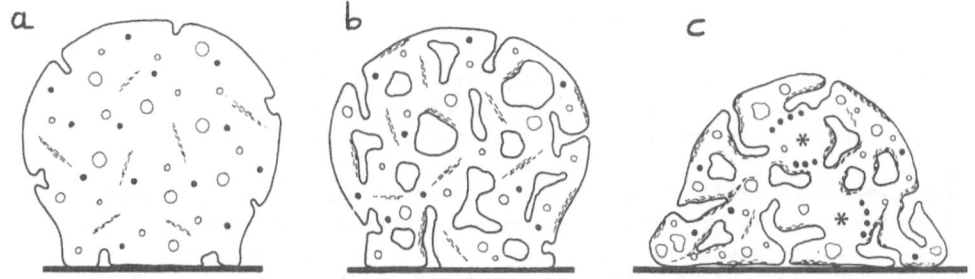

Figure 3: Schematic representation of sections through a protoplasmic drop of age (a) 5 min,(b) 10 min and (c) in an older stage. From [2]:Fig.10 .

A second typical example of cell plasma activity is the migration of leukocytes, fibroblasts and other similar cells on surfaces. During locomotion,they often polarize their cell shape into a "leading edge", where membrane protrusions (pseudopods, lamellipods or filopods) are extended and retracted in an oscillatory manner, and into an opposite "tail", where a relatively stable knob-like formation (uropod) appears. See the schematic view of a leukocyte in Fig.4. Apparently the uropod is a "center of contraction": A high concentration of polymerized actin is observed (compare e.g. [19]) , and the cell leaves retraction fibers,or sometimes even small cell fragments, behind it (see Fig.4: 3 and 4).

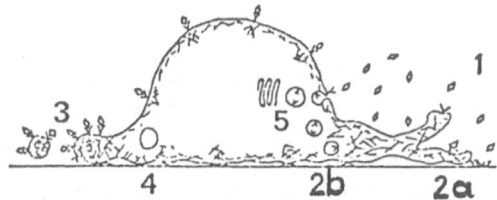

Figure 4: Schematic view of a leukocyte moving to the right on a glass surface. From [18]. See text.

In contrast to this, the front region shows more variable dynamic activities: In a more or less regular temporal and spatial order, lamellipods (flat, broad membrane extensions) are protruded along the substrate (Fig.4:2a). They partially adhere , while their front part usually erects into the medium (Fig.4:1) and moves backwards to the cell body like a wave or "ruffle" (for fibroblasts see [1]).

The ruffles , together with other particles in the upper cell membrane,are actively transported backwards [8,10], probably by an intracellular contractive system acting on membrane proteins. The same system could also induce membrane invaginations and internal vesicles in that region (Fig.4:5). Moreover, there is strong evidence that actin filaments are attached to the lower membrane at sites of adhesion (see Fig.4:2b and [19]). These different observations suggest that, as in *physarum*, a contractive actomyosin system serves as one "motor" for various motions of leukocytes, fibroblasts and similar cells. For a more detailed review see [7].

Fibroblasts treated with cytochalasin B show the same 5 min cyclic protrusion and retraction process as observed in lamellipodial ruffling, but now of thin filopods extending from a zig-zag shaped cell surface, while cells treated with colchicine show the extension and retraction of larger pseudopods with a period of about 10 minutes [15].

Moreover, even small cytoplasmic fragments (microplasts), treated mechanically or biochemically as above, continue to produce the different membrane projections and move them in a stereotypical and autonomous way [3]. This, and the knowledge of the effects which the mentioned drugs have on various intracellular filaments, again favor an actomyosin "motor" to be responsible for driving the oscillatory movements. They seem to obtain their different appearence only because of different chemical and physical side conditions.

For the purpose of mathematical modelling, which might help to interpret these observations or to perform new quantitative experiments with cell plasma motion, let us summarize the desirable properties of such an actomyosin "motor":

1. Formation of stable local contraction centers (e.g. at the uropod)
2. Induction of membrane invaginations and vesicle formation
3. Fusion of different membrane fragments (vacuoles, plasmalemma)
4. Tendency to oscillate in connection with parts of cell membranes and possibly also a
5. Coordination of contractile oscillations at different parts of the cell.

From the biochemical viewpoint, all these functions are realizable since the required molecules have been localized in the cytoplasma of physarum, leukocytes, fibroblasts and similar cells (compare e.g. [17]): Monomeric G-actin, polymeric F-actin, actin-binding proteins (leading to nucleation, filament or bundle formation) and oligomeric myosin (serving as links contracting different actin filaments towards each other).

2. Mathematical model of a viscous actomyosin system

In 1981, M.Dembo, F.Harlow and I [7], proposed a mathematical model which in an idealized way regards the distributed and interconnected system of actin filaments or bundles ("actin nodes") as a viscous, reactive fluid. (See [14,16], where also elasticity is included.) The modelling equations are based on the following hypotheses concerning the physico-chemical interactions between the actin nodes (see Fig.5 and compare [9,12] for an experimental background):

a. Assembly: Nucleation of monomeric G-actin at actin-binding proteins and subsequent polymerization
b. Contraction: Mutual attraction of actin nodes via binding to myosin filaments
c. Viscosity: Mutual friction between moving actin nodes by shearing forces
d. Disassembly: Steric inhibition of actin filaments in high densities and subsequent dissolution of the actin nodes

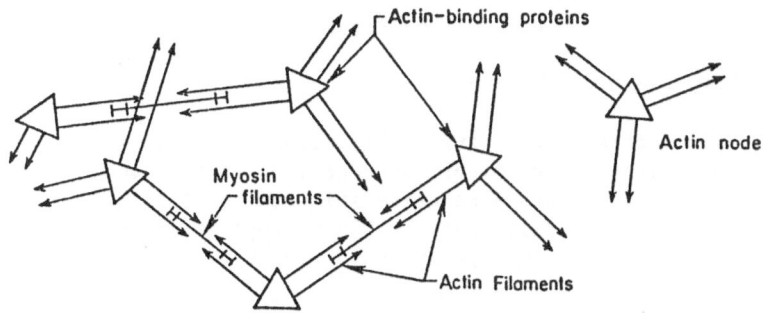

Figure 5: Actin node hypothesis: Actin-binding proteins serve as "nuclei" for the poly-merization of actin filaments. Myosin filaments cause their mutual attraction. From [7].

Denoting by G(t) the region filled with cytoplasma at time t and by

u(t,x)   the mean density
v(t,x)   the mean velocity   } of actin nodes at location x ∈ G(t) ,

then the mass balance equations are

$$\partial_t u \; + \; \nabla_x \cdot (uv) \;\; = \;\; N_+(u) - u \, N_-(u) \;\; = \;\; f(u) \; . \tag{1}$$

Here N_+ and N_- denote the assembly and disassembly rate of actin nodes, which might depend on their actual density $u=u(t,x)$. The resulting reaction kinetics f(u) is supposed to have exactly one stable equilibrium, at u=1 say. Take for example Fig.6:

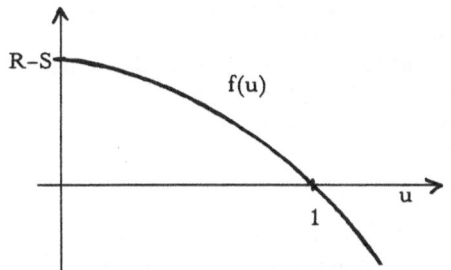

Figure 6: Particular assembly-. disassembly kinetics for actin nodes $f(u) = R(1-u) - S(1-u)^2$. The negative slope $R = -f'(1)$ is called the "reactivity" of the actomyosin system.

This model is based on the simplifying assumption that there are constant reservoirs of G-actin, of actin binding protein and of regulatory molecules (ATP,Ca^{++}) equally available in any part of the cell. Variations or inhomogeneities of these controlling substances can easily be built into the model (compare the model in [16] with Ca^{++}-regulation). However, our point is to demonstrate that already a very simple reaction kinetics (as in Fig.6) , in combination with the attractivity and viscosity properties described below can induce contractive oscillations.

Under the assumption that attraction and shearing forces dominate any inertia effects, and that they are proportional to the gradient $\nabla_x u$

and the density   u   respectively,   the resulting force balance equations
are

$$A \ u\nabla_x u + \nabla_x \cdot (u\nabla_x v) = B \ u \ v \qquad\qquad (2)$$

Here   A   denotes   the   quotient   of   the   attraction   coefficient
and   the   viscosity   coefficient,   and   might   be   called   the   local
"contractivity",   whereas   B   represents   a   relative   friction coefficient.
In all subsequent investigations   I will   assume   B = 0   neglecting   any
friction between actin filaments and the cytoskeleton (see Fig.7).

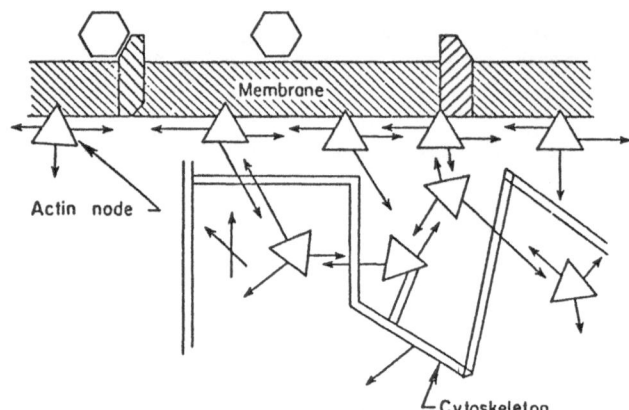

Actin node

Figure 7: Schematic view of
possible interactions of actin
nodes with each other, with
the membrane and the
cytosceleton. Comp. also [7].

   On   the   other   hand , I   want   to   include   interactions   of   the
actomyosin system with the plasma membrane,which is represented   by   the
boundary   B(t)   of   the   domain   G(t).   Let   us   assume   that   the   actin
binding   proteins   anchored   in   the   membrane   (see   Fig. 7)   are
uniformly   distributed.   Clearly,   this is not realistic, since   in   all
processes like cell migration, membrane invagination   or   cap   formation
actin filaments become concentrated at   "centers   of   action"   (see [7]).
However, models   for contraction phenomena   along   the   membrane surface
itself would require 2-dimensional analoga of equations (1) and   (2)   on
the boundary B(t).

   Instead of this   quite complex   mathematical system , let us   attempt
a simple model which considers the balance equation only for   forces   in
the   normal   direction n(t,x)   at   a   boundary   point $x \in B(t)$.   Again
neglecting   any   inertia   effects , we   arrive   at   two   free   boundary
conditions for the <u>normal speed of the membrane surface</u>

$$S'(t,x) = \tfrac{1}{2} \ M \ \{ \ P(V) + T(x) - A \ u(t,x) \ \} \qquad\qquad (3)$$

and the mean <u>velocity of the actin fluid</u>

$$v(t,x) \ = \ S'(t,x) \ n(t,x) \qquad \text{for } \ x \in B(t) \qquad\qquad (4)$$

which just follows   the   displacement   of   the   membrane.   The   membrane
speed S' in (3) is proportional to the sum of
a. the <u>pressure</u> P depending on the volume   V   of the cell G(t) , e.g.

$$P(V) = p \ (1/V - 1/V_{max})$$

with a maximal volume V_{max},

b. the <u>surface tension</u> T depending on the local curvature or other properties of the membrane at location x, and

c. the <u>inward attraction</u> by intracellular actin nodes attaching the point x of the membrane with density $u(t,x)$.

The coefficient M is related to the mobility of the membrane in the corresponding medium.

Although equations (1)-(4) represent a mathematically well–posed system, at the moment it seems to be too difficult to analyze, even numerically. Therefore, I want to restrict the present investigations to the 1-dimensional case.

3. Analysis of the 1-dimensional problem

Let us consider an experimental situation of a cylindrical glass tube containing a cytoplasma fragment

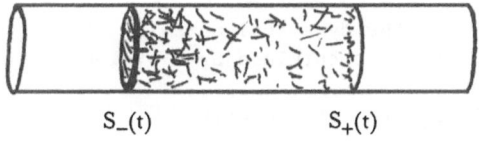

S_(t)　　　　　　　S_+(t)

Figure 8: Plasma fragment within a cylindrical tube having two free membrane surfaces at $x=S_-(t),S_+(t)$ possibly changing with time t.

having a length $L(t) = S_+(t) - S_-(t)$.

Our aim is to describe the dynamical behavior of the free boundaries S_+ and S_- as well as of the interior density distribution $u(t,x)$, $S_-(t) < x < S_+(t)$ in dependence on the system parameters, namely the "reactivity" R, the "contractivity" A and the "boundary parameters" in equation (3).

It is appropriate to consider the stream curves (or characteristics) $S=S(t)$ defined by $S'(t)=v(t,S(t))$, which emanate from initial points on the interval $G(0)=[S_-(0),S_+(0)]$. Note that the actin fluid contracts at places where the stream curves converge towards each other, see [6].

After transforming the domain $G(t) = [S_-(t),S_+(t)]$ to the unit interval [0,1] by

$$y = {}^{1}/_{L(t)} \ \{x - S_-(t)\}$$

we deduce from the 1-dimensional version of system (1)-(4) that the <u>change of actin density along the stream curves</u> is given by

$$\partial_t u + v \ \partial_y u = f(u) + {}^{1}/_{2} \ A \ u^2 - K(t) \tag{5}$$

where the <u>velocity of stream curves</u> is

$$v(t,y) = \int_0^y \frac{K(t)}{u(t,\bullet)} - {}^{1}/_{2} \ A \ u(t,\bullet) - L'(t)/_{L(t)} \tag{6}$$

with the averaging functional

$$K(t) = \{ L'(t)/L(t) + 1/2 \; A \int_0^1 u(t,\bullet) \} \Big/ \int_0^1 \frac{1}{u(t,\bullet)} \tag{7}$$

The <u>cell length</u> itself satisfies the following differential equation

$$L'(t) = M \{ p \; (1/L(t) - 1/L_*) + A \; (1 - \frac{u(t,1)-u(t,0)}{2}) \} \tag{8}$$

Here the critical length

$$L_* = (\; 1/L_{max} + (A-T)/p \;)^{-1} < L_{max} \qquad , \text{ provided } T < A$$

characterizes the unique homogeneous equilibrium of system (5)-(8), namely $u = 1$, $v = 0$. Decomposing the density $u(t,y) = U(t) + z(t,y)$ into the total mass $U(t)$ of actin nodes in the cell, and the deviation $z(t,y)$ from the homogeneous distribution, we obtain the equation

$$\partial_t z = (A - R) \; z + (\text{higher order terms in } z).$$

This proves that under the <u>stability condition</u>

$$A < R \; , \qquad \text{meaning "reactivity", dominates "contractivity",}$$

all spatial disturbances are homogenized. Thus the dynamical behavior of the asymptotically homogeneous plasma fragment is determined by a system of ordinary differential equations for the <u>polymerized actin mass</u> $U(t)$ and the <u>cell length</u> $L(t)$, namely

$$\left. \begin{array}{l} U' = f(U) - L'/L \; U \\[2mm] L' = M \{ p(1/L-1/L_*) + A \; (1-U)\} \end{array} \right\} \tag{9}$$

The equilibrium $U = 1$, $L = L_*$ is stable if $A < p + L_*/M \; R$, and no activity of the contractile system persists.
In the other case, the negative feedback system (9) exhibits oscillations in such a way that all trajectories converge to a stable limit cycle (Fig.9), which periodically goes through the following three stages

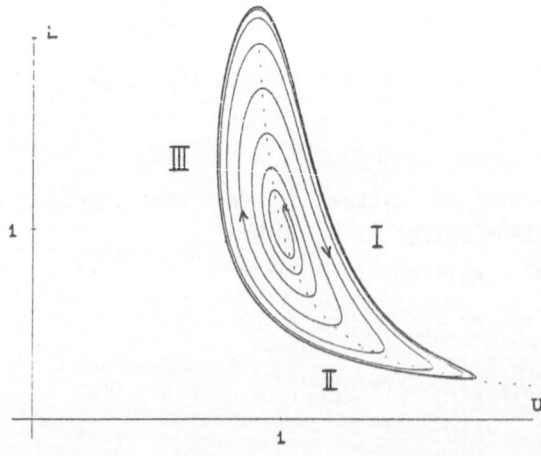

Figure 9: Contraction–relaxation limit cycle as solution of (9) with A=5, R=4, S=3, M=3, p=1 and L_*=1. See text below.

I.   Contraction (U increasing, L decreasing)
II. Disassembly (U decreasing)
III. Relaxation (L increasing)
This contraction-relaxation cycle seems to represent the basic process
which drives the observed plasma oscillations of plasmodial drops [2]
or freely suspended (isotonic) plasmodial strands [11] of *physarum* .

   In the reactivity-contractivity parameter diagram (Fig.10) we have
discussed so far the two upper regions, namely that of no activity (I)
and that of "homogeneous" oscillations (II) with a uniform   spatial
distribution of actin filaments in the plasma.

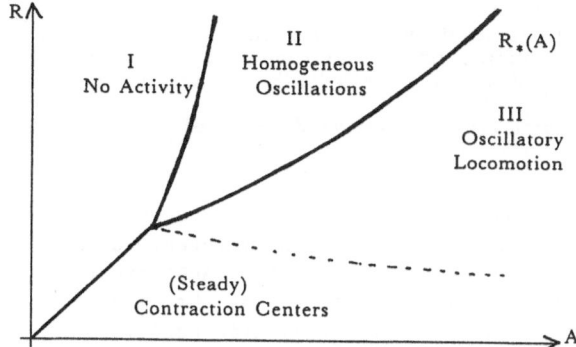

Figure 10: Parameter diagram R (reactivity) versus A (contractivity) for system (5)–(8).
The bifurcation curve $R=R_*(A)$ is implicitly given by $R = A\underline{U} + 2S(1-\underline{U})$,
where $\underline{U}$ is the time average of U(t) over the limit cycle in Fig.9 .

   When the contractivity parameter A increases and   eventually
crosses one of the two "bifurcation curves" $R = A$, or $R = R_*(A)$ given
in the caption of Fig.10, then the model system (5)-(8) shows a rich
variety of nonhomogeneous contraction and oscillation phenomena. The
simplest is the formation of steady contraction centers at one side of
the interval. For fixed boundaries (M = 0) this has been extensively
discussed in [5]. Actin nodes are transported by attraction towards
the contraction center, where they are dissolved due to the high
density. Secondary local contraction centers generally are attracted
and absorbed by the dominating center (see Fig.5 in [5]) in a way
which resembles the observed waves of ruffles moving backwards to the
cell body (see section 1).

   In cases where the secondary center is marginally localized at the
opposite   boundary , we   (numerically)   observe   the   onset   of
spatio-temporal oscillations between a strong and a weak contraction
center (Fig.11). A stable contraction-relaxation cycle develops , which
at places between the two centers periodically goes through the
following two stages
I. Attractive transport away to the dominating center and
       depletion of polymerized actin filaments
II. De novo assembly of actin filaments

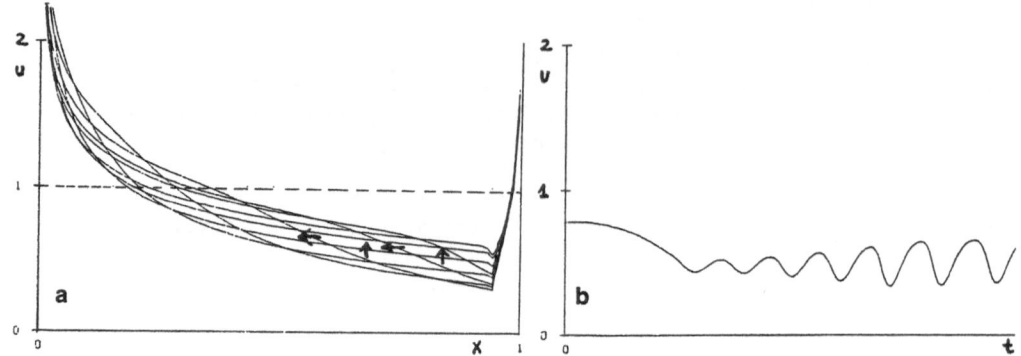

Figure 11: Spatio–temporal oscillations in a fixed cell. Parameters as in Fig.9, but A=5.5 and M=0. (a) Density u(t,x) for different times t with contractile waves (⬅) from a smaller marginal center (right) to the dominating center (left), and subsequent de novo assembly of actin nodes (⬆).
(b) u(t,x) for fixed x=0.95, approaching a limit cycle with period of 2.6 [min].

During this cycle, the plasmatic flow towards the dominating contraction center varies by a factor of about 2 . In different cases where two equally strong contraction centers compete with each other, we even observe a reversal of this flow, but only in the middle region, and the amplitude of oscillations is quite small (compare[7]).

These spatio-temporal contractive oscillations in the simple 1-dimensional model with fixed boundaries are, on one hand, too special or too weak to explain the complex phenomena like shuttle streaming or lamellipodial protrusion. On the other hand, they demonstrate that already a viscous, contractile polymer fluid with a simple reaction kinetics (as in Fig.6) is capable of producing regular oscillations.

Moreover, when both the formation of local contraction centers and the interaction with a movable membrane surface (M positive) is allowed, then the mathematical "model plasmodium" represented by system (5)-(8) is able to show its "first steps" in locomotion. For example , a contraction center at the left margin "pulls" the left membrane surface (comp. Fig.8) stronger than the right one, so that due to the internal pressure the latter is "pushed" and the whole plasma fragment is displaced to the right. In cases where R is large enough, so that (A,R) lies in region III of Fig.10 , the resulting "protrusion" to the right and the subsequent "retraction" at the left proceed in a regular oscillatory manner (see Fig.12a).

If, on occasion (e.g. after long numerical error addition), a secondary contraction center develops at the other margin, this usually becomes equally strong and induces a competitive contraction at both ends of the cell. This finally results in a regular for- and backward displacement of the whole cell due to the oscillatory properties described above (see Fig.12b).

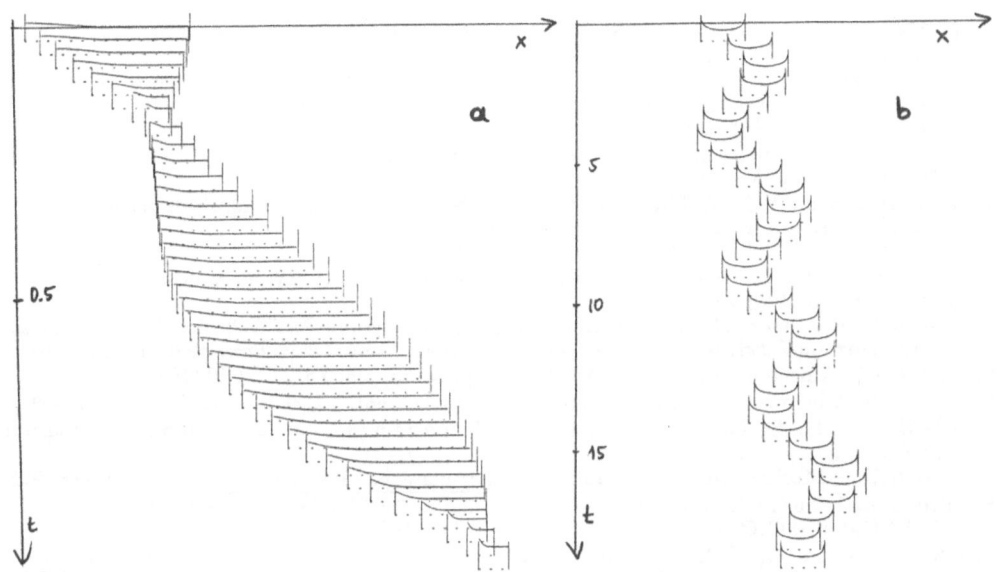

Figure 12: Locomoting "model plasmodium" as solution of the 1–dimensional system (5)–(8) with parameters as in Fig.9, but M=4. For description see the text above.

4. Summary

The contractile intracellular actomyosin system has, in a simplified way, been modelled as a highly viscous reactive fluid with density dependent disassembly rate of polymerized filaments. Their interaction with the membrane, together with the intracellular hydrostatic pressure leads to spatio-temporal contractive oscillations of the cell plasma, according to a contraction - disassembly - relaxation - assembly cycle. The presented mathematical model could be one elementary step towards more detailed explanations of the mechanism of cell locomotion.

References

1. M. Abercrombie,J.E.M. Heaysman and S.M. Pegrum: The locomotion of fibroblasts in culture: II."Ruffling." Exp.Cell Res. **60**, 437–444 (1970)
2. F. Achenbach and K.E. Wohlfarth–Bottermann: Morphogenesis and disassembly of the circular plasmalemma invagination system in *physarum polycephalum*. Differentiation **19**,179–188
3. G. Albrecht–Buehler: Autonomous movements of cytoplasmic fragments. (1981) Proc.Natl.Acad.Sci.USA **77**, 6639–6643 (1980)
4. R.D. Allen and D.L. Taylor: The molecular basis of amoeboid movement. In:"Molecules and Cell Movement". Eds.S.Inoue,R.E.Stephens. Raven Press, New York 1975, pp.239–258
5. W. Alt: Contraction patterns in a viscous polymer system. In: Proc. Workshop "Modelling of Patterns in space and time". Heidelberg 1983. Eds.W.Jäger,J.Murray To appear in Springer Lect.Notes in Biomath. Vol.**55**
6. W. Alt and M. Dembo: A contraction–disassembly model for intracellular actin gels. In: Proc. EQUADIFF 82, Würzburg. Eds.H.W.Knobloch,K.Schmitt Springer Lect.Notes in Math. Vol.**1017**, Berlin 1983, pp.2–9
7. M. Dembo, F. Harlow and W. Alt: The biophysics of cell surface motility. In:"Cell Surface Dynamics. Concepts and Models" Eds.A.Perelson,Ch.DeLisi,F.Wiegel. M.Dekker,New York 1984, pp.495–542

8. M. Dembo and A.K. Harris: The motion of particles adhering to the leading lamella of crawling cells. J.Cell Biol. **91**, 528–536 (1981)
9. K. Götz von Olenhusen and K.E. Wohlfarth–Bottermann: Evidence for actin transformation during the contraction–relaxation cycle of cytoplasmic actomyosin.
Cell Tiss.Res. **196**, 445–470 (1979)
10. A.K. Harris: Cell surface movements related to cell locomotion.
Ciba Found.Symp. Vol.**14** (NS), 3–26 (1973)
11. N. Hülsmann and K.E. Wohlfarth–Bottermann: Spatio–temporal relationships between protoplasmic streaming and contraction activities in plasmodial veins of *physarum polycephalum*. Cytobiol. **17**, 317–334 (1978)
12. G. Isenberg and K.E. Wohlfarth–Bottermann: Transformation of cytoplasmic actin.
Cell Tissue.Res. **173**, 495–528 (1976)
13. K. Kuroda: Movement of cytoplasm in a membrane free system. In:"Cell Motility. Molecules and Organization" Eds.S.Hatano,H.Ishikawa,H.Sato. Univ.Tokyo Press 1979, pp.347–361
14. G.M. Odell: Amoeboid motions. (AMS) Lect.Appl.Math. Vol.**16**, 191–220 (1977)
15. R. Ohnishi: Cinematographic analysis of ruffling movement of L cells. As ref.13, pp.251–261
16. G.F. Oster and G.M. Odell: A mechanochemical model for plasmodial oscillations in *physarum*. As ref. 5
17. P. Sheterline: Mechanisms of Cell Motility. Molecular aspects of contractility. Acad.Press 1983
18. Th.P. Stossel: The mechanism of leukocyte locomotion. In:"Leukocyte Chemotaxis" Eds.J.I.Gallin,P.G.Quie. Raven Press 1978, pp.143–160
19. J.A. Sullivan and G.L. Mandell: Motility of human polymorphonuclear neutrophils: Micro–scopic analysis of substrate adhesion and distribution of F–actin. Cell Motility **3**, 31–46 (1983)
20. S. Zigmond, H.I. Levitsky and B.J. Kreel: Cell polarity: An examination of its behavioral expression and its consequences for PMN chemotaxis. J.Cell Biol. **89**, 585–592 (1981)

Circle Maps and the Periodic Forcing of Limit Cycle Oscillators

J. Bélair

Département de mathématiques et de statistique, Université de Montréal
Montréal, Québec, Canada H3C 3J7

L. Glass

Department of Physiology, McGill University, 3655 Drummond
Montréal, Québec, Canada H3G 1Y6

1. Introduction

Transformations of an interval into itself have been studied extensively in the last few years [1]. We briefly review results we have found on a related but more complex class of one-dimensional maps, namely maps of the circle into itself [2,3]. We assume that a biological rhythm (in our case, the spontaneous oscillation of cardiac cells) is associated with a globally attracting limit cycle. Under the hypotheses that a perturbation of this cycle is followed by a rapid return to it, one is led to the family of maps

$$\phi_{i+1} = f(\phi_i) = \tau + g(b;\phi_i) \qquad (1)$$

where ϕ_i is the phase of the oscillator just before application of the i-th stimulus, and b and τ are parameters corresponding, respectively, to the intensity and the (normalised) period of the perturbation. The map g is called a PTC (Phase-Transition Curve), and is experimentally obtained by application of one stimulus at different phases of the cycle. (A detailed derivation is given in [3]).

We consider in particular maps of the form

$$g(b;\phi_i) = M\phi_i + bh(\phi_i) \qquad (2)$$

where h is a function "close to" $\sin 2\pi\phi_i$. The cases M = 0,1 are relevant, in a topological sense, to the actual experiments performed on aggregates of embryonic chicken ventricular cells, submitted to intracellular electrical stimulation [2, 3, 4].

2. Rotation number and the existence of periodic orbits

A fundamental concept in these studies is the rotation number, as introduced by Poincaré [5]. The sequence of iterates ϕ_0, $\phi_1 = f(\phi_0),\ldots,\phi_N = f(\phi_{N-1}) = f^N(\phi_0)$ is used to define the auxiliary function

$$\Delta_i(\phi_0) = f(\phi_{i-1}) - \phi_{i-1}$$

and the rotation number

$$\rho = \limsup_{N\to\infty} \frac{1}{N} \sum_{i=1}^{N} \Delta_i(\phi_0).$$

When $\phi_N^\star = \phi_0$ and $\phi_i^\star \neq \phi_0^\star$, $1 \leq i < N$, $\phi_0^\star$ is called a periodic point of period N, and $\{\phi_0^\star,\phi_1^\star,\ldots,\phi_{N-1}^\star\}$ is a periodic orbit of period N. If $\left|\frac{\partial f^N}{\partial^N \phi_i}\Big|_{\phi_0^\star}\right| < 1$, this orbit is stable and if, furthermore, $\rho = \frac{K}{N}$, there is N : K phase-locking. If a critical point of F is a point on a periodic orbit,the orbit is called superstable.

In the case when M = 1 and f is a homeomorphism, the rotation number of a map is unique, and thus, finding phase-locked solutions is equivalent to having a map with rational rotation number [6]. In the more general case, when M is not necessarily one and f is not a homeomorphism, the rotation number is no longer sufficient to completely describe the dynamics.

A useful concept used in conjunction with the rotation number is the lift. Consider the continuous map $F : \mathbb{R} \to \mathbb{R}$ with the symmetry $F(x+1) = f(x) + M$ for all $x \in R$. Then,by considering $f = F$ (mod 1),we restrict the function to the cycle. The map F is called the lift of f and M is the degree of f. Now consider the nth iterate of the lift. Simple arguments [7] can be used to show that

$$F^n(x_0;\tau_0+1) = F^n(x;\tau_0) + 1 + M + M^2 + \ldots + M^{n-1}$$

$$F^n(x_0+1,\tau_0) = F^n(x_0;\tau_0) + M^n$$

as schematically shown in Fig. 1. Now fix x_0 and let τ vary from τ_0 to $\tau_0 + 1$. There must be at least $\sum\limits_{j=0}^{n-1} M^j$ different values of τ which give rise to periodic orbits for x_0 fixed. An immediate consequence is that, for x_0 fixed for M = 1. at least one value of τ can be found such that x_0 is a fixed point with any given rational rotation number [7, 8].

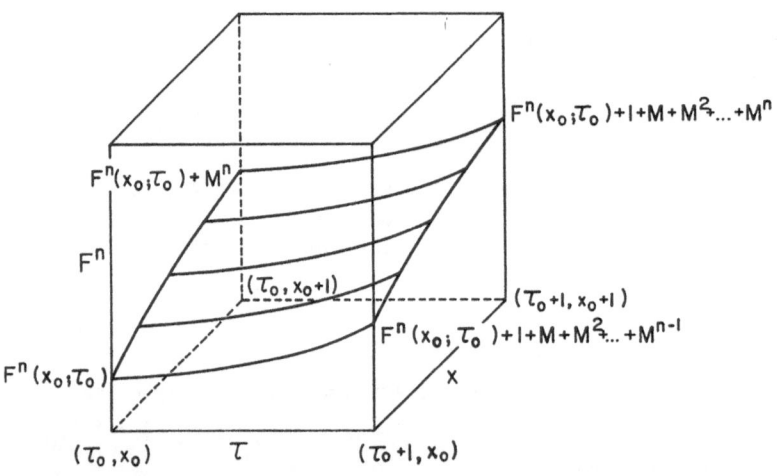

Fig. 1. Three-dimensional representation of the n-th iterate of the lift F of f, for any value of M.

Now consider the family of maps given by (1) and (2),where M = 1 and the map $g(b;\phi_i)$ is a homeomorphism. There is a locus of points in τ,b parameter space along which any given point x* is a fixed point with given rotation number (dotted line in Fig. 2). The projection of all such continuation lines, associated with any given rotation number, on the (b,τ) plane is called the Arnol'd tongue [6]. The Arnol'd tongues are ordered so that the rotation number of the Arnol'd tongues increases and τ increases keeping b fixed. Difficulties involved in extending these results to maps with degree different from 1 are discussed in [7].

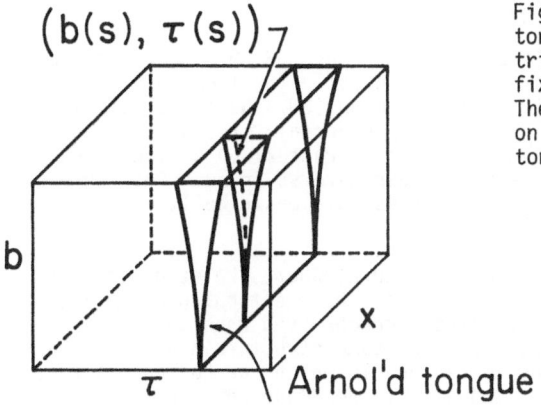

Fig. 2 The construction of an Arnol'd tongue. The dotted line shows in parametric representation b and τ for which a fixed value of x is on a periodic orbit. The closure of all such lines projected on the x = 0 phase is called the Arnol'd tongue.

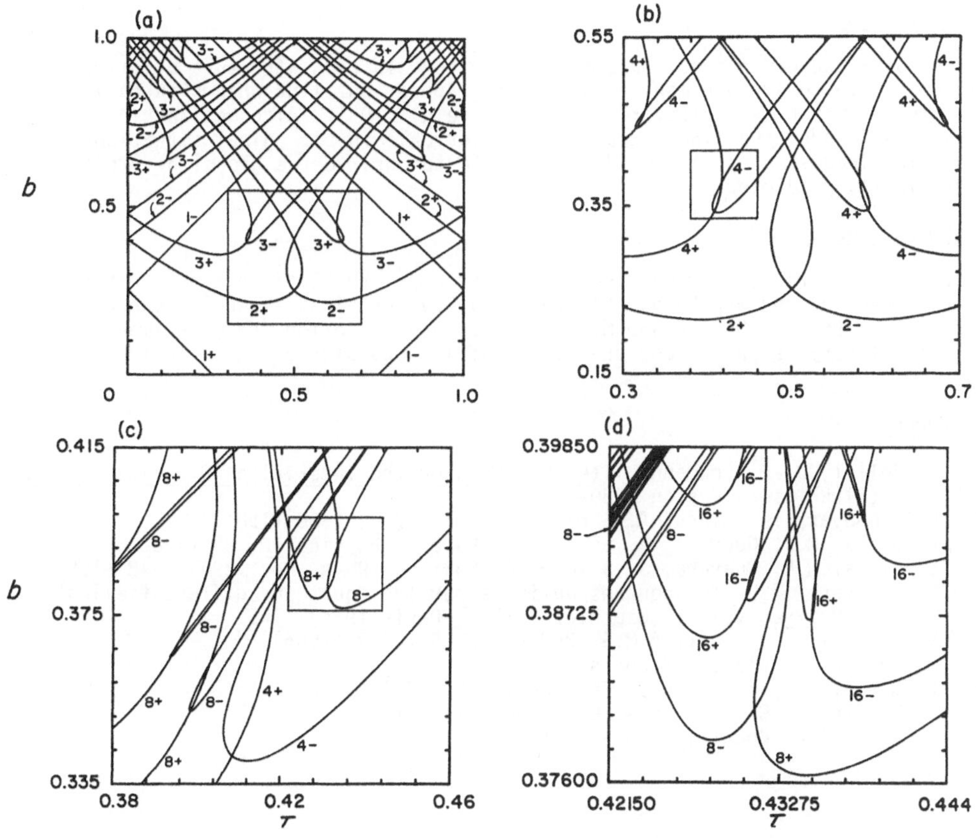

Fig. 3. Skeleton for a map of degree 0; k + (k-) represents a cycle of period k associated with the maximum (minimum).
(a) cycles of periods 1,2,3; not all cycles of period 3 are labeled;
(b) enlargement of the square in (a); periods 2 and 4 are shown;
(c) enlargement of the square in (b); periods 4 and 8 are shown;
(d) enlargement of the square in (c); showing period 8 and 16.
Note the topological equivalence.

3. Bifurcations of periodic orbits: The skeleton

We have performed a global analysis of the parameter space [8, 9], searching for the values of b and τ for which superstable orbits exist. The locus of super-stable orbits in parameter space is called the skeleton [8]. Numerical simulations have led to the observation of a self-similar and universal structure in the organization of the skeletons. This self-similarity is characterized by the repeated presence of a fundamental "cell" in the space of parameters, in numerous regions and at numerous scale-lengths. Figure 3 illustrates this basic cell, computed for a map of degree 0 [9].

   Successive blow-ups reveal a self-similar structure, where skeleton lines corresponding to periods 2 and 4, 4 and 8, 8 and 16 etc. have a topologically invariant position. Identical structures have also been observed when the basic period is higher (i.e. periods 3, 6, 12,...). The same structure has been observed in the extensions of the Arnol'd tongues to regions in which the Poincaré map is non-monotonic in degree one maps [4, 8], and also in non-monotonic regions of circle maps with degree greater than 1 [7]. This structure is a generalization of the period-doubling bifurcations in maps with one extrema and one parameter, to maps with two extrema and two parameters.

3. Conclusion

The use of a two-dimensional parameter space reveals rich structure of bifurcations which may help to explain many results which appear puzzling when bifurcations,as a single parameter is varied,are considered (e.g. [10]). Although intrinsic "noise" may render the finer structure unobservable in biological situations, in favorable circumstances it should be possible to observe the global bifurcation theoretically predicted.

Acknowledgements

The Canadian Heart Foundation, the Natural Sciences and Engineering Research Council (Canada) and the Fonds FCAC (Québec) have provided financial support for this study. Drs. A. Shrier and M.R. Guevara have performed the experiments on which the theory is based, and provided useful discussion of the main ideas.

References

1. P. Collet, J.-P. Eckmann: Iterated Maps on the Interval as Dynamical Systems
        (Birkhauser, Boston 1980).
2. M.R. Guevara, L. Glass, A. Shrier: Science 214, 1350 (1981).
3. L. Glass, M.R. Guevara, A. Shrier, R. Perez: Physica 7D, 89 (1983).
4. L. Glass, M.R. Guevara, J. Bélair, A. Shrier: Phys. Rev. 29A, 1348 (1984).
5. H. Poincaré: "Sur les courbes définies par les équations différentielles",
        in Oeuvres vol. I (Gauthier-Villar, Paris 1954), 145.
6. V.I. Arnol'd: Transl. A.M.S. 2nd series 46, 213 (1965).
7   J. Bélair, L. Glass: Preprint (1984).
8. L. Glass, R. Perez: Phys. Rev. Lett. 48, 1772 (1982).
9. J. Bélair, L. Glass: Phys. Lett. 96A, 113 (1983).
10. M. Bier, T.C. Bountis: Phys. Lett. 104A, 239 (1984).

Cellular Metabolism and Transport

Dynamic Coupling and Time Patterns in Biochemical Processes

Benno Hess and Mario Markus

Max-Planck-Institut für Ernährungsphysiologie, Rheinlanddamm 201
D-4600 Dortmund 1, Fed. Rep. of Germany

1. Introduction

Biological systems are characterized by elaborate catalytic reaction
networks, which are organized in time and space. In order to understand these systems, biochemists have isolated subsets of the reaction networks, and studied them in detail as homogeneous open chemical systems. The simplest and most intuitive dynamic behaviour that
an open system can display is a stationary state. However, due to
the non-linearity of the processes involved, phenomena may appear
which are unexpected by intuition. Thus, a paradigm was broken when
observations of self-sustained oscillations were reported in chemical
[1] and biochemical [2,3] systems (for reviews see [4,5]). A second
counterintuitive phenomenon appeared with the observation of chaotic
oscillations in a chemical system [6] and in a single-enzyme system [7].

As a next step in the investigation of the dynamics of reaction
networks, the effect of coupling between oscillatory systems was investigated. The simplest result to be expected in such coupling phenomena is entrainment of one oscillation by the other. In fact,
Huygens already had observed that two mechanical clocks having slightly different periods become synchronized when fixed together on a
thin wooden board [8]. In biology, entrainment is observed at all levels of organization: cells, organs and higher organisms (see references in [9]). Recently, the oscillatory coupling between two prominent cellular functions could directly be demonstrated, as shown in
Fig. 1. In Fig. 1a we see a 1:1 synchronization between the yeast
membrane potential, as indicated by Rhodamin-6G fluorescence, and
glycolysis, as indicated by NADH fluorescence. Fig. 1b shows a cycle
obtained by plotting one oscillatory signal from Fig. 1a versus the
other [10,11].

In a first effort to understand coupling phenomenae at the cellular
level, periodic forcing of self-sustained oscillations were investigated in electrochemical [12-14] and biochemical [15-20] systems. Measurements of NADH flurescence in glycolyzing yeast extracts at perio-

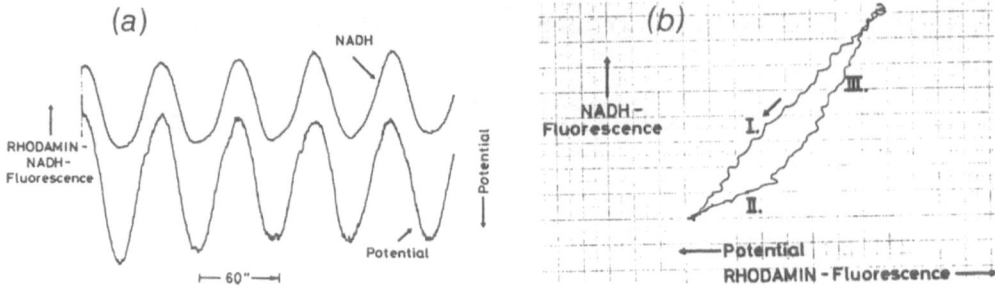

Fig. 1 Coupling of oscillations in yeast cells: a) Oscillations of NADH and Rhodamin fluorescence, as indicators of glycolysis and of membrane potential, respectively. b) Cycle determined by the two oscillations, showing three quasi-linear relationships I, II and III [10,11]

dic glucose input flux yielded entrained oscillations with periods 1,2 and 3 times the input period [15]. Later on, entrainment with periods 4,5,7 and 9 times the input period, as well as quasiperiodicity (frequency mixing) and chaos, were found [16,20].

In this report, we present a numerical description of oscillatory phenomenae in periodically forced glycolysis, partially covering the experimental observations, and partially giving an outlook to possible observable phenomena.

2. Model and Methods of Display

Simulations of chemical and biochemical systems under periodic forcing have been performed with relatively simple models, like the Sel'kov model (see [21] and references therein) and the Brusselator [22,23]. Due to their simplicity, these models have the advantage that some calculations can be performed analytically, that the numerical integration of the differential equations is fast, and that the number of control parameters is low, permitting a relatively easy evaluation. Therefore, these models can be useful tools to obtain a general view of the possible dynamic features that the system might display. However, the results obtained with these models are mostly qualitative and are not sensitive to all relevant biochemical control parameters. In contrast, we aimed in our model towards a quantitative comparison with the experimental results i.e. towards the closest possible simulation of natural phenomena [16-20]. Therefore, we chose the simplest model of glycolysis which represents the properties of this process in accordance with biochemical experiments. This model includes detailed rate laws for the two dynamically relevant enzymatic processes: the phosphofructokinase (PFK) reaction [24], and the pyruvate kinase (PK)

reaction [25]. These rate laws had been derived as two-substrate mechanisms after extensive kinetic measurements. The rate law of PK includes the effect of potassium and magnesium ions, as described in [26].

Our model (see [17,18]) describes the time-evolution of the concentrations of phosphoenolpyruvate (PEP), fructose-6-phosphate (F6P),
ATP and ADP, by the following equations:

$$\frac{d[ADP]}{dt} = V_{PFK} - V_{PK} \tag{1}$$

$$\frac{d[PEP]}{dt} = V_{in} - V_{PK} \tag{2}$$

$$V_{in} = \bar{V}_{in} + A \sin\omega t \tag{3}$$

$$[PEP] + [ATP] - [F6P] = C \tag{4}$$

$$[ADP] + [ATP] = ADN . \tag{5}$$

Here, $\bar{V}_{in}$, A and ω are the parameters that characterize the input flux
rate V_{in}. C and ADN are constants. V_{PFK} and V_{PK} are complicated functions of the concentrations [PEP], [F6P], [ADP] and [ATP]. We have chosen
here [ADP] and [PEP] as phase variables, while [F6P] and [ATP] depend
linearly on [PEP] and [ADP] by virtue of the conservation laws (4-5).

Equations (1-3) can be written as a system of three autonomous
differential equations:

$$\frac{d[ADP]}{dt} = V_{PFK} - V_{PK} \tag{6}$$

$$\frac{d[PEP]}{dt} = \bar{V}_{in} + A \sin\varphi - V_{PK} \tag{7}$$

$$\frac{d\varphi}{dt} = \omega . \tag{8}$$

We thus describe our system by three independent variables. We have
shown from experimental data [27] that this number of variables is,
in fact, sufficient to describe the dynamics of glycolysis, in spite
of being composed of a much larger number of reacting substances.
The equations (6-8) are invariant under the transformation $[PEP] \rightarrow [PEP]$,
$[ADP] \rightarrow [ADP]$, $\varphi \rightarrow \varphi + 2\pi$. Consequently, (6-8) are defined in $R^2 x S^1$,
where R^2 is the [PEP]-[ADP]-plane and S^1 is the unit circle. Thus,
we can represent a solution by plotting [PEP] versus [ADP] on R^2 and
rotating this plane around the [PEP]-axis with angular velocity ω,
i. e. one rotation is completed after φ has increased by 2π. Alternatively, one may plot any other pair X_1, X_2 of independent phase variables, instead of [ADP] and [PEP] in this manner. The resulting rectangular coordinates are X_2, $X_1 \cos(\omega t)$ and $X_1 \sin(\omega t)$. As a further
display method in this work we use stroboscopic plots on the plane at
times $t = (\varphi_o + 2\pi n)/\omega$, where φ_o is the initial phase and $n = 1, 2, \ldots$

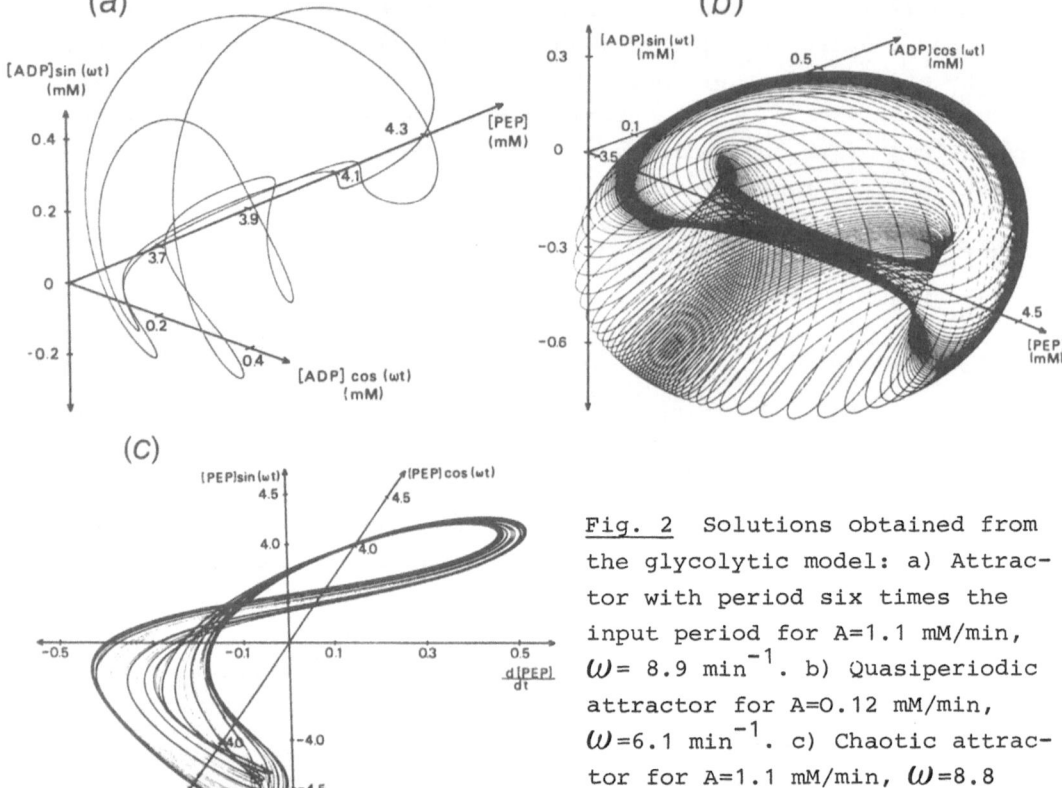

(a)

(b)

(c)

Fig. 2 Solutions obtained from
the glycolytic model: a) Attractor with period six times the
input period for A=1.1 mM/min,
$\omega = 8.9$ min^{-1}. b) Quasiperiodic
attractor for A=0.12 mM/min,
$\omega =6.1$ min^{-1}. c) Chaotic attractor for A=1.1 mM/min, $\omega =8.8$
min^{-1}

Our differential equations were solved numerically with the subroutine DC01AD from the Harwell Subroutine Library [28] . The conditions
for the calculations are those given in reference [18], unless indicated otherwise. Examples of solutions, plotted three-dimensionally as
described in the last paragraph, are given in Fig. 2. We have chosen
X_1=[ADP], X_2=[PEP] for Figs. 2a and 2b, and X_1=[PEP], X_2 = d[PEP]/dt
for Fig. 2c. The chaotic solution shown in Fig. 2c is reached on increasing ω by a series of period-doubling bifurcations, as described
in [29].

In order to describe quantitatively the computed solutions, we determined the Liapunov exponents using the method given in [30]. With
these exponents we determined the Liapunov dimension using the Kaplan-
Yorke formula [27,31,32]. For periodic and quasiperiodic solutions,
the Liapunov dimensions are 1 and 2, respectively. For chaotic solutions, we obtained values slightly above 2, indicating that the attractors are almost two-dimensional. For the chaotic attractor shown in
Fig. 2c, for example, we obtained a Liapunov dimension equal to 2.17 .

Entrained, quasiperiodic and chaotic oscillations are found experimentally in the theoretically predicted ranges of control parameters by measurements of NADH fluorescence (F) in yeast extracts [16,20]. Fig. 3a shows a typical case of measured entrainment. This figure demonstrates how glycolysis, indicated by F, follows the input oscillation, even though the self-sustained oscillation of the system has a 1.4 times larger period. Fig. 3b shows a plot of one oscillation versus the other, revealing a cycle of triangular shape, as in Fig. 1b. However, other cycle shapes are obtained for different experimental conditions. In Fig. 4 we show an experimental chaotic attractor constructed in an analogous way as the attractors shown in Fig. 2, but setting here $X_1=F$ and $X_2=dF/dt$. Clearly, the attractor in Fig. 4 consists of an almost two-dimensional strip twisting in phase space, just as our theoretical chaotic attractors, exemplified in Fig. 2c. For comparison of our theoretical and experimental results, we want to point out that the oscillations of [PEP] are highly correlated to those of [NAD], as shown experimentally in [33], and that [NAD] is correlated to $-F$, since [NADH]+[NAD] remains constant.

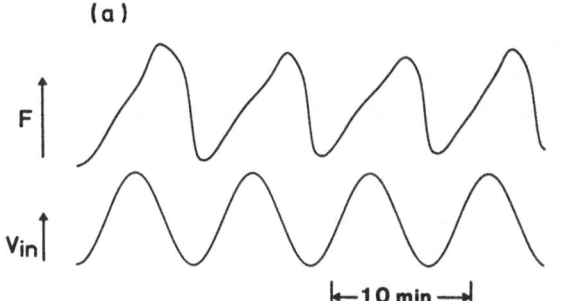

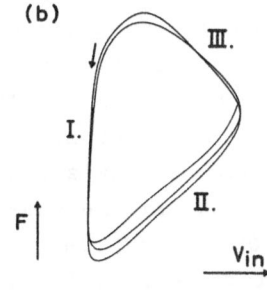

Fig. 3 Measurement of entrainment of glycolytic oscillations in yeast extracts: a) Record of NADH fluorescence F, as indicator of glycolysis, at sinusoidal glucose input flux V_{in}. b) Cycle determined by F and V_{in}, showing three quasi-linear relationships I, II and III as in Fig. 1b

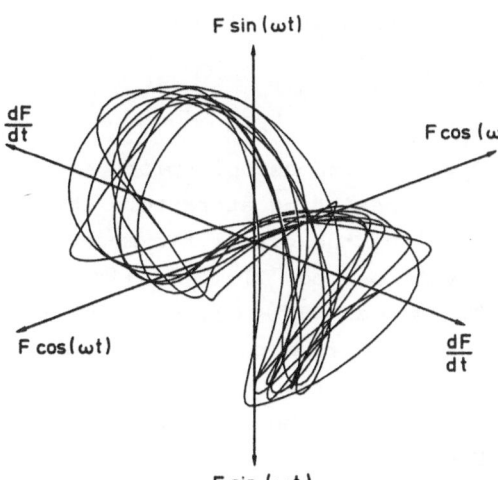

Fig. 4 Chaotic attractor obtained from a measurement of NADH fluorescence (F) in yeast extracts under sinusoidal glucose input (unpublished results). ω: input frequency

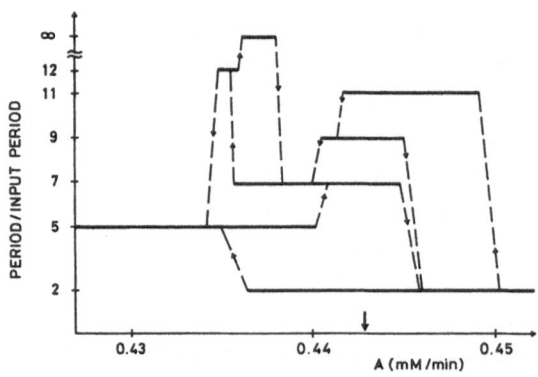

Fig. 5 Complex hysteretic response of the glycolytic model system depending on the input flux amplitude A (ω = 6.5 min^{-1}, period ∞ means chaos)

In a nonlinear system as the one described here, different attractors may c o e x i s t in phase space for a given set of control parameter values. In other words, the system of equations may have more than one solution, depending on the initial conditions. Each coexisting solution has its basin of attraction, that is, its manifold of initial conditions leading asymptotically to the corresponding solution. In the system described by (1-5), we found up to four coexisting attractors ("t e t r a r h y t h m i c i t y" [18]), as exemplified in Fig. 5. This figure was obtained by slowly changing A back and forth, holding the other parameters constant, at a rate $\Delta A/A =$ 10^{-5} per input period.Fig. 5 shows an extension of a complex hysteretic response scheme given in [18]. At the value of A indicated by the arrow pointing to the abszissa in Fig. 5 (A=0.443 mM/min), four periodic attractors coexist, having periods 2,7,9 and 11 times the input flux period. Fig. 6 shows stroboscopic plots of these attractors for a phase φ_{o} = (5/9)x2π .

If attractors coexist in phase space as described above, one of them may be s w i t c h e d into the other by resetting initial conditions through pulsed additions or substractions of metabolites.

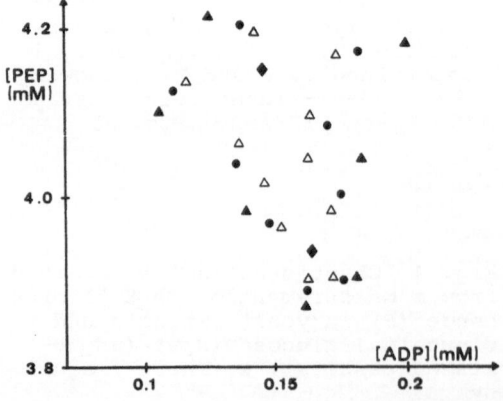

Fig. 6 Stroboscopic plots of four coexisting attractors obtained from the glycolytic model, having periods 2T ($\blacklozenge$), 11T ($\triangle$), 9T ($\bullet$) and 7T ($\blacktriangle$), where T is the input flux period. The conditions correspond to the arrow pointing to the abszissa in Fig. 5

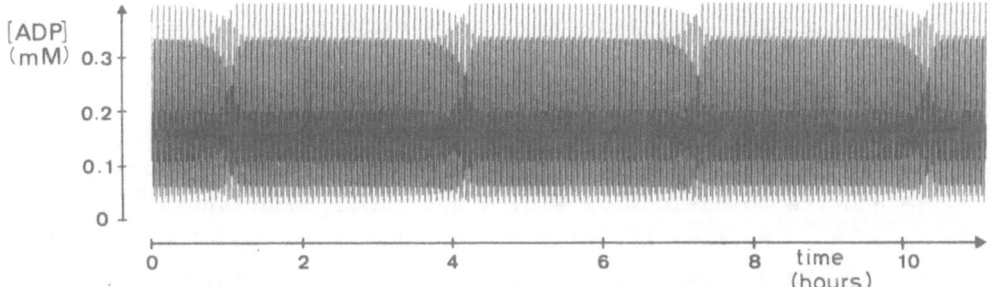

Fig. 7 Spontaneous modulation of a periodic oscillation calculated
at sinusoidal input flux with A=0.451 mM/min and $\omega = 6.7$ min^{-1}

In such a procedure, transients can be avoided by the right choice
of the phase and the magnitude of the pulses, so that a sudden transition from one time pattern into another is obtained [17,18]. A change from one time pattern into another can also occur spontaneously,
as shown in Fig. 7. This figure shows an oscillation having a period
equal to five times the input flux period, which is endogenously
modulated by short quasiperiodic interruptions. The modulating period
in this example is in the order of hours, while the period of the
carrier oscillation is in the order of minutes.

The discovery that an oscillatory system is capable of displaying a
spontaneously modulated periodic time pattern, and the knowledge of
coupling of oscillatory systems, led us to carry out the model analysis under modulation of the sinusoidal input flux. Allowing for amplitude and frequency modulation (AM and FM) we write:

$$V_{in} = \bar{V}_{in} + A\left[1 + A_{AM}\sin(\omega_{AM}t)\right]\sin\left[\omega t + A_{FM}\sin(\omega_{FM}t)\right]. \quad (9)$$

Note that replacing (3) by (9) in our model still permits one to describe the system by three homogeneous differential equations by setting again $d\varphi/dt = \omega$ and

$$V_{in} = \bar{V}_{in} + A\left[1 + A_{AM}\sin\left(\frac{\omega_{AM}}{\omega}\varphi\right)\right]\sin\left[\varphi + A_{FM}\sin\left(\frac{\omega_{FM}}{\omega}\varphi\right)\right] \quad (10)$$

We found that the numerical analysis of the system, including AM or
FM input flux, leads to a considerable enrichment of the diversity of
glycolytic time patterns [19,34]. We want to point out here that it
is important to distinguish between "slow modulation", defined by
$\omega_{AM} \ll \omega$, $\omega_{FM} \ll \omega$, and "fast modulation", defined by ω_{AM}, ω_{FM}
in the order of ω. In the first case, the effect is quite simple:
the system runs periodically through the time patterns it would display without modulation. In the second case, however, the modulation

parameters A_{AM}, A_{FM}, ω_{AM} and ω_{FM} act as bifurcation parameters, just as A and ω act as bifurcation parameters in the system described by (1-5).

The effect of "slow modulation" becomes especially interesting when the system displays hysteresis. In order to demonstrate such a case, we use the conditions of Fig. 5 and slowly modulate A sinusoidally, in such a way that A varies between 0.427 and 0.447 mM/min. In terms of (9), this means ω_{FM} =0, A_{FM}=0, A=0.437 mM/min, ω= 6.5 min^{-1}, A_{AM}=0.01 mM/min. We choose ω_{AM}=0.001ω. We start the integration at the far left of Fig. 5 (A=0.427 mM/min) on the attractor with period five times the input period. Under these conditions, the periods 9, 11, 12 and ∞ are not reached, and the system alternates between the periods 2,5 and 7 times the input period, with transients in between. This behaviour is illustrated in the upper part of Fig. 8. The lower part of this figure shows the input flux rate. Certain regions appear completely black in Fig. 8 because the oscillations there are too fast to be graphically resolved. A remarkable aspect of this figure is the demonstration of a system "m e m o r y", in the sense that the state of the system does not depend only on the values of the control parameters, but also on its history. In fact, the time patterns in Fig. 8 are different for equal values of V_{in}, depending on whether V_{in} is decreasing or increasing.

The effect of "fast modulation" is exemplified in Figs. 9 and 10. We have chosen here the conditions corresponding to Fig. 6, replacing (3) by (9), and setting A_{AM}=0, ω_{AM}=0, ω_{FM}=0.4ω, variable A_{FM} in Fig. 9, and A_{FM}=0.01 in Fig. 10. The choice of ω_{FM} implies that the input has a period equal to five times T=2π/ω. The words "input period" at the ordinate of Fig. 9 thus mean 5T. Fig. 9 was obtained, analogously as Fig. 5, by slowly varying A_{FM} back and

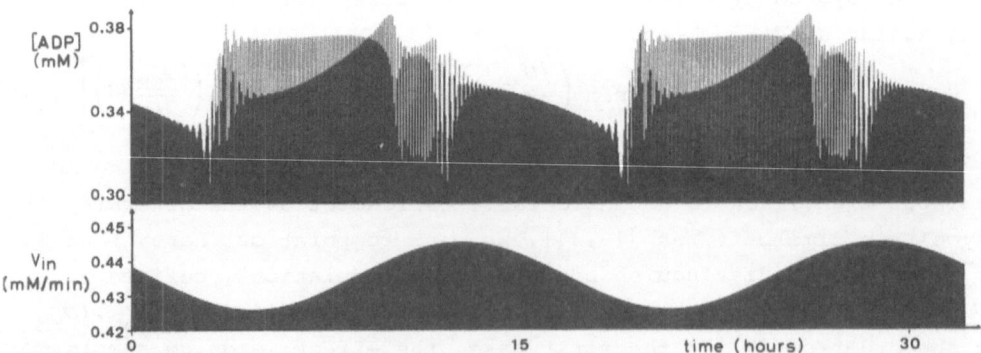

Fig. 8 Response calculated from the glycolytic model (top) under slow modulation of the amplitude of the input flux V_{in} (bottom)

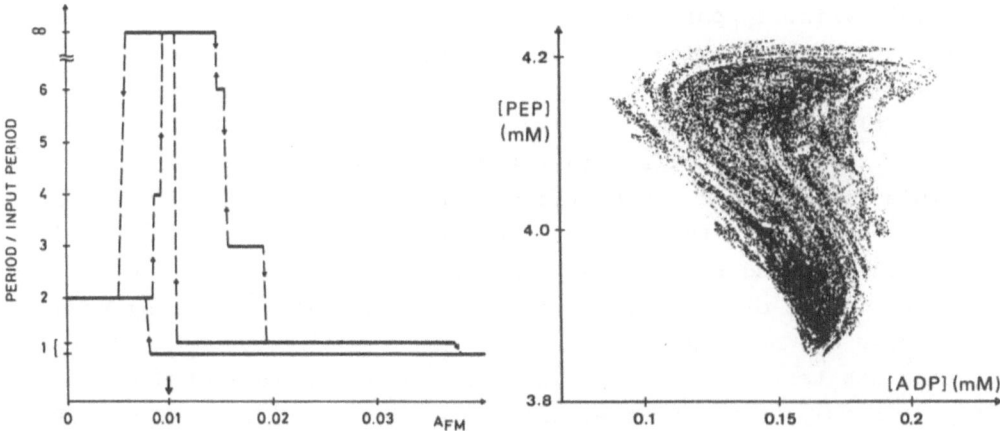

Fig. 9 Complex hysteretic response of the glycolytic model system depending on the modulating amplitude of FM input flux. Period ∞ means chaos. The conditions at A_{FM}=0 are those of Fig. 6

Fig. 10 Stroboscopic plot obtained from the glycolytic model in a "highly chaotic" dynamic state. The conditions correspond to the arrow pointing to the abszissa of Fig. 9

forth ($\Delta A_{FM}/A_{FM}$=5x10^{-6} per time T) starting on the attractor having period 2T at A_{FM}=0. (A start from the attractors with periods 7T, 9T and 11T at A_{FM}=0 leads to transitions to period 2T for A_{FM} smaller than 10^{-3}, which are not shown in Fig. 9). Fig. 9 shows the coexistence of two attractors having a period equal to the input period (number 1 on the ordinate). In Fig. 10 we show a stroboscopic plot with phase φ_o=(5/9)x2π of the chaotic attractor corresponding to the arrow pointing to the abszissa in Fig. 9. A remarkable feature of this attractor is that it covers a broader region of phase space than any attractor under unmodulated periodic input. In fact, the attractor corresponding to Fig. 10 does not cover the usual quasi-two-dimensional subset of phase space, but a region approximately containing the loci of all four periodic attractors that coexist at A_{FM} = 0 (Fig. 6). This broadening of the attracting region can be quantified by determining the Liapunov dimension, which results to be 2.94 for the example shown in Fig. 10. This value of the Liapunov dimension is almost equal to the phase dimension, which is 3, as compared to the values slightly above 2 for unmodulated sinusoidal input. The scenario leading to this "h i g h l y   c h a o t i c" solution is currently under investigation.

3. Discussion

In the present work, we have given examples illustrating the diversity of time patterns obtained from different types of forcing of oscillatory glycolysis. In particular, we have pointed to the existence of chaotic attractors having an unusually high degree of randomness. A broadening of the "excursion region" of a trajectory in phase space has recently been reported for a simpler oscillator [35], where two periodic attractors coexist and external noise is applied to the system. This noise causes stochastic switching between the two attractors. In contrast to this effect, the enhancement of the degree of randomness reported here comes from purely deterministic laws. In fact, the effect of "numerical noise" caused, for example, by round-off errors, can be excluded here.

Chaos has tentatively been associated with a number of diseases [36,37]. Lately, particular attention has been paid to its possible role in neural disorders [38-40].

In addition, we would like to draw the attention to some possible physiological roles of the glycolytic clock, keeping in mind that each role may apply as well to other biochemical or biological oscillators:

a) An "o r d e r i n g   p r i n c i p l e" for the organization of organisms: This may well apply to intracellular organisation, as well as to the organization of cell populations. The latter concerns, for example, the aggregation of the slime mould Dictyostelium discoideum, being organized by oscillation of cAMP (see [41]).

b) A "d i s s i p a t i o n   r e d u c e r": It has been shown that the dissipation of free energy per unit time at fixed average reactant concentrations tends to be lower in an oscillatory state than in the corresponding unstable steady state, and lower still when the oscillation is entrained by an external periodic force (see [21] and references therein).

c) A "f r e q u e n c y   g e n e r a t o r" [42]: Just as we have considered here the effect of an oscillator on glycolysis as "frequency acceptor", enriching considerably the dynamic possibilities of the system, one may conceive glycolysis as the source of forcing of another oscillator. However, it may in general be difficult to properly distinguish between "generator" and "acceptor".

d) A "r a n d o m   g e n e r a t o r": This function may allow the system in a chaotic mode to adapt to new external conditions by trial and error in processes like taxis, signal reception or even evolution. The "highly chaotic" state, illustrated in Fig. 10 gives the system a greater choice of combinations of phase-variables than a chaotic state with low Liapunov dimension, as the one shown in Fig. 2c.

e) A "m e m o r y   d e v i c e": If attractors coexist in phase space for the same set of control parameters, the system dynamics is not only determined by the values of the control parameters, but also by its previous history, as in Fig. 8.

f) a "c a r r i e r   o s c i l l a t i o n" for an endogenous, low frequency modulation: This function allows the system to display two different clocks in a single dynamic process (see Fig. 7). Periodic oscillations that are modulated may serve as "frequency donors" in the sense of point c), giving rise to "highly chaotic" states in the sense of point d).

Acknowledgements

We thank Miss Heike Kessel for her valuable programming assistance and Mrs. Bettina Plettenberg for efficient typing of the manuscript.

1. A. M. Zhabotinskii: Dokl. Nauk. SSSR 157, 392 (1964)
2. B. Chance, B. Hess and A. Betz: Biochem. Biophys. Res. Commun. 16, 182 (1964)
3. B. Hess, B. Chance and A. Betz: Ber. Bunsenges. Phys. Chem. 68, 768 (1964)
4. B. Hess and A. Boiteux: Annual Rev. Biochem. 40, 237 (1971)
5. Biological and Biochemical Oscillations, ed. by B. Chance, E.K. Pye, A.K. Ghosh and B. Hess (Academic Press, N.Y., London, 1973)
6. R. H. Simoyi, A. Wolf and H.L. Swinney: Phys. Rev. Lett. 49, 245 (1982)
7. L.F. Olsen and H. Degn: Nature 267, 177 (1977)
8. N. Minorsky: Nonlinear Oscillations (R.E. Krieger Publ., Huntington, N. Y., 1974) p. 438
9. J. Grasman: Bull. Math. Biol. 46, 407 (1984)
10. B. Hess, A. Boiteux and D. Kuschmitz: in Biological Oxidations, ed. by H. Sund and V. Ullrich (Springer, Berlin, 1983) p. 249
11. B. Hess, D. Kuschmitz and M. Markus: in Dynamics of Biochemical Systems, ed. by J. Ricard and A. Cornish-Bowden (Plenum Press, N.Y.), in press
12. H. Hayashi, S. Ishizuka, M. Ohta and K. Hirakawa: Phys.Lett. 88A 435 (1982)
13. H. Hayashi, M. Nakao and K. Hirakawa: Phys. Lett. 88A, 265 (1982)
14. H. Hayashi, S. Ishizuka and K. Hirakawa: Phys. Lett. 98A, 474 (1983)
15. A. Boiteux, A. Goldbeter and B. Hess: Proc. Natl. Acad. Sci.USA 72, 3829 (1975)
16. M. Markus, D. Kuschmitz and B. Hess: FEBS Lett. 172, 235 (1984)
17. B. Hess and M. Markus: in Synergetics - from Microscopic to Macroscopic Order, ed. by E. Frehland (Springer, Berlin, 1984) p.6

18. M. Markus and B. Hess: Proc. Natl. Acad. Sci. USA 81, 4394(1984)

19. M. Markus and B. Hess: Arch. Biol. Med. Exp., in press

20. M. Markus, S.C. Müller and B. Hess: in Non Equilibrium Dynamics in Chemical Systems, ed. by A. Pacault and C. Vidal (Springer-Verlag, Berlin, Heidelberg, New York), in press

21. P.H. Richter: Physica 10D, 353 (1984)

22. T. Kai and K. Tomita: Progr. Theor. Phys. 61, 54 (1979)

23. K. Tomita and T. Kai: J. Stat. Phys. 21, 65 (1979)

24. D. Blangy, H. Buc and J. Monod: J. Mol. Biol. 31, 13 (1968)

25. A. Boiteux, M. Markus, Th. Plesser, B. Hess and M. Malcovati: Biochem. J. 211, 631 (1983)

26. M. Markus, Th. Plesser, A. Boiteux, B. Hess and M. Malcovati: Biochem. J. 189, 421 (1980)

27. M. Markus and B. Hess: this volume

28. Harwell Subroutine Library: A Catalogue of Subroutines (Theoretical Physics Division, A.E.R.E., Harwell, England, 1973)

29. M. Feigenbaum: J. Stat. Phys 19, 25 (1978) and 21, 669 (1979)

30. I. Shimada and T. Nagashima: Progr. Theor. Phys. 61, 1605(1979)

31. J. Kaplan and J. Yorke: in Functional Differential Equations and Approximation of Fixed Points, ed. by H.O. Peitgen and H.O. Walther (Springer-Verlag, Berlin, 1978) p.228

32. J.D. Farmer, E. Ott and J.A. Yorke: Physica 7D, 153 (1983)

33. B. Hess, A. Boiteux and J. Krüger: in Advances in Enzyme Regulation, Vol. 7 (Pergamon Press, Oxford, New York, 1969) p. 149

34. M. Markus, H. Becher and B. Hess: in Proc. of the 8th Int. Biophysics Congress, Bristol, England (1984) p.80

35. F.T. Arecchi and A. Califano: Phys. Lett. 101A, 443 (1984)

36. M.C. Mackey and L. Glass: Science 197, 287 (1977)

37. L. Glass and M.C. Mackey: Ann.N.Y.Acad.Sci. 316, 214 (1979)

38. M.R. Guevara, L. Glass, M.C. Mackey and A. Shrier: IEEE Trans. on Systems, Man and Cybernetics 13, 790 (1983)

39. G.B. Ermentrout, SIAM J. Appl. Math. 44, 80 (1984)

40. R. King, J.D. Barchas and B.A. Huberman: Proc. Natl. Acad. Sci. USA 81, 1244 (1984)

41. B. Hess: 8. Fritz-Lippman-Vorlesung, Hoppe-Seyler's Z. Physiol. Chem. 364, 1 (1984)

42. B. Hess: Nova Acta Leopoldina 46, 103 (1977)

Dimension and Liapunov Exponents of a Strange Attractor from Biochemical Data

Mario Markus and Benno Hess

Max-Planck-Institut für Ernährungsphysiologie, Rheinlanddamm 201
D-4600 Dortmund 1, Fed. Rep. of Germany

Chaotic oscillations of glycolysis in yeast extracts under periodic substrate input were recently observed by registering the fluorescence of NADH. These observations are within the range of control parameters predicted by a two-enzyme glycolytic model [1,2].

A three-dimensional phase space can be constructed by plotting dF/dt versus F on a plane and rotating this plane around the dF/dt-axis with an angular velocity equal to the input angular frequency [3]. This representation yields strange attractors, having the shape of closed, almost two-dimensional strips in phase space. In this report, we will describe such an attractor by determining its dimension, as a measure of its "strangeness", and its Liapunov exponents, as a quantification of its average stability and instability properties (see [4-7]).

If one wants to determine the dimension of a strange attractor, one first encounters the difficulty of an appropriate definition of dimension. In fact, an infinite number of generalized dimensions D_q (q=0,1, 2,...), obeying $D_{q-1} \geq D_q$ can be defined for the description of strange attractors [8]. So far, such attractors have been characterized especially by D_0 (fractal dimension), D_1 (information dimension), D_2 (correlation exponent) and D_∞. For the present investigation, we decided to make use of the information dimension $D_I = D_1$, since it has recently been shown [4] that, excluding some exceptions not met here, D_I is related to the Liapunov exponents by the equation $D_I = D_L$, where D_L is the so called "Liapunov dimension" and is given by the Kaplan-Yorke formula

$$D_L = j + \left(\sum_{i=1}^{j} \lambda_i \right) \Big/ |\lambda_{j+1}| \qquad (1)$$

(see [5-7]). The λ_i (i=1,2,...,M) are the Liapunov exponents. M is the phase-space dimension, i.e. the number of independent variables needed to specify the state of the system at any time. j is the largest integer such that $\lambda_1 + \lambda_2 + ... + \lambda_j \geq 0$, assuming that the λ_i have been ordered so that $\lambda_1 \geq \lambda_2 \geq \lambda_3 ...$. On the other hand, the information dimension is defined by

$$D_I = \lim_{\epsilon \to 0} dI / d[\ln(1/\epsilon)] \quad , \tag{2}$$

where the information I is given by

$$I = -\sum_{i=1}^{N} P_i \ln P_i \quad . \tag{3}$$

Here, the phase space is partitioned into N equal cubes of side length ϵ, and P_i is the probability of finding the trajectory in the cube i.

We determined D_I from the experimental attractor by the definition (2-3). It has been found that this method ("box counting") may cause convergence difficulties, especially for $N \geq 3$ [9]. However, the box-counting method has proven successful for other three-dimensional systems [7]. In Fig. 1 we show I, calculated with (3), as a function of $\ln(1/\epsilon)$. We normalized each rectangular coordinate by dividing through the length of the interval covered by the attractor. The points for M=3 were obtained for the three-dimensional space defined above. For M=4, we added a dimension defined by $d^2 F / dt^2$. For M=2, we used the dimensions dF/dt and F. For M=1, we used only F. The slope of the linear region of each plot in Fig. 1 yields D_I, according to (2). The equivalence of the slopes for M=3 and M=4, yielding $D_I \approx 2.22$, demonstrates that three phase variables are enough to characterize the attractor, in spite of the much larger number of reactants involved in glycolysis. We confirmed this result with fluorescence measurements for different control parameters. A similar reduction of dimensionality has been reported for other systems [5,6,10].

Having shown the sufficiency of a three-dimensional phase space (M=3), we now determine the three Liapunov exponents λ_1, λ_2 and λ_3. The maximum Liapunov exponent λ_1 can be estimated by considering data

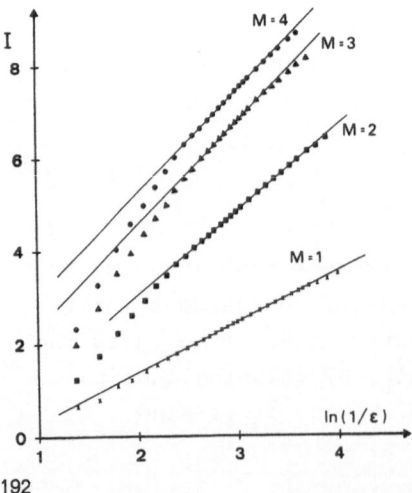

Fig. 1 Information I computed from a measurement of chaos in glycolysis of yeast. ϵ is the length of a box in a partitioned phase space. M is the dimension of the phase space. The slopes of the linear regions are equal to the information dimension of the chaotic attractor.

points F_i $(i=1,\ldots,n)$ taken at times spaced by the input period, then plotting F_{i+1} versus F_i (stroboscopic transfer function), interpolating a smooth function $y(x)$ through the resulting points, and calculating

$$\lambda_1 \approx \frac{1}{n} \sum_{i=1}^{n} \log_2 \left|\frac{dy}{dx}\right|_i \tag{4}$$

(see [10] and references therein). A stroboscopic transfer function of the attractor analyzed here is given in [1]. Application of (4) to that function yields a positive $\lambda_1 \approx 0.93$ per input period. The input period is 4.4 min, and thus $\lambda_1 \approx 0.21$ min^{-1}. Since the average divergence $\delta = \lambda_1 + \lambda_2 + \lambda_3$ is negative (the system is dissipative) [5,6], and since the Liapunov exponent corresponding to the direction of the trajectory is 0, we have $\lambda_2=0$, $\lambda_3 < 0$, and thus $D_L=2-\lambda_1/\lambda_3$. This equation can now be solved for λ_3 since λ_1 and D_L (equal to D_I) are known. We obtain $\lambda_3 \approx 0.95$ min^{-1}. The resulting average divergence δ is ≈ 0.74 min^{-1}.

Chaos is known to have features of both instability and stability. λ_1 and δ may be used as figures of merit of these two features for any complex chaotic system with phase dimension 3. The quantitative meaning of λ_1 and of δ for the example studied here is: the distance of nearby trajectories is amplified in the average by a factor $2^{\lambda_1} = 2^{0.21}=1.16$ per minute (instability) and a volume defined by a set of nearby points of the attractor is contracted in the average by factor $2^{-\delta} = 2^{0.74}=1.67$ per minute (stability).

References

1. M. Markus, D. Kuschmitz and B. Hess: FEBS Lett. 172, 235 (1984)
2. M. Markus and B. Hess: Proc. Natl. Acad. Sci. USA 81, 4394 (1984)
3. B. Hess and M. Markus: in this volume
4. J.D. Farmer, E. Ott and J.A. Yorke: Physica 7D, 153 (1983)
5. P. Grassberger and I. Procaccia: Physica 9D, 189 (1983)
6. J.D. Farmer: Physica 4D, 366 (1982)
7. D.A. Russell, J.D. Hanson and E. Ott: Phys.Rev.Lett. 45,1175(1980)
8. H.G.E. Hentschel and I. Procaccia: Physica 8D, 435 (1983)
9. H.S. Greenside, A. Wolf, J. Swift and T. Pignataro: Phys.Rev. A 25, 3453 (1982)
10. J.C. Roux, R.H. Simoyi and H.L. Swinney: Physica 8D, 257 (1983)

Coupling of Glycolytic Oscillations and Convective Patterns

Stefan C. Müller, Theo Plesser, and Benno Hess

Max-Planck-Institut für Ernährungsphysiologie, Rheinlanddamm 201
D-4600 Dortmund 1, Fed. Rep. of Germany

When a spatially extended light beam passes through a thin layer of yeast extract placed in a petri dish, a wavelength specific for the absorption of reduced
nicotinamide adenine dinucleotide (NADH) is selected, and when glycolysis in the cytoplasmic medium proceeds in an oscillatory manner, spatial patterns are observed on
a length scale comparable to the layer depth (1 to 2 mm). They form and disappear
periodically in intervals of several minutes [1]. Further insight into the coupling of the oscillatory degradation of glycolytic substrates such as glycogen and
transport processes in a liquid layer is gained with a two-dimensional UV sensitive spectrophotometer, which uses computer-based video techniques for fast recording of spatially resolved light intensities [2]. The transmission experiments are
combined with a dark-field type optical method for detecting spatial distributions
of refractive index gradients.

The cytoplasma was extracted from yeast cells (S. Carlsbergensis) grown under aerobic conditions [3]. The typical yield of 50 mg/ml protein content corresponds to
about one third of the concentration in an intact yeast cell. After adding glycogen and cleaning the sample with 0.44 um Millipore filter, the intensity distribution in an extract layer was observed with the transmission and dark-field techniques. Images, consisting of 512 x 512 picture elements with an intensity resolution
of 256 digital units, could be collected at a frequency up to 30 per min.

Transmission patterns photographed from the TV observation screen are presented
in Fig. 1. In the initial stage, faint bright patches begin to evolve around thin
dark lines (Fig. 1A). Switching to the dark-field method (at arbitrary wavelength)
shows that the lines coincide with lines of strong refractive index gradients. In
transmitted light, these gradients cause shadows due to scattering effects. Thus,
the dark lines represent the momentary morphology of a refractive index pattern
which, after an induction period of a few minutes, permanently exists in the sample.

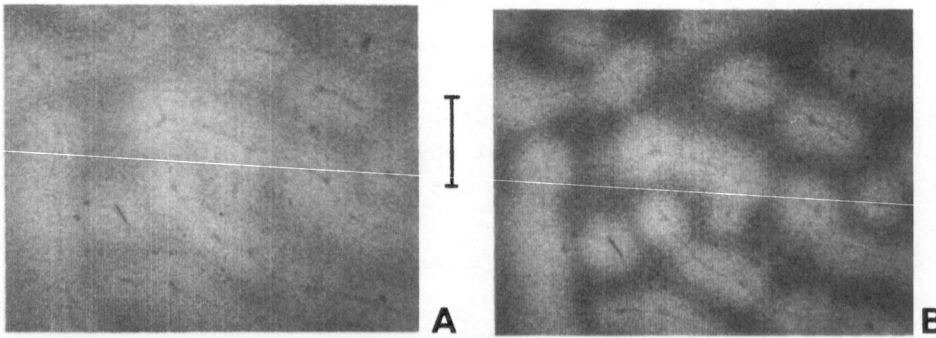

A   **B**

Fig. 1: Images of pattern formation during oscillating glycolysis in a 1.8 mm layer
of yeast extract. Intensity modulation of transmitted light at 380 nm at the beginning of patterning (A) and with maximum contrast 30 sec later (B). Scale bar: 2 mm.

The initially rod-like pattern, which later breaks up into a polygonal line network, evolves independently of the glycolytic activity in the extract and indicates the spatial arrangement of convection cells [4]. The cells form because of gradients in surface tension at the open liquid/gas interface, due to inhomogeneities in temperature as well as in biochemical composition (Marangoni convection [5,6]). Fig. 1B shows how the faint bright patches of picture A evolve to maximum contrast with respect to the dark background. The dark lines are somewhat blurred in the transmission image but still equally well detectable with the dark-field method. Thus, we observe a strong spatial correlation between NADH specific transmission patterns and the spatial distribution of convective flow.

The occurrence of these transmission patterns is coupled to the time course of the glycolytic oscillations in the bulk of the layer according to Fig. 2. The shaded stripes indicate those time intervals during which patterns are detected by visual inspection of the images. These intervals occur twice for each oscillation (Fig. 2A) and overlap with the peaks in the standard deviation (STD) of the intensity distribution in the images (Fig. 2B). Thereby the STD serves as a heuristic means for quantifying the distinctness of patterns. The extrema of the time derivative of the absorbance curve (Fig. 2c) fall into exactly the same time intervals. Consequently, during each glycolytic half-period, transmission patterns build up and decay just when the turnover of NADH passes through maximum values.

A remarkably regular, transient pattern is observed at the beginning of an experiment. Other than the periodically occurring structures just described, this pattern appears only once and can be detected by transmitted light of any wavelength

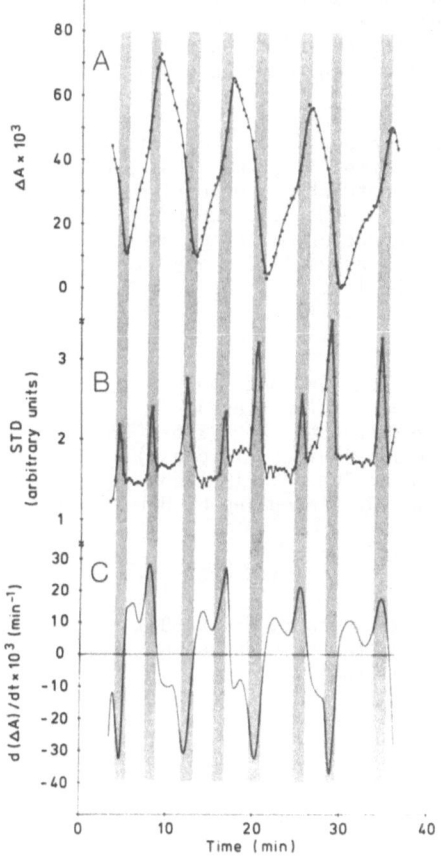

Fig. 2: Transmission patterns observed at 380 nm forming twice during each glycolytic cycle. (A) Oscillations of NADH absorption in a 1.8 mm layer of yeast extract. Absorbance changes ΔA (arbitrary reference intensity) are derived from averaged image data of transmitted light taken at 15 sec intervals. Measured points are fitted by splines (solid curve). (B) Standard deviation (STD) of the spatial intensity distribution of each image of A. (C) Derivative with respect to time of the fitted ΔA-curve in graph A. The shaded stripes indicate the time intervals of visual detection of patterns on the TV images.

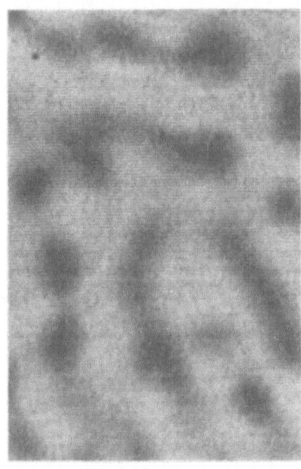 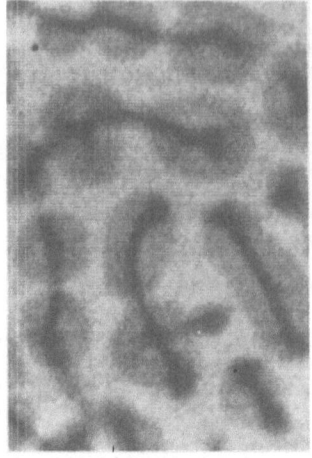 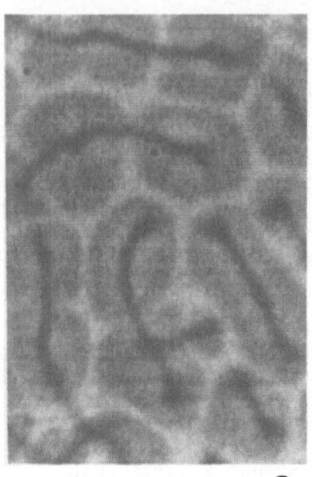

├─────────────┤   **A**   **B**   **C**

Fig. 3: Evolution of the initial transient pattern in a 1.8 mm layer of yeast extract containing 0.02% methyl orange, observed at 430 nm 1.75 min (A), 2 min (B),
and 2.25 min (C) after the start of the experiment. Scale bar: 2 mm

provided that a suitable (biochemically inert) dye is added to the solution. An
example is shown by the time sequence of pictures in Fig. 3. It reflects the effect
of unbalanced forces at the open layer surface during an induction period of several minutes which precedes the completion of the evolving rod-like convective
structure.

Our experiments show that pattern formation during oscillating glycolysis in an
open layer of yeast extract as observed with transmitted light at NADH specific
wavelengths is strongly correlated in space with a network of convection cells and,
in time, with the periodically occuring phases of high metabolic turnover. The glycolytic system thus generates chemical reaction-convection patterns, a class of
patterns in which symmetry is broken by interfacial instabilities due to inhomogeneities in temperature or in chemical composition [7-11].

References

 1. A. Boiteux and B. Hess, Ber. Bunsenges. Phys. Chem. 84, 392 (1980)
 2. S.C. Müller, Th. Plesser and B. Hess, Anal. Biochem. (in press)
 3. B. Hess and A. Boiteux, Hoppe-Seyler's Z. Physiol. Chem. 349, 1567 (1968)
 4. S.C. Müller, Th. Plesser and B. Hess, Naturwissenschaften (in press)
 5. J.C. Berg, A. Acrivos, and M. Boudart, Adv. Chem. Eng. 6, 61 (1966)
 6. Convective Transport and Instability Phenomena (J. Zierep and H. Ortel, eds.)
    G. Braun, Karlsruhe (1982)
 7. K. Showalter, J. Chem. Phys. 73, 3735 (1980)
 8. J.-C. Micheau, M. Gimenez, P. Brockmans, and G. Dewel, Nature 305, 43 (1983)
 9. D. Avnir and M. Kagan, Nature 307, 727 (1984)
10. K.I. Agladze, V.I. Krinsky, and A.M. Pertsov, Nature 308, 834 (1984)
11. S.C. Müller and Th. Plesser, in: Modelling of Patterns in Space and Time
    (W. Jäger and J. Murray, eds.) Springer, Berlin (1984).

Computation of Bifurcation Diagrams for Selkov's Model of Glycolytic Oscillations

Bruno Eckhardt and Peter Richter

Forschungsschwerpunkt "Dynamische Systeme" Fachbereich Physik, Universität Bremen
D-2800 Bremen 33, Fed. Rep. of Germany

Consider a system of N first order differential equations

$$\frac{d}{dt} x_i = f_i(x_1, \ldots, x_N, \lambda, t)$$

where $f_i(t+\tau) = f_i(t)$ and λ is an additional parameter. For certain ranges of λ and τ the long-time behaviour of the solution may be stationary or periodic, for others it may be quasi-periodic or chaotic. We study bifurcations of periodic orbits by using the following mapping (stroboscopic map; Poincaré section):

Let $x_i(x^{(o)}, \lambda, t)$ be the solution to initial condition $x^{(o)}$. Then

$$T_\lambda : x^{(o)} \to x(x^{(o)}, \lambda, \tau)$$

The Jacobian $DT_\lambda = \partial x_i / \partial x_j^{(o)}$ gives information on the stability of the orbit:

| all eigenvalues | < 1 : stable |
| at least one eigenvalue | > 1 : unstable |

Different bifurcations arise from different ways in which eigenvalues of DT_λ cross the unit circle:

Type D: one eigenvalue = +1; saddle-node bifurcation: a stable and an unstable orbit collide and annihilate each other; intermittency (Pomeau-Manneville)

Type I: one eigenvalue = -1; period doubling: an orbit of period n becomes unstable while a stable orbit of period 2n appears (Feigenbaum)

Type H: two complex conjugate eigenvalues $e^{\pm i\varphi}, \varphi \neq 0, \pi$; a new frequency shows up in the motion (Hopf)

The bifurcation diagram is obtained with an algorithm due to Kawakami [1] (see flowchart in Fig. 1). In order to find a fixed point of T_λ or any of its higher iterates, one has to solve the nonlinear equation

$$x^{(o)} = x(x^{(o)}, \lambda, \tau). \qquad (1)$$

This is done using a Newton method. To find a fixed point with a given eigenvalue of T_λ one has, in addition, to solve for

$$\det(DT_\lambda - \mu) = 0 \qquad (2a)$$

where $\mu = +1, -1$, or $\exp(\pm i\varphi)$. Eliminating φ in the last case, the conditions becomes (for N=2)

$$\det(DT_\lambda) = 1 \qquad (2b)$$

197

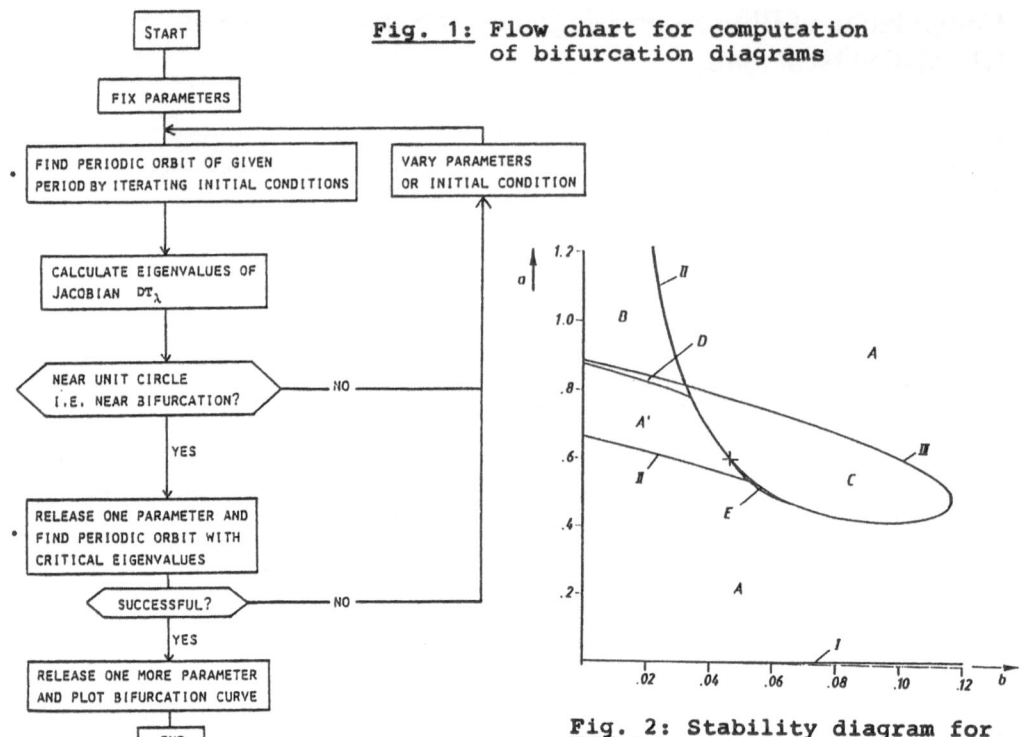

Fig. 1: Flow chart for computation of bifurcation diagrams

START

FIX PARAMETERS

FIND PERIODIC ORBIT OF GIVEN
PERIOD BY ITERATING INITIAL CONDITIONS

VARY PARAMETERS
OR INITIAL CONDITION

CALCULATE EIGENVALUES OF
JACOBIAN DT_λ

NEAR UNIT CIRCLE
I.E. NEAR BIFURCATION?   NO

YES

RELEASE ONE PARAMETER AND
FIND PERIODIC ORBIT WITH
CRITICAL EIGENVALUES

SUCCESSFUL?   NO

YES

RELEASE ONE MORE PARAMETER
AND PLOT BIFURCATION CURVE

END

*USING NEWTON'S METHOD

Fig. 2: Stability diagram for the unperturbed system

The N+1 constraints (1) and (2) require another degree of freedom besides the N initial conditions $x^{(0)}$. This could be λ or τ, for example.

In order to apply Newton's method to find the solution of (1) and (2) in the Poincaré section, we need derivatives of the mapping with respect to the variables $x^{(0)}$, λ, and τ. These are computed by integration of new ODE's obtained from (1) and (2) via differentiation. - For N = 2 a total of 24 ODE's has to be integrated.

Application:

Extended Selkov model for glygolytic oscillations [2,3]

$$A \rightleftharpoons S$$
$$S + 2P \rightleftharpoons 3P$$
$$P \rightleftharpoons B \qquad \text{A and B are reservoirs.}$$

After scaling, one obtains the following equations

$$s = a - ks - sp^2 + Kp^3$$
$$p = b - p + sp^2 - Kp^3$$

For the following we fix k = 0.1 and K = 1.

For a,b constant, the bifurcation diagram can be studied analytically (Fig. 2) [2]:

198

Line I   : equilibrium a = kb, s = p = b

Line II : boundary between parameter regions with one and three
           stationary states

Line III: onset of limit-cycle oscillations

Region A: one stable fixed point

Region A':one excitable stable fixed point + two unstable fixed
          points

Region B: two stable fixed points + one unstable fixed point

Region C: one stable limit-cycle + one unstable fixed point

Region D: stable limit-cycle + stable + unstable fixed point

Region E: stable fixed point + stable + unstable limit-cycle

We now pick a working point a = 0.05, b = 0.09 inside region C and
study the effect of a small oscillatory perturbation of a:

$$a \rightarrow a + \delta a \cos \omega t$$

The external oscillation competes with the stable limit-cycle oscillation of frequency ω_0 = 0.28 and for rational values of ω/ω_0 we
obtain resonance and frequency locking.

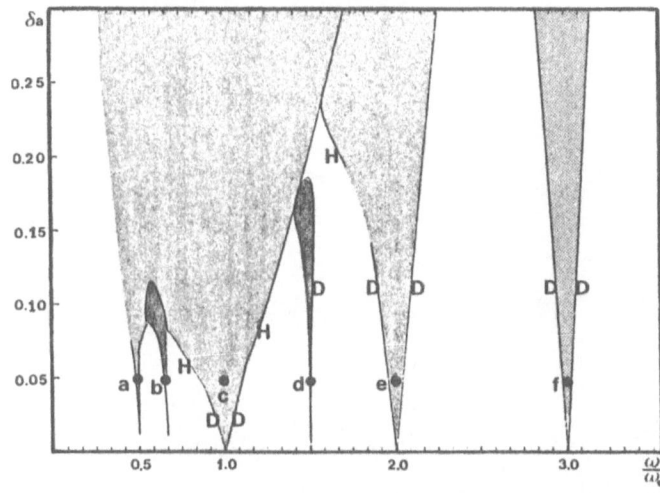

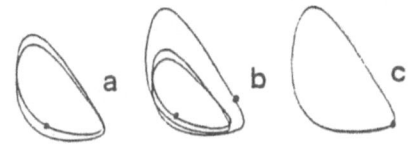

Fig. 3a: Bifurcation diagram for a =
0.5, b = 0.9. Capital letters indicate the type of bifurcation bounding
the resonance horns. For small δa it
always is D. Small letters indicate
the parameter values for the orbits
shown in b).

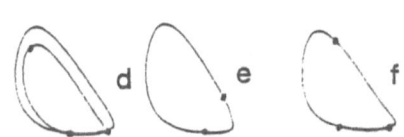

Fig. 3b: Some representative orbits
for the parameter values as shown in
a). The small dots along the orbits
identify initial points and values
after one period. (For a resonance of
order n:m we have n = number of dots,
m = number of loops).

For small δa, some results can be obtained analytically [3], but the full resonance diagram as shown in Fig. 3a requires techniques such as the above program. In Fig. 3b some representative orbits are shown.

Acknowledgements:

The authors would like to express their sincere gratitude to Prof. Kawakami for discussions on his algorithm.
This research was supported by Stiftung Volkswagenwerk.

References:

[1] H. Kawakami, IEEE Circuits and Systems, 31, 248-260 (1984)

[2] P.H. Richter, P. Rehmus, J. Ross, Prog. Theor. Phys. 66, 285 (1981)

[3] P.H. Richter, Physica 10D, 353 (1984)

Comparison of the Linear and Non-Linear Temperature Properties and Model Simulations of the In Vivo Glycolytic System

R. Kerutt, K. Rinast, W. Martin, and K. Brinkmann
Botanisches Institut der Universität, Kirschallee 1
D-5300 Bonn, Fed. Rep. of Germany

Although the known models of glycolysis do imply temperature as a possible parameter, an investigation of the influence of temperature on the experimental as well as on the model system is still missing. Using sinusoidal and step-like temperature signals as defined inputs, we examined the responses of NADH fluorescence. Comparing the experimental results with model simulations under a variety of options, our special interest was the identification of linear ranges.

Saccharomyces carlsbergensis was grown in aerobic conditions (1) and starved at least 4 hrs in 50 mM KH_2PO_4 (pH 4.6) before starting the experiments. The experimental setup and data sampling were described earlier (2), in data processing we used the program system TIMESDIA (3). The periodically excited system was analyzed with respect to phase differences and the amplitude gain by means of the Bode-plot technique; step responses were analyzed using a modified adaptive least square algorithm to obtain the parameters of a finite non-recursive filter displayed as impulse response and Bode plot (2). Numerical simulations were performed by means of the program system MISS (4).

Up to a temperature level of 19° C the experimentally evolved dynamics of the stationary part of glycolysis are most appropriately modelled by a finite linear non-recursive filter with pass-stop characteristics and a weak resonance at 0.46 min^{-1} (2). The low pass filter behaviour may have been selected in the direction of protecting the energy balance of the poikilothermic system against short time disturbances by temperature fluctuations. The stationary system responses to temperature steps with a strongly damped oscillatory transition. The modelled filter derived from this result as impulse reaction, fits very well with the results independently obtained with the sinusoidal excitations (2).

In the simulation studies, we introduced the parameter temperature into the non-linear models of Goldbeter (5) and Richter (6) assuming an Arrhenius-type dependency of all velocity constants. Within this frame the following options were used: 1. All reaction constants have identical temperature-dependence, 2. In addition to option 1 the allosteric constant of the PFK reaction varies with temperature, 3. Different temperature-dependencies of individual reaction constants.

The experimentally shown exponential temperature-dependence of period length is qualitatively simulated by both models and all options. However, in both models option 1 is insufficient to meet the experimentally observed temperature-dependence of amplitude. In simulating the experimental results with sinusoidal temperature excitement of the non stationary part of glycolysis, the two models differ strongly. Using Goldbeters model, none of the defined options is sufficient to simulate the experimental result of a very weak coupling between the glycolytic oscillation and the periodic temperature input, as shown in

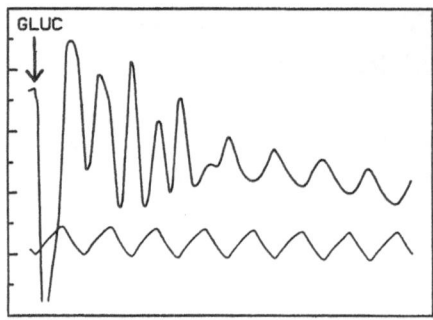

Fig. 1

Response of NADH fluorescence
(upper part) to a periodic
temperature input (lower part).
Temperature period 220 sec,
Temperature range 12-18° C.

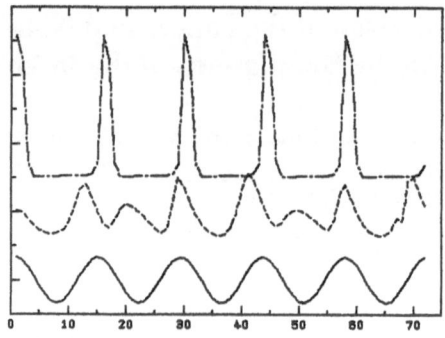

RELATIVE TIME

Fig. 2

Periodic temperature input
(lower curve) and time course
of PFK product in Richters
model (middle curve) and Goldbeters model (upper one).
Q_{10} of PFK velocity 1.76 and
of all other velocities 2.4 Q_{10}
of the allosteric constant
0.27. Temperature periods relative to the eigenperiod in
both models, amplitude 5° C.

fig.1 ; the model always produces a rigid coupling of a pulse-like
output (fig. 1, upper curve). Using Richters model, which is an extension of the Goldbeter model completed by ATP consuming processes,
adenylate kinase and storage connections, a simulation of weak coupling is possible, (fig. 2, middle curve) if the reaction constants
describing the exchange with sinks and sources are more sensitive to
temperature than the reactions of the chain.

1 Gosh, A. et al.: J. Biol. Chem. 235, 2522-2528 (1960)
2 Rinast, K. et al.: Ber. Deutsch. Bot. Ges.   (1984)
3 Martin, W. et al.: EDV in Biol. Med.:8, 90-94 (1977)
4 Gottwald, B.A.: Proc. 1st Europ. Simulation Congr. 574-579, Aachen
  1983
5 Goldbeter, A., Nicolis: Progr. Theor. Biol. 4, 65-160 (1976)
6 Richter, O. et al.: Chronobiologia 5, 56-65 (1978)

Circadian Rhythms

High Molecular Weight Protein is Presumably Essential for the Circadian Clock

H.G. Schweiger, R. Hartwig, G. Neuhaus, G. Neuhaus-Url, and M. Li-Weber

Max-Planck-Institut für Zellbiologie, Rosenhof,
D-6802 Ladenburg, Fed. Rep. of Germany

M. Schweiger

Institut für Biochemie, Universität, A-6020 Innsbruck, Austria

Many functions in living organisms are known to oscillate. The frequency spectrum ranges from fractions of a second up to at least a year (1). Among these oscillations, those with a period of about 24 hours, the so-called circadian rhythms, play a major role (2, 3). For a number of years, the extent to which oscillations in living systems can be compared with those occurring in physico-chemical systems has been discussed (4). A critical question is whether the insight gained in recent years into the mechanism of physico-chemical oscillations can be of help in understanding oscillations in living systems. Before answering this question, one should recall that a basic difference between the oscillations of physico-chemical and living systems is that living systems have been, and still are, subjected to evolution. This implies that the oscillations under discussion should have a selective advantage. In contrast to other oscillations in livings systems, most circadian rhythms allow a convincing answer to be given to this question (3, 5).

Diurnal fluctuations have been known to man throughout history. The interest in these rhythms, however, has changed, because for the last two decades the emphasis has no longer been on the purely descriptive aspect, instead the intention has been to learn more about the underlying physiological and molecular mechanisms (6).

Studies of the underlying molecular mechanisms of circadian rhythms favour the use of unicellular systems, and it is obvious that individual cells deserve particular attention due to their lower degree of complexity (7). The unicellular and uninucleate marine green alga Acetabularia is a suitable choice, and this giant cell has been thoroughly studied during the last two decades (8). A major advantage of this unicellular organism is the size which, in the case of the preferentially studied species Acetabularia mediterranea, may reach a length of up to 50 mm. Other advantages are its resistance against cell surgery and the relatively simple culture of this alga in the laboratory (9).

Methods have been developed which enable different parameters in individual cells to be monitored over several weeks. One of the parameters studied is the oxygen evolution due to photosynthesis. The oxygen concentration in the medium is monitored after it has passed the cell (10).

Another possibility takes advantage of a recently described phenomenon (11). The Acetabularia cell contains a high number of chloroplasts, which can exceed 10^6 (12). These chloroplasts tend to accumulate during the day, preferentially in the apical part of the cell, while during the night they prefer to accumulate in the basal part of the cell. This chloroplast-migration phenomenon is also retained under constant temperature and constant light. Under constant conditions, the period of these oscillations is approximately 24 hours (13). Chloroplast migration is monitored by means of a light-beam passing through the basal or apical part of the cell, and the absorption of the light indicates where the chloroplasts preferentially lie in the cell.

A third parameter that can be monitored over long time periods is the electropotential. This is done in the following way (14): A petri-dish is subdivided into three compartments by plastic walls. The three compartments are interconnected only by the cell through small holes in the plastic walls. The outer compartments contain electrodes which enable the potential to be monitored. An oscillation with a period of about 24 hours is again observed, even under constant conditions (14). These three parameters, oxygen evolution, chloroplast migration and electropotential, exhibit endogenous oscillations, and a number of features point to the conclusion that these oscillations represent circadian rhythms. The availability of methods to monitor these three different parameters enables the question of whether or not these three rhythms are independent from each other to be answered, or, in other words, whether an individual cell has one or several clocks.

An apparatus has been developed which allows two rhythms to be monitored simultaneously, e.g. oxygen evolution and chloroplast migration or chloroplast migration and electropotential (15, SCHWEIGER and BRODA, unpublished data). Using these two devices, investigations have been performed revealing that the periods of the two parameters measured were identical. This was true even if the period changed during the experiment. Such changes in period are observed quite frequently in individual Acetabularia cells. The phase-relationship between the two parameters was retained even if the periodicity was perturbed by pulses, e.g. by dark-pulses resulting in a shift of the phase (SCHWEIGER, unpublished data).

In a very few cases, however, pulse experiments have shown that the phase-relationship is not always retained (SCHWEIGER, unpublished data). In some cases dark-pulses did affect the phase-relationship, if only for a short time. After the shift, a fixed phase-relationship was again established. This leads to the conclusion that in one cell several oscillators may occur, and that these oscillators may be firmly coupled to each other; it also means that there is no master-clock - slave-clock relationship, as was suggested some time ago (16).

Other experiments performed on individual Acetabularia cells have answered questions about the location of the clock. Experiments in which cells were fragmented have demonstrated that the rhythmicity is retained regardless of whether the fragment was nucleate or anucleate (17). This result excludes the possibility of the cell nucleus being the site of the clock or, in other words, it clearly shows that the cytoplasm is capable of autonomously expressing circadian oscillations. Another answer obtained from these experiments was that, within the cytoplasm, no defined localization can be attributed to the clock. One has to assume instead that even very small anucleate fragments are capable of performing circadian oscillations, and this could mean that a circadian oscillation is an inherent feature of the cytoplasm.

The conclusion that the cytoplasm is autonomous in expressing a circadian rhythm is apparently contradictory to results which were obtained by a nuclear exchange experiment (18). In this experiment, a nucleus shifted the phase after transplantation into an anucleate cell whose phase was different from the nucleus donor cell. This meant that the nucleus was capable of affecting the phase.

The nuclear exchange experiment, moreover, indicated that gene expression is involved in the central molecular mechanism of the clock. Such a hypothesis was corroborated by experiments using inhibitors of gene expression. Experiments performed with specific inhibitors of gene expression showed that translation on 80S ribosomes may play a central role in the expression of a circadian rhythm. Oxygen evolution decreased in the presence of cycloheximide, a specific inhibitor of translation on 80S ribosomes, and the oscillatory behaviour was lost. The oscillations were reestablished after removal of the inhibitor (19).

More recently, an interesting experiment has been performed underlining the role of antibiotics. Besides cycloheximide, a number of

other antibiotics which inhibit translation on 80S ribosomes have
also been shown to have an effect on the circadian rhythm (20, 21,
22, 23). One of these antibiotics is G 418, which affects translation on 80S and 70S ribosomes (24). This antibiotic loses its inhibitory effect on translation if it is phosphorylated. Such a phosphorylation reaction is catalysed by the aminoglycoside-3'phosphotransferase II. While Acetabularia apparently does not contain this
enzyme, it is found in bacteria. The bacterial gene coding for the
phosphotransferase is available and originates from a transposon
(25).

Recently this gene was shown to be a useful selection marker for
transformation of eukaryotic cells (25, 26, 27, 28). The experiment
started with the amputation of the rhizoid of an Acetabularia cell
and isolation of the nucleus (29). The isolated nucleus was microinjected with a construction containing the gene (neo gene) for the
phosphotransferase and regulatory sequences from Antirrhinum and
parsley (KREUZALER and KAULEN, unpublished data). The microinjected
nucleus was implanted into an enucleated cell and after a couple of
days the cell was tested for sensitivity against the antibiotic
G 418. While 100 % of the control cells, which did not contain the
neo gene, died in the presence of the antibiotic, more than 50 % of
the microinjected cells survived. After the first period of exposure
to the antibiotic the surviving cells were again treated with G 418
until in a simultaneously G 418-treated control group all the cells
had died. Essentially all the microinjected cells survived the second G 418 treatment. In such a G 418-resistent cell the oxygen evolution was monitored. Non-microinjected control cells showed a
strong reaction to a 24 hour G 418 pulse. The oxygen evolution decreased, and during exposure to the antibiotic no oscillation could
be observed. The G 418 resistent cells, however, were not at all
affected by the antibiotic. It can therefore be concluded that the
microinjected gene was expressed in the Acetabularia cell, and that
the presence of a phosphotransferase, which phosphorylates the antibiotic, prevents a response of the circadian rhythm to a pulse of
the antibiotic G 418.

The involvement of translation on 80S ribosomes in the expression
of a circadian rhythm has been proven by experiments in which the
phase of oxygen evolution was shifted by pulses of cycloheximide
(30). Depending on the time in the cycle when the cell was exposed
to cycloheximide, a substantial or even no shift in the phase was
observed. It is generally accepted that the phase-shifting effect

reflects a disturbance of the central mechanism of the circadian
clock (6).

From these and other experiments, the phase-shifting effect of cycloheximide pulses has prompted the proposal of the coupled-translation-membrane-model (31). This postulates the following steps: One
or only a few polypeptides essential for the circadian clock are
synthesized on 80S ribosomes. This polypeptide is integrated into a
membrane. A membrane which contains the specific polypeptide changes
its features and this eventually results in a suspension of the supply of this specific protein due to inhibition of either the synthesis or the transport of the protein. The concentration of the polypeptide in the membrane is subjected to turnover, and when a certain
threshold is passed, the suspension of the supply with the polypeptide is abolished, i.e. the cycle starts from the beginning.

The coupled-translation-membrane-model postulates the existence of
one or a few essential polypeptides which are translated in a cyclic
manner. Moreover, the cyclic synthesis should be phase-shifted in a
characteristic way by cycloheximide pulses. A polypeptide has recently been identified in our laboratory that meets these requirements (HARTWIG, unpublished data). After in vivo incorporation of
radioactive methionine at different times of the day and preparation
of a chloroplast fraction, a number of bands can be separated elec-
trophoretically on a polyacrylamide gel. Autoradiography of such
gels reveals that among all these bands one corresponding to an
apparent molecular weight of 230 000 exhibits pronounced oscillations. The oscillations are retained even under constant light. The
labelling of this band is inhibited by cycloheximide, and cycloheximide pulses given at different times in the cycle affect the phase
of the synthesis rhythms in a characteristic way (HARTWIG, unpublished data).

It has been shown in other experiments that it is the different rates of synthesis rather than rates of degradation which are responsible for the oscillations (HARTWIG, unpublished data).

Another prominent feature is that oscillations in the rate of synthesis of this polypeptide are retained after enucleation (HARTWIG,
unpublished data). From the fact that the essential polypeptide is
synthesized on 80S ribosomes, one has to conclude that the coding
site of this polypeptide is in the nuclear genome. This means that
the diurnal oscillations in the anucleate cell can not be due to
different rates of transcription of the corresponding genes. Therefore, at least in an anucleate cell, the regulatory mechanism has to
be exclusively in the cytoplasm. This is in full agreement with the

experiments on the circadian rhythm of the oxygen evolution in anucleate cells.

The cytoplasmic regulation of the oscillatory synthesis of the 230 000 molecular weight polypeptide suggests that the cytoplasm is capable of selecting and translating a distinct message at a distinct time of the day. This suggestion raises the question as to whether there is any evidence that the cytoplasm is capable of selecting messages. In fact, recent evidence suggests that the Acetabularia cell has such capabilities. In this experiment (LI-WEBER, unpublished data) polyadenylated RNA was isolated from Acetabularia cells. The polyadenylated RNA was converted to double-stranded DNA via single-stranded DNA, and was integrated into a plasmid. The plasmid was propagated in bacteria. The plasmid containing the DNA sequence complementary to the polyadenylated RNA was isolated after propagation. A number of such specific DNA sequences were used for hybridization experiments. In these experiments it was shown that if total RNA was isolated from Acetabularia, and was separated by gel electrophoresis, then specific complementary DNA sequences from Acetabularia hybridized with different RNA bands to different extent. A comparison of total RNA isolated from nucleate and anucleate cells revealed that in both cases approximately the same amount of hybridization signal was obtained, although in anucleate cells a certain decrease was observed. If, however, polysomal RNA, which is supposed to be actively involved in the translation process, was studied, pronounced differences were found.

A comparison between total RNA and polysomal RNA revealed that, in contrast to total RNA in anucleate cells, immediately after removing the nucleus, certain DNA-hybridizing polysomal RNA species were decreased, and that these increased later above the level prior to enucleation. This, however, means that at the beginning of the experiment, a sufficient amount of message was there but was not bound in polysomes. At a later time, in a more progressed developmental stage, specific RNA sequences are selected from the total RNA pool and bound to ribosomes, where they actively participate in translation. This demonstrates that the cell is capable of selecting and translating specific messages.

In further studies it will be interesting to discover whether or not a similar process happens during circadian oscillations. If so, this would be another step in the direction of proving that the coupled-translation-membrane-model is in agreement with occurrences in the cell.

References:

1. J.E. Pauly: In: Chronobiology: Principles and applications to shifts in schedules, L.E. Scheving and F. Halberg, eds. (Nato advanced study institutes series, Alphen aan den Rijn, 1980) p. 33

2. J. Aschoff: Circadian clocks (North Holland Publishing Co., Amsterdam 1965)

3. E. Bünning: The physiological clock, circadian rhythms and biological chronometry (Springer Verlag, Berlin 1973)

4. A.M. Zhabotinsky: In: Biological and biochemical oscillators, B.C. Chance, E.K. Pye, A.K. Ghosh, B. Hess, eds. (Academic Press, New York, London 1973) p. 89

5. K. Hoffmann: In: Circadian rhythmicity, proceedings of the international symposium on circadian rhythmicity (Centre for Agricultural Publishing and Documentation, Wageningen 1972) p. 175

6. J.W. Hastings and H.G. Schweiger: The molecular basis of circadian rhythms (Dahlem Konferenzen, Berlin 1976)

7. B.M. Sweeney: In: Circadian rhythmicity, proceedings of the international symposium on circadian rhythmicity (Centre for Agricultural Publishing and Documentation, Wageningen 1972) p. 137

8. H.G. Schweiger: In: Chronobiology: Principles and application to shifts in schedules, L.E. Scheving and F. Halberg, eds. (Nato advanced study institute series, Alphen aan den Rijn, 1980) p. 239

9. H.G. Schweiger and S. Berger: Int. Rev. Cytol., Suppl. 9, 12 (1979)

10. D. Mergenhagen and H.G. Schweiger: Exp. Cell Res. 81, 360 (1973)

11. H.U. Koop, R. Schmid, H.H. Heunert and B. Milthaler: Protoplasma 97, 301 (1978)

12. H.G. Schweiger, S. Berger, K. Kloppstech, K. Apel and M. Schweiger: Phycologia 13, 11 (1974)

13. H. Broda, G. Schweiger, H.U. Koop, R. Schmid and H.G. Schweiger: In: Developmental biology of Acetabularia, S. Bonotto, V. Kefeli and S. Puiseux-Dao, eds. (Elsevier/North-Holland Biomedical Press, Amsterdam 1979) p. 163

14. H. Broda and H.G. Schweiger: Eur. J. Cell Biol. 26, 1 (1981)

15. H.G. Schweiger, H. Broda, D. Wolff and G. Schweiger: In: Polarographic oxygen sensors, aquatic and physiological applica-

tions, E. Gnaiger and H. Forstner, eds. (Springer Verlag, Berlin 1983) p. 190

16. C.S. Pittendrigh: In: Handbook of behavioral neurobiology 4 (1981), biological rhythms, J. Aschoff, ed., p. 57

17. D. Mergenhagen and H.G. Schweiger: Exp. Cell Res. 92, 127 (1975)

18. E. Schweiger, H.G. Wallraff and H.G. Schweiger: Science 146, 658 (1964)

19. D. Mergenhagen and H.G. Schweiger: Exp. Cell Res. 94, 321 (1975)

20. J. Jacklet: Science 198, 69 (1977)

21. J.C. Dunlap, W. Taylor and J.W. Hastings: J. Comp. Physiol. 138, 1 (1980)

22. N. Stahr, G. Holzapfel and R. Hardeland: J. Interdiscipl. Cycle Res. 11, 277 (1980)

23. H. Nakashima, J. Perlman and J.F. Feldman: Ann. J. Physiol. 241, R31 (1981)

24. S. Bar-Nun, Y. Shneyour and J.S. Beckmann: Biochim. et Biophys. Acta 741, 123 (1983)

25. A. Jimenez and J. Davies: Nature 287, 869 (1980)

26. F. Colbère-Garapin, F. Horodniceanu, P. Kourilsky and A.-C. Garapin: J. Mol. Biol. 150, 1 (1982)

27. P.J. Southern and P. Berg: J. Mol. Appl. Genet. 1, 327 (1982)

28. L. Herrera-Estrella, M. de Block, E. Messens, J.-P. Hernalsteens, M. van Montagu and J. Schell: EMBO J. 2, 987 (1983)

29. G. Neuhaus, G. Neuhaus-Url, P. Gruss and H.G. Schweiger: EMBO J. 3, 2169 (1984)

30. M.W. Karakashian and H.G. Schweiger: Exp. Cell Res. 98, 303 (1976)

31. H.G. Schweiger and M. Schweiger: Int. Rev. Cytol. 51, 315 (1977)

Circadian Time-Dependent Effects of Plant Growth Regulators on Morphogenesis in *Acetabularia*

Thérèsa Vanden Driessche

Département de Biologie moléculaire, Université Libre de Bruxelles
Rue des Chevaux 67, B-1640 Rhode St. Genèse, Belgium

1. Introduction

Morphogenesis, a highly complex process, appears to require a particular balance between the major plant growth regulators (PGRs) (1). During cap morphogenesis and development in Acetabularia, a unicellular and uninucleate alga, the content of auxin has been found to increase (2). Some PGR undergo changes in level under the influence of light, although very little is understood about the mechanism of this regulation.

On the other hand, morphogenesis in many plant species depends on a particular ratio between the light and the dark periods in the 24 h cycle. This is the foundation of BUNNING's theory of photoperiodism (see 3). Whatever the mechanism involved (the external or internal coincidence model according to PITTENDRIGH,4),the circadian organization of organism is implicated in some way. We have recently evidenced that morphogenesis in Acetabularia is sensitive to changes in the light-dark schedule,and that this sensitivity is modulated in a circadian way (5).

Since auxin has a role to play in the development of Acetabularia, we examined whether steps of auxin could exert a differential effect on this process, according to the time of the day. The highly differential effects exerted by both the auxin indole acectic acid (IAA) and its competitive antagonist morphactin show that at least one pathway required for cap formation and modulated by IAA is temporally regulated (6).

The next question is whether other PGRs also exert a differential effect when given as a step. In this paper, we present the experimental results of a screen of various PGRs for the time-dependency of their effects. Although the results should be considered as preliminary for each of the individual PGR, the results of the general screening clearly show that their physiological effects are dependent on the circadian time organization of the cell.

2. Material and Methods

Acetabularia mediterranea was cultivated as previously described (7) in L-D 12-12.

Treatment with the various PGRs was begun at different times in the 24 h cycle and lasted until at least half of the caps were formed by the controls. For practical reasons, experiments comprising both steps during the light period and steps during the dark period were carried out in parallel on algae cultivated in two culture rooms in inverted L-D schedule: therefore, the algae, although as similar as possible, did not belong to the same batch (L-D schedule perturbations bringing about modifications in the rate of cap formation,5). The PGRs used were: abscisic acid (ABA), gibberellic acid (GA) A3, kinetin (6-furfurylaminopurine), 6-benzylaminopurine (BAP) (N -benzyladenine), and etephon, 2-chloroethanephosphonic acid, an ethylene releaser all from Sigma.

3. Results

The results obtained with ABA (2 exp.), etephon (1 exp.), GA (4 exp.), and cytokinins of both the kinetin type (2 exp.) and BAP (2 exp.) by following the appearance of the caps are schematically represented in fig.1. In addition, experiments carried out in order to determine the appropriate concentrations have revealed "chrono-toxicity" as well. Even though the exact timing of the circadian modulation of effectiveness of PGRs steps cannot as yet be inferred, it is clear that they are most effective at different circadian phases: ABA during the L period (whereas it is known that IAA acts from 11 to 22 h), etephon at the beginning of the L period, GA during the dark period and cytokinins during the L one.

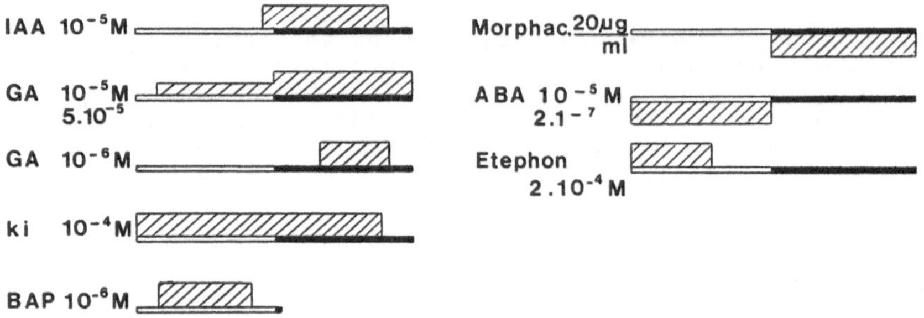

IAA 10^{-5} M

Morphac. $\dfrac{20\mu g}{ml}$

GA 10^{-5} M 5.10^{-5}

ABA 10^{-5} M 2.1^{-7}

GA 10^{-6} M

Etephon 2.10^{-4} M

ki 10^{-4} M

BAP 10^{-6} M

Fig.1. Circadian time-dependency of the effect of different PGRs steps on cap formation in Acetabularia. The white bar represents the light period and the black one, the dark one. The effects are only qualitatively represented either above the time axis, when it is stimulating, or below, when it is inhibiting. Only for GA 10^{-5} and GA 5.10^{-5} M refers to a stimulation of different strength.

4. Conclusions

All PGRs accelerate or slow down cap formation in Acetabularia in a circadian time-dependent way. Since this modulation by the different PGRs is differently phased, morphogenesis appears to result from the combination of several oscillating pathways, possibly pertaining to a metabolic network. The results are at least compatible with a population of oscillators. Both the external and the internal coincidence models are possible hypotheses. It should also be kept in mind that energy-supplying processes may be affected by PGRs, such as photosynthesis by IAA (8).

References

(1) J. MacMillan (ed.) Hormonal regulation of development I. Encyclopedia of Plant Physiology 9. (Springer Verlag 1980)
(2) E. Kof and V. Kefeli. In Developmental Biology of Acetabularia. S. Bonotto, V. Kefeli and S. Puiseux-Dao eds. 45 (Elsevier/North Holland Biomedical Press 1979)
(3) E. Bünning. The Physiological Clock (Springer Verlag 1967)
(4) C. S. Pittendrigh. Proc. Nat. Acad. Sci USA 69, 2734 (1972)
(5) T. Vanden Driessche (I) Chronobiology Internat. (in press)
(6) T. Vanden Driessche (II) Chronobiology Internat. (in press)
(7) L. Lateur Rev. Algologique 1, 26 (1963)
(8) T. Vanden Driessche. In Developmental Biology of Acetabularia. S. Bonotto, V. Kefeli and S. Puiseux-Dao eds. 195 (Elsevier/North Holland Biomedical Press 1979)

Phase and Period Effects of Physical and Chemical Factors.
Do Cells Communicate?

J. Woodland Hastings, Hellmuth Broda[*], and Carl H. Johnson

Department of Cellular and Developmental Biology, Harvard University
Cambridge, MA 02138, USA

It was early appreciated (1,2) and more recently explicitly argued (3) that insight concerning the nature of the underlying circadian oscillator might be sought by examining phase and/or period alterations as a consequence of some environmental condition or perturbation - be it physical or chemical. In this brief review we will consider effects of three classes of perturbations on phase and/or period: temperature, light and chemical, administered as either steps or pulses.

Parenthetically, it may be noted that effects of such conditions on amplitude have been largely ignored by most workers, though they surely should not be. In part, the lack of attention to this aspect may be attributed to the fact that interpretations of amplitude effects seem more complex and may require a more explicit formulation of a postulated mechanism.

Period Stability

Depending on the model chosen, permanent effects on phase and period are expected for perturbations or conditions which affect state variables or parameters of a circadian oscillation (3). Yet starting with the discovery of the unexpected but physiologically relevant "temperature-independence" in the Drosophila circadian system (4), and including lack of effects by many drugs and inhibitors (2), the circadian oscillator has gained a reputation for stability of period (frequency) under different physical and chemical environments. With drugs, this might be explained by arguing that they fail to react with the crucial components of the oscillator. But temperature affects the rate constants of all cellular reactions. The existence of negative Q10 values for period in some circadian systems led to the proposal, now generally accepted, that this is due to temperature overcompensation in a temperature compensation mechanism (5). But its cellular and molecular nature remains elusive. In the Njus et al. (6) model for the circadian oscillator, period was proposed to be related to membrane fluidity, which could be self-regulatory. But the chemical reagent, A2C, synthesized by Kosower and colleagues (7), which alters membrane mobility, has not been found to affect the period in our Gonyaulax polyedra test system (unpublished data). Another possibility is that auto-temperature compensation might occur if the period were governed by the ratio of two rate constants. However, no model system has been proposed with such a feature.

The concept of period stability in the face of diverse environmental changes was advanced by Pittendrigh and Caldarola (8). They rallied evidence from several systems, indicative of a general homeostatic conservation of the frequency of circadian oscillations, irrespective of the changes they are likely to encounter in the cell; temperature compensation was then viewed as a special case of this principle, but again, the possible mechanism remains elusive. In fact, the comfort taken by appreciating the physiological significance may have diverted attention from the need to understand the mechanism.

---

[*] Present address: Fachbereich Biologie, Universität Konstanz
D-775 Konstanz, FRG.

Circadian rhythms may fail to be expressed under certain conditions, most notably bright light and low temperature (9,10). By analogy with temperature conditional mutants, such conditions have been called nonpermissive (9).

Conditional rhythmicity due to both bright light and low temperature appear similar in that normal rhythmicity may be restored by simply switching back to permissive conditions in a single step. Even after weeks or years of nonpermissive (and constant) conditions, no information on period is required. It may be compared to holding a pendulum and then releasing it, for in both cases the phase is determined by the time at which the step occurred; the rhythm appears to start from CT 12. This was earlier viewed as restarting a stopped clock, as in the pendulum analogy. However, a newer interpretation due to Peterson and Jones (11) is that an underlying rhythm continues but without expression, and that expression is rephased to CT 12 at the time of the return to permissive conditions. Simple exposures to low temperature for a few hours will not reset the Gonyaulax clock in the same way; pulses or flashes of light, however, are very effective.

Unlike the usual temperature-sensitive mutants, the property of circadian rhythmicity is lost at low rather than high temperatures. Although the effect of low temperature is presumably not due to thermal denaturation of a protein (enzyme), as is usually the case in temperature-sensitive mutants, it does exhibit a similarly high temperature coefficient: in Gonyaulax rhythmicity is normal at 14 [C] but is lost at 12.5 [C] (10). The effect could be due to subunit dissociation. An understanding of the molecular basis for conditionality of circadian systems could provide important insight concerning mechanism.

Phase-shifting by light: phase response curves (PRC)

Phase-shifting by single light pulses has been documented in numerous studies with many different organisms. The subject is too extensive to be fully considered here, but in general bright light pulses (usually minutes or a few hours in duration) during the day (or subjective day phase), have little or no effect, pulses during early night phase cause delays, and those during the second half of the night phase cause advances. The consequent phase response curve has been viewed as a highly relevant indication of the nature of the underlying driving oscillation. For example, in Gonyaulax the different rhythms of cell division, luminescence and photosynthetic capacity have different phases, yet the phase response curves for the different rhythms are the same (12,13). More studies of this kind with other systems would be highly desirable.

The perceived importance of phase response curves is such that an atlas for different organisms is being prepared for reference and comparison purposes. Unfortunately, partly due to the large amount of work required to determine a PRC, most such determinations are for but a single set of conditions. It would be preferable to have many measurements on a single organism with environmental conditions (e.g., light, temperature, medium composition) varied in a systematic way. It has been shown, for example, that the light PRC for Drosophila is the same at different temperatures (14), but published data of this kind with other systems is scanty. Such measurements might lead to a better understanding of the resetting surfaces (15) under different conditions and perhaps new insights concerning the nature of the fundamental driving oscillator.

A good example of how the PRC may depend on conditions was reported by Christianson and Sweeney (16). The PRC for phase-shifting of the glow rhythm in Gonyaulax by 3 hour light pulses (800 foot candles) has both advances and delays (12), but this is with cells maintained in the dark. Cells maintained instead in dim white or red light exhibit a PRC in which the phase-shifts for such light pulses are largely confined to advances. While we are not prepared to interpret this, there is a large difference in the PRCs due only to background light intensity (16). An understanding of the PRC must include an explanation of such effects.

214

Knowledge concerning the nature of the photoreceptor in phase-shifting by light could provide important clues concerning the mechanism of the clock. To date, knowledge concerning the action spectrum for phase-shifting is restricted to only a few organisms. The action spectrum for Gonyaulax (17) was determined with pulses given in the latter half of the subjective night - thus causing advances. No similar measurements have been made with pulses in the early part of the night phase, which give delays. The action spectrum for phase-shifting the activity rhythm in the hamster has recently been found to be similar to the absorption spectrum for rhodopsin (18). The results suggest, however, that there are important differences between the photoreceptive system involved in entrainment and that mediating vision.

An equally large gap in our knowledge lies in the area of dose-response measurements. For most organisms there do not exist published data concerning the effect of light intensity upon the extent of phase-shifting. It would thus be important to determine PRCs at different light intensities.

Drugs and inhibitors

It is evident that much attention has been paid to effects upon circadian rhythms of physical conditions, notably light and temperature. Though not ignored, far less work has been concerned with effects of chemical factors as probes for the identity of the putative state variables or parameters of circadian oscillations. This experimental approach relies on the specificity of the drug or inhibitor for some key component of the circadian oscillation.

In early work with Gonyaulax (19,20) an apparent abolition of rhythmicity was found both with an inhibitor of m-RNA synthesis (actinomycin D) and an inhibitor of protein synthesis (puromycin). Interestingly, inhibition by puromycin was prompt, whereas actinomycin required a full cycle (day) to take effect, as though the putative message for the expression of the rhythm was formed well in advance. Small but significant phase-shifts were reported for puromycin. Unfortunately, the action of actinomycin D was irreversible, so possible phase-shifts could not be determined. In the meantime, there has been little additional definitive information concerning the possible role of m-RNA in this or other systems.

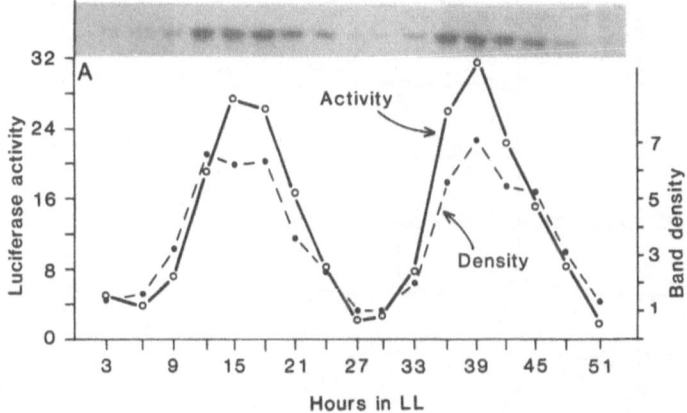

Figure 1. Rhythm of luciferase activity and its amount (density) extractable from cells maintained in constant dim light (LL), intensity about 1000 lux. Extracts were made every three hours, assayed for luciferase activity and prepared for sodium dodecyl sulfate electrophoresis, immunoblotting onto nitrocellulose and autoradiography. Above the graph is an immunoblot of total soluble Gonyaulax proteins reacted against an antibody to luciferase.

Nevertheless, the type of postulate which emerged from these and other studies was one in which the hypothetical oscillator involves the synthesis of m-RNA and possibly protein synthesis as well. Specific models were advanced both by Goodwin (21) and by Ehret and Trucco (22); in the latter, it was proposed that some part of the DNA of eucaryotic cells comprises functional subunits termed "chronons", each corresponding to a daily scanning cycle; a chronon is regulated so that its transcription occurs in a linear and temperature-compensated fashion once each day. Although it was not specified how many such chronons may occur in a given genome, a considerable number were apparently envisioned. More important, a major prediction of the chronon model is that a complete (or substantial) turnover of some specific components would occur on a daily basis. This has recently been experimentally verified in the <u>Gonyaulax</u> system, where it has been shown that luciferase is turned over in each circadian cycle (23,24; Fig. 1). At the same time, we have no reason to believe that luciferase itself is involved directly in the biochemical mechanism of the clock. The cellular clock itself could be transcriptional in nature, but it could also be translational. Or the mechanism might include neither; the oscillatory system might involve other components while protein synthesis and m-RNA synthesis might be simply hands of the clock (3).

The strongest evidence that protein synthesis is involved as a key component of the oscillator itself is that pulses of inhibitors of protein synthesis on 80s ribosomes cause large and permanent phase-shifts in a variety of different systems (25,26). This conclusion is strengthened by the observations that different types of inhibitors of protein synthesis, having different modes of action, all cause phase-shifts, and that a ribosomal mutant in <u>Neurospora</u> resistant to inhibition of protein synthesis by cycloheximide was also immune to phase-shifting by cycloheximide (27).

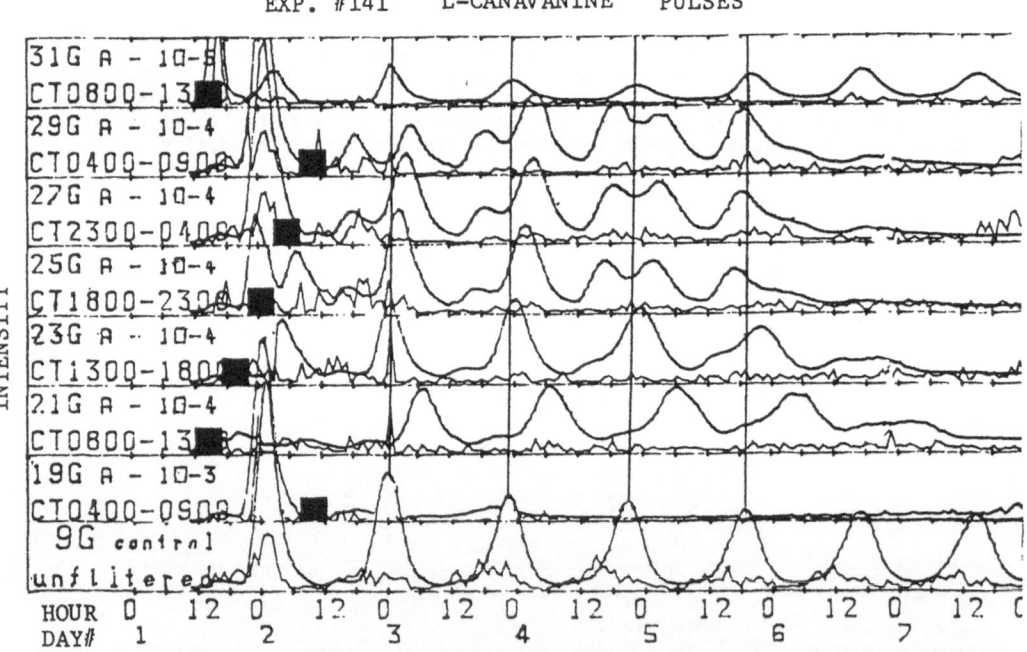

EXP. #141   L-CANAVANINE   PULSES

Figure 2. Effect of five-hour pulses of L-canavanine (given at different times in the cycle, as shown) on the glow rhythm of <u>Gonyaulax</u>. Cultures were kept at 19 [C] and constant illumination (white fluorescent lights, 35 uE/m^2/s). The phase of a control culture (bottom), marked by a vertical line, indicates that 10^{-5} M canavanine (top) had little or no effect; 10^{-3} M (second from bottom) abolished rhythmicity, while 10^{-4} M had major but different effects when administered at different times in the cycle as shown.

We have recently found that both pulses of and continuous exposure to certain amino acid analogues will cause major phase changes in the Gonyaulax rhythm (Fig. 2). In this case, it may be not simply because of the lack of protein synthesis but could be the result of the synthesis of defective or inactive protein(s). This could relate to the suggestion from the experiments of Leong and Schweiger (28) that there is a key "clock protein" involved in the mechanism.

Chemical communication between cells?

Influences on phase and period by light and temperature are readily interpreted in relation to environmental fluctuations of these parameters. Might chemical substances in the environment also have significant and physiologically relevant effects on phase and period?

Several authors (29,30,31,32) have considered the possibility that some medium-mediated chemical communication might occur between unicells. One idea, for example, is that such communication might help cells remain in better synchrony while being maintained under constant conditions. In none of these cases, however, including a recently published study with Gonyaulax (33), has evidence indicative of communication been found. But more recently, with improved instrumentation and analysis, we have measured the circadian glow of luminescence in mixed cultures for 14 days and do find some evidence for medium-related effects; these effects become evident only during the second week.

The shape of glow rhythm in Gonyaulax provides a good measure of the degree of synchrony of cells in a population (34). In a population of cells kept in the dark, the first glow peak after release from a light-dark (LD) cycle has a shape similar to that of a single isolated cell under the same conditions (Fig. 3). In a population kept for several days under constant conditions of dim light, the period can be determined to within a few minutes; during this time the wave form of the glow peak becomes broader, indicative of the expected desynchronization between individual cells (35). The consequent damping (Fig. 4) is presumed (though not yet demonstrated) not to occur at the level of the individual cell. A single cell is thus postulated to maintain its rhythmicity. In an asynchronous population, a single light perturbation is then sufficient to resynchronize the cells.

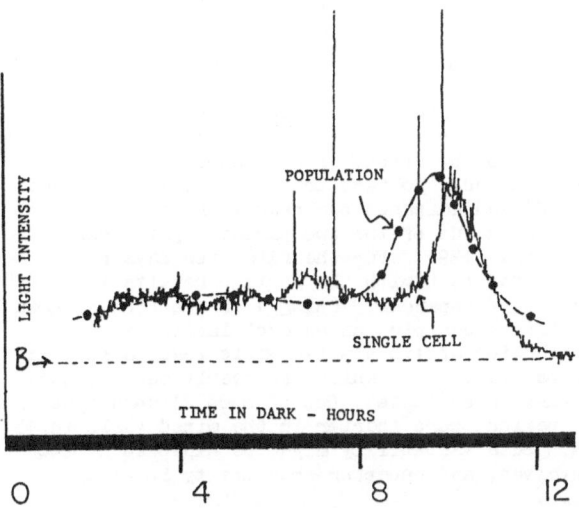

Figure 3. Comparison of time courses of glow peaks measured in continuous dark from a single cell (jagged line with four flashes superimposed) and a population (10,000 cells; solid dots, flashes not shown). Ordinate, light intensity, arbitrary units; abscissa, time in hours from beginning of dark. B = baseline. Redrawn from Krasnow et al. (34).

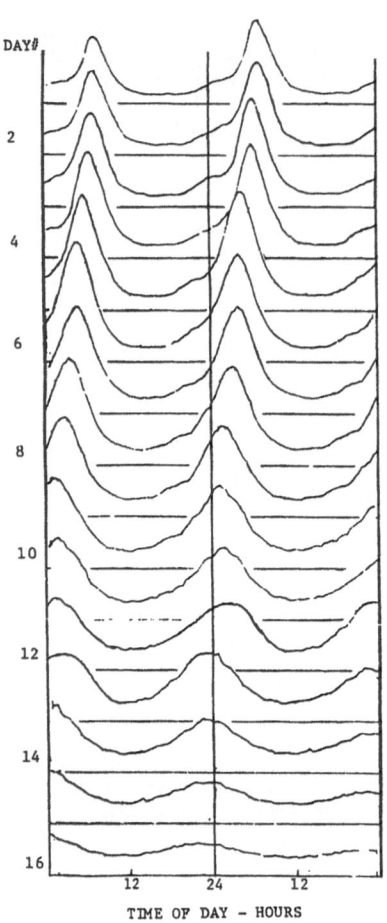

DAY#

TIME OF DAY - HOURS

Figure 4. Progressive increase with time in width of the glow peaks of a population of cells maintained in constant dim light (LL) at 19 (C). This may be due to a gradual drifting out of phase with one another of the individual cells in the population. Cultures had been in constant dim light for 10 days prior to the recordings shown, but were resynchronized by an eight hour dark period given one day before records shown here.

Is it possible that the cells in a population desynchronize more slowly than would be predicted on the basis of individual cell behavior? The answer to this is not known, but if it does occur, it might be by virtue of some sort of medium-mediated mutual entrainment phenomenon. To test this possibility, we mixed two out-of-phase populations (Fig. 5a,d) and compared the resulting bioluminescent glow output (Fig. 5b,c) with the calculated sum of the two parent populations maintained under the same conditions (Exp. #139). For the first few days no differences were apparent, this being consistent with the earlier results (33). But during the second week differences were apparent. Using a subtraction program with amplitude normalized data, the differences plot as an oscillation with an increasing amplitude (Fig. 5e). Subtraction of two similar vials results in a line with only small or no amplitude ripples (Fig. 5f). While the result seems clear, its statistical significance is not easy to evaluate. Out of some 21 such cases in two experiments (#139 and #152), the periods were shorter in the mixed vials in 15 cases, longer in 2 and the same in 4. Such variability might be expected if two "competing" signal substances are involved, and receptor sensitivity is also varying as a function of time in cycle.

The typical differences between an experimental mix and the corresponding calculated trace may be better visualized in phase plots of the maxima (Exp. #152; Fig. 6). In the plots of the summation of the separate cultures, the two peaks remain distinct and parallel one another over the course of the experiment (vial 62); in the cultures where the cells were actually mixed, the peaks corresponding to

218

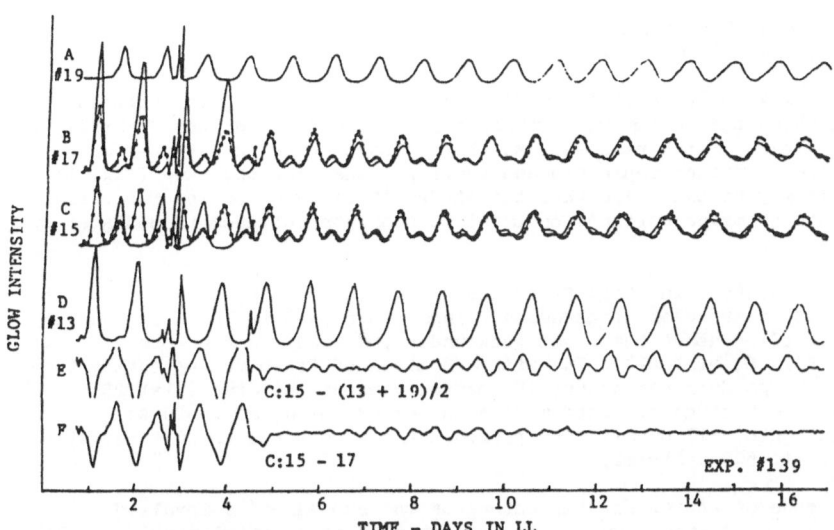

Figure 5. (Exp. #139) Comparison of circadian rhythm of luminescence in a culture comprising a mixture of two out-of-phase cultures (referred to as X and Y) with the expected luminescence based on the behavior of parent cultures maintained under the same conditions. For the first 4 days traces in panels A and C contained Y cells, while those of B and D contained X cells. On day five the contents of the vials (#15, #17) were evenly mixed and divided back into the two vials. The corresponding calculated trace [C:(13 + 19)/2], representing the addition of traces A and D, is plotted with solid dots and superimposed on both traces B and C. Panel E shows the result of subtracting the actual mix (#15) from the calculated trace. The increasing amplitude after day 9 is indicative of the growing phase-difference. This is not the case with the traces from panels B and C (#15, #17); as can be seen by the subtraction shown in panel F, there are only small differences.

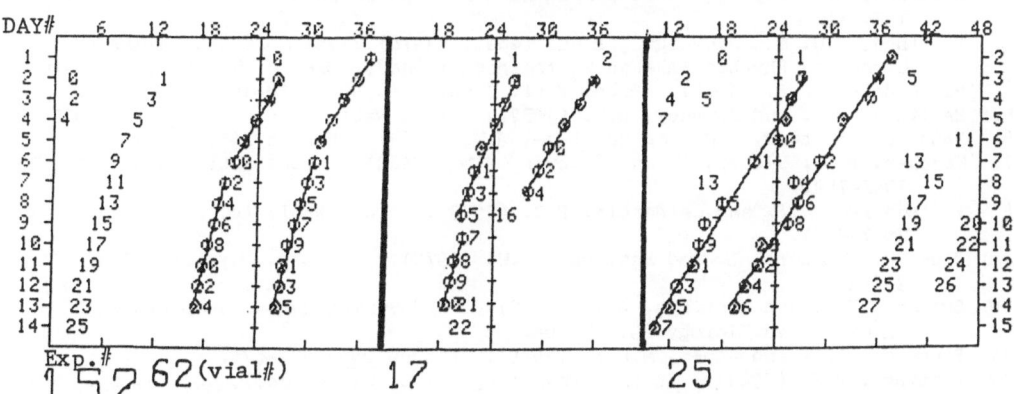

Figure 6. Data from mixed cultures (Exp. #152) presented in the form of plots of peak times on sequential days. The first panel at left shows the calculation for the two parent cultures maintained separately (vials #15 and #19; calculation given assignment #62). The behavior of the actual mix (vial #17) is shown in the center panel, while the right panel shows a similar actual mix (vial #25) in which the medium was replaced every 2 days during the first week.

the two cultures had different periods and merged to form a single peak in the course of the experiment (vial 17).

In order to check the role of the medium in such experiments, we included several mixes in which the medium was exchanged every two days for the first week. The results were clear. With fresh medium, the two components in cultures where the cells were mixed remained separated and their periods were the same (Fig. 6; vial 25). Thus it may be concluded that the different behavior of the two cellular rhythmic components in mixed cultures compared to separate cultures is mediated by the medium.

In these experiments with added fresh medium, an interesting additional effect was seen. There is an apparent spontaneous change (increase) in the free running period in cultures after about the first week under the conditions of our experiments (see Fig. 6; vial 62). But with additions of fresh medium during the first week, this change does not occur; the period remains the same (vial 25). Thus, there is an effect of conditioned medium on period which, as would be expected, is related to cell density: no period changes occurred in cultures with low cell densities (< 500 cells/ml).

Neither the mode of action nor the nature of the substance(s) involved in causing period changes is known, but it is evident that it is attributable to the action of the cells themselves. A realistic pursuit of the problem and the isolation of the putative phase-affecting substance will require a much better bioassay. This might be possible if one high density population is established as the dominant effector, and a second (low density) culture as the responder. A nonluminous but still circadian population of cells might be used to exert effects on a low density of out-of-phase luminous circadian responder cells.

This research was supported in part by a NIH grant (GM 19536) and the Deutsche Forschungsgemeinschaft. We are grateful to Nick Aiuto for his assistance with data analysis.

REFERENCES

1. Pittendrigh, C.S. and Bruce, V.C. (1957). In Rhythmic and Synthetic Processes in Growth, pp. 75-109 (D. Rudwick, ed.) Princeton University Press.
2. Hastings, J.W. (1961). Cold Spring Harbor Symposia on Quantitative Biology 25: 131-144.
3. Hastings, J.W. and Schweiger, H.-G. (eds.) (1976). The Molecular Basis of Circadian Rhythms, Abakon Verlagsgesellschaft, Berlin, Germany, 462 pages.
4. Pittendrigh, C.S. (1954). Proc. Natl. Acad. Sci. 40: 1018-1029.
5. Hastings, J.W. and Sweeney, B.M. (1957). Proc. Natl. Acad. Sci. 43: 804-811.
6. Njus, D., Sulzman, F.M. and Hastings, J.W. (1974). Nature 248: 116-120.
7. Kosower, N.S., Glaser, T. and Kosower, E.M. (1983). Proc. Natl. Acad. Sci. 80: 7542-7546.
8. Pittendrigh, C.S. and Caldarola, P.C. (1973). Proc. Natl. Acad. Sci. 70: 2697-2701.
9. Njus, D., McMurry, L. and Hastings, J.W. (1977). J. Comp. Physiol. 117: 335-344.
10. Sweeney, B.M. and Hastings, J.W. (1960). Cold Spring Harbor Symposia on Quantitative Biology 25: 87-104.
11. Peterson, E.L. and Jones, M.D.R. (1979). Nature 280: 677-679.
12. Hastings, J.W. (1964). In Photophysiology (A.C. Giese, ed.), Vol. I, pp. 333-361, Academic Press, New York.
13. Sweeney, B.M. (1969). Rhythmic Phenomena in Plants, Academic Press, New York.
14. Pittendrigh, C.S. (1981). In Biological Clocks in Seasonal Reproductive Cycles, pp. 1-35 (B.K. and D.E. Follett, eds.), Bristol, Wright.
15. Winfree, A.T. (1980). The Geometry of Biological Time, Springer Verlag, New York.
16. Christiansen, R. and Sweeney, B.M. (1972). Int. J. Chronobiol. 1: 95-100.
17. Hastings, J.W. and Sweeney, B.M. (1960). J. Gen. Physiol. 43: 697-706.

18. Takahashi, J.S., DeCoursey, P.J., Bauman, L. and Menaker, M. (1984). Nature 308: 186-188.
19. Karakashian, M. and Hastings, J.W. (1962). Proc. Natl. Acad. Sci. 48: 2130-2137.
20. Karakashian, M.W. and Hastings, J.W. (1963). J. Gen. Physiol. 47: 1-12.
21. Goodwin, B. (1963). Temporal Organization in Cells, Academic Press, New York.
22. Ehret, C.F. and Trucco, E. (1967). J. Theoret. Biol. 15: 240-262.
23. Dunlap, J. and Hastings, J.W. (1981). J. Biol. Chem. 256: 10509-10518.
24. Johnson, C.H., Roeber, J.F. and Hastings, J.W. (1984). Science 223: 1428-1430.
25. Jacklet, J.W. (1981). Biol. Bull. 160: 199-227.
26. Taylor, W.R., Dunlap, J.D. and Hastings, J.W. (1982). J. Exp. Biol. 97: 121-136.
27. Nakashima, H., Perlman, J. and Feldman, J.F. (1981). Science 212: 361-362.
28. Leong, T.Y. and Schweiger, H.-G. (1979). Eur. J. Biochem. 98: 187-194.
29. Hastings, J.W. and Sweeney, B.M. (1958). Biol. Bull. 115: 440-458.
30. Brinkmann, K. (1966). Planta 70: 344-389.
31. Edmunds, L.N. (1971). In Biochronometry, pp. 594-611 (M. Menaker, ed.) NAS, Washington D.C.
32. Mergenhagen, D. and Schweiger, H.-G. (1974). Plant Sci. Lett. 3: 387.
33. Sulzman, F.M., Gooch, V.D., Homma, K. and Hastings, J.W. (1982). Cell Biophysics 4: 97-103.
34. Krasnow, R., Dunlap, J., Taylor, W., Hastings, J.W., Vetterling, W. and Haas, E. (1981). In Bioluminescence: Current Perspectives, (K.H. Nealson, ed.) pp. 52-63, Burgess Publishing Company, Minneapolis, MN.
35. Njus, D., Gooch, V. and Hastings, J.W. (1981). Cell Biophysics 3: 223-231.

Singularity in a Unicell: Can Pulses of Protein Synthesis Inhibitors Stop the Biological Clock?

H. Broda* and J.W. Hastings

The Biological Laboratories, Harvard University, Cambridge, MA 02138, USA

The unicellular marine dinoflagellate Gonyaulax polyedra exhibits a pronounced circadian rhythm in its bioluminescence (for reference see (1)). This rhythm can be phase-shifted by light pulses as well as by biochemical agents. Extremely effective are the inhibitors of protein synthesis on 80s ribosomes.

Dose and phase response curves with one of these drugs (anisomycin) were studied in great detail (1,2,3). A 3-D-plot of the results of over 500 such experiments yields a spiral surface for the new phase plane (1). The center of this surface (axis of a spiral "staircase") corresponds to the new phases following weak pulses of the drug at an intermediate concentration and one specific circadian time (fig.1).

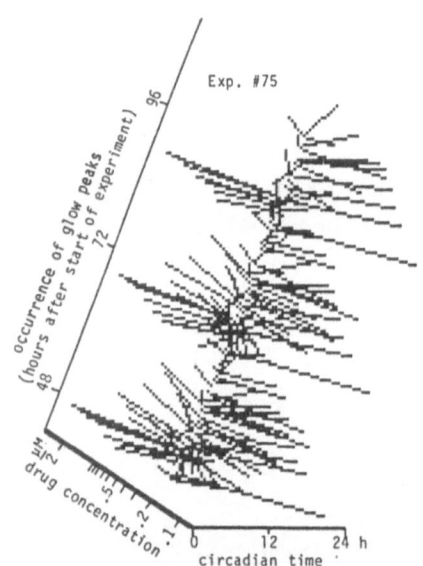

Figure 1
Spiral surface of the new phase after one-hour pulses of anisomycin with different concentrations and at different circadian times. To help visualize the surface, experimental points are connected to the resulting axis of the spiral.

This axis may be interpreted as the singularity (4) that coincides with the transition from a weak to a strong phase response curve at the circadian time of the strongest achievable phase-shift. Drug pulses in the close vicinity of this point may result in double oscillations of the rhythm under investigation (fig.2) or even an apparent abolishment or "paralysis" of that rhythm. The double peaking may be interpreted as a population phenomenon,which could be explained if there were slight differences in the effective concentration of the drug for different individual cells. In the neighborhood of the singularity this could lead to a splitting into two different subpopulations, one of which will have received a

*Present address: Fachbereich Biologie, Universität Konstanz,
D-7750 Konstanz, Fed. Rep. of Germany

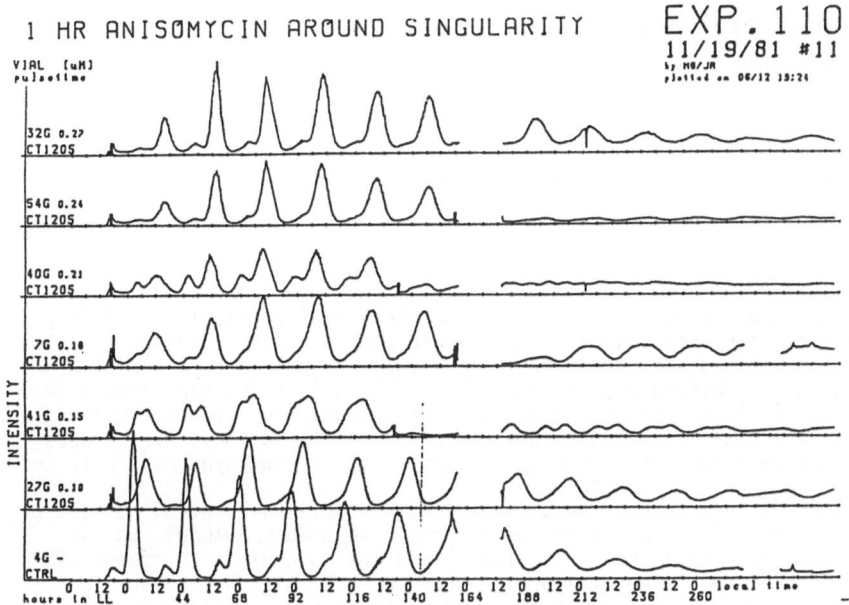

1 HR ANISOMYCIN AROUND SINGULARITY EXP.110
11/19/81 #11
by MB/JR
plotted on 06/12 15:24

Figure 2. Double peaks caused by pulsing close to the singularity point. In some of these vials a second pulse (marked with a bar) of the same concentration was added. Note that e.g. in vial 40 the subsequent glow shows no more detectable rhythm.

strong, the other a weak pulse. Furthermore, the double peaking populations can be brought into apparent arrhythmia by administering a second weak drug pulse at the specific time estimated for the predicted singularity of the major peak in the double oscillation (fig.2). The question remains to be resolved, whether the arrhythmia reflects a property of each individual cell,or is due to still rhythmic individual cells with random phase-distribution.

The oscillator itself which drives the circadian clock is believed to have a frequency of about 24 hrs (shape of the phase response curves; only one point of singularity within one day). Presently we are not yet able to specify the biochemical mechanism producing such a long period and precise oscillations (5). We presume that the biochemical reactions which form the components of this clock all have time constants that are several orders of magnitude smaller. How such a system achieves the unusual temperature compensation of the period remains still a puzzle. The investigation of the mechanisms responsible for this complex clockwork can be viewed as one of the most exciting fields of cybernetics in biological systems.

References

(1) W. Taylor, R. Krasnow, J.C. Dunlap, H. Broda, J.W. Hastings: Critical pulses of anisomycin drive the circadian oscillator in Gonyaulax towards its singularity. J.comp.Physiol. 148, 11-25 (1982).

(2) W. Taylor, J.W. Hastings: Minute-long pulses of anisomycin phase-shift the biological clock in Gonyaulax by hours. Naturwissenschaften 69, 94-96 (1982).

(3) W. Taylor, J.C. Dunlap, J.W. Hastings: Inhibitors of protein synthesis on 80s ribosomes phase-shift the Gonyaulax clock. J.exp.Biol. 97, 121-136 (1982).

(4) A.T. Winfree: The geometry of biological time. Springer, Berlin, Heidelberg, New York (1980).

(5) J.W. Hastings, H. Broda, C.H. Johnson: Phase and period effects of physical and chemical factors. Do cells communicate? (This book).

Circadian Control of Protein Synthesis Rate in Cell-Free Extracts of *Gonyaulax polyedra*

A. Schroeder-Lorenz and Ludger Rensing

Biology Department, University of Bremen, D-2800 Bremen, Fed. Rep. of Germany

Protein synthesis, or rather the concentration changes of particular proteins, may be essential for the functioning of the circadian clock (1). Evidence for this statement has been derived from phase-shifting effects by pulses of antibiotics that inhibit the 80 S-ribosome dependent protein synthesis of the dinoflagellate, Gonyaulax polyedra (2,3) and other organisms. From this type of evidence, working hypotheses about the role of proteins in a circadian oscillator have been proposed that rely on negative feed back of a protein species on its own synthesis rate (4, 5).

The simplest known feed back systems involving translational control consist of two elements: the rate of translation of a certain messenger molecule on ribosomes (for example, ribosomal mRNA of bacteria) and the concentration of one of the synthesized proteins (a ribosomal protein species). Simulations based, for example, on Goodwin's model oscillator (6) require three or more elements for a stable limit cycle oscillation (7) which, however, need not change the basic assumption of translational feed back control.

Growing cultures of Gonyaulax polyedra show circadian changes in the synthesis rate of total protein (7,8,9). Since we were interested in the question whether or not these changes are based on translational control, we first tested the translational capacity of a cell-free system. This system consisted of a cell extract

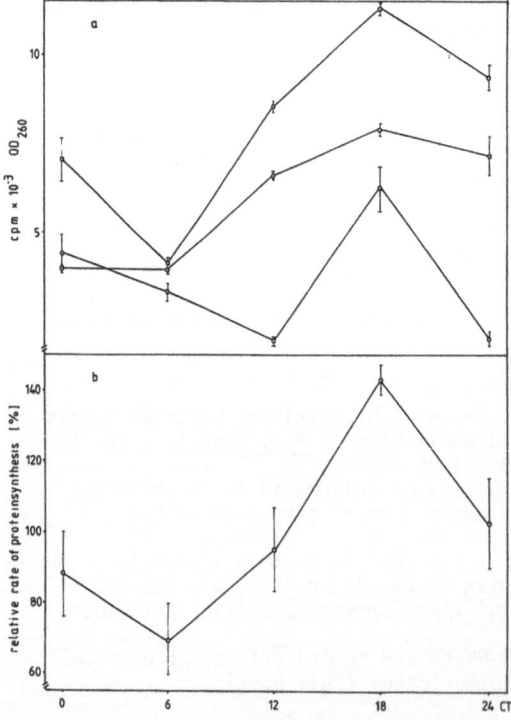

Fig. 1. In vitro translation system of Gonyaulax polyedra. S30 extracts were prepared at different phases of cultures in constant dim light (1800 lx) and the ability to incorporate 35S-methionine into TCA-insoluble material measured with reference to OD 260
a. shows the data of 3 series of experiments,
b. summarizes the data expressed as percent of the mean value (ordinate). Abscissa: time in h one day after beginning of constant light conditions. The previous day time (="subjective"day) corresponds to 0-12, the previous night time (="subjective" night) to 12-24, circadian time (CT). As the free-running period length of these cultures is approximately 24h, the circadian time is equal to the hours in constant light. Vertical bars: 1SD.

of growing cultures (supernate of centrifuging the cell homogenate for 15 min at 30 000 xg (S30 extract)), cold amino acids and 35 S-methionine, ions and energy metabolites (10). The translational capacity of the cell-free system shows almost exactly the same circadian changes as the cells in vivo (fig. 1).

Because this result might be due to circadian differences either of the ribosome and/or mRNA content of the extract or a translational control mechanism,we centrifuged the extract for 2 h at high speed (100 000xg) in order to eliminate the ribosomes (S100 extract). We added S100 extracts from the phase of high capacity ("subjective" night phase S100N) to S30 extracts of the low capacity phase ("subjective" day phase, S30D) and vice versa. The results showed that S100N increases the translational capacity of S30D in a concentration-dependent way (fig.2, upper panel) whereas the S100D inhibited the S30N (fig. 2, lower panel).

In order to distinguish between protein and RNA components in the S100 extracts, we exposed them to heat of 95°C for 5 min. As no activating or inhibitory effects of the extracts were observed after this treatment,we conclude that proteins - possibly kinases and/or phosphatases - play a role in controlling the circadian changes of total protein synthetic capacity. We cannot answer the question, however, whether this control system is part of the clock mechanism or whether it is one (of many) dependent functions of the clock.

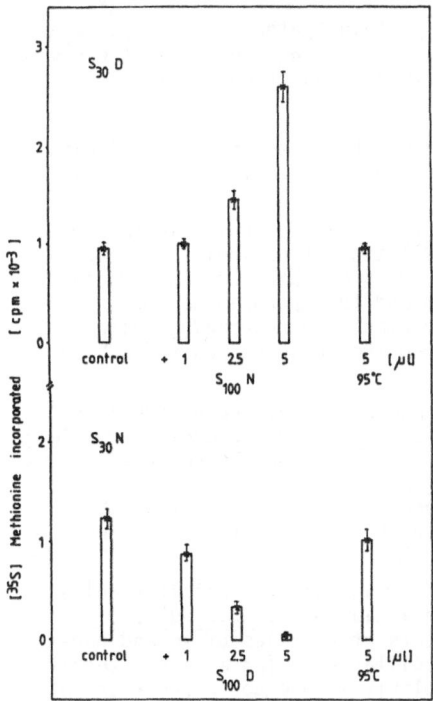

Fig. 2. Influence of postribosomal supernatant (S100) on the translational capacity of the S30 extracts. Postribosomal extracts of the high capacity phase (S100N) were added to extracts of the low capacity phase (S30D, upper panel) and postribosomal extracts of the low capacity phase (S100D) were added to extracts of the high capacity phase (S30N, lower panel). The amounts of S100 are given in µl underneath each column. Right columns give the results with S100 extracts heated for 5 min. Ordinate: Incorporation of 35S-methionine into TCA precibitable proteins. Vertical bars: 1SD.

1. G. Cornelius, L. Rensing: Biosystems 15, 35-47 (1982)
2. J.C. Dunlap, W. Taylor, J.W. Hastings: J. Comp. Physiol. 138, 1-8 (1980)
3. W. Taylor, R. Krasnow, J.C. Dunlap, H. Broda, J.W. Hastings: J. Comp. Physiol. 148, 11-25 (1982)
4. H.G. Schweiger, M. Schweiger: Int. Rev. Cytol. 51, 314-342 (1977)
5. K. Drescher, G. Cornelius, L. Rensing: J. theor. Biol. 94, 345-353 (1982)
6. B.C. Goodwin: Adv. Enzyme Regul.3,425 (1965)
7. L. Rensing, W. Schill, this volume
8. G. Cornelius, A. Schröder-Lorenz, L. Rensing, submitted
9. W. Volknandt, R. Hardeland: Comp. Biochem. Physiol. 77B, 493-500 (1984)
10. A. Schroeder-Lorenz, ms. in preparation.

Perturbation by Single and Double Pulses as Analytical Tool for Analyzing

Ludger Rensing and Walter Schill

Biology/Mathematics Department, University of Bremen
D-2800 Bremen, Fed. Rep. of Germany

1. Introduction

The functional structure of the circadian oscillator and its molecular compounds has frequently been analyzed by applying pulsed perturbations of various kinds and registering the response of the oscillator - mainly the phase shifting response [1,2]. This approach has been extensively exploited in the case of the circadian rhythm of Gonyaulax.

On the basis of previous data on the role of protein synthesis in the circadian oscillator (1) and earlier mathematical models [3,4], we used a limit cycle oscillator to simulate the response to sudden perturbations of one of its variables [5].

We focussed our interest on comparing the dynamic features of the model with those of the circadian rhythm. We, therefore, exposed both oscillators to similar perturbing treatments.

a) The phase-shifting effects of single pulses of protein synthesis inhibitors applied at different phases of the circadian oscillation of Gonyaulax have been analyzed previously [6,7]. We measured the effect of the inhibitor (anisomycin) on the rate of protein synthesis in this organism and the approximate course of recovery from this perturbation [8]. We then simulated a corresponding pulse perturbation with the model oscillator and compared the phase responses in both oscillators.

b) We determined the phase relation between the phase response curve (PRC) to inhibitor pulses and the perturbed variable. In Gonyaulax the perturbed variable (possibly the concentration of a protein species) is not known. Therefore, we measured the rate of total protein synthesis in growing cultures [9,10], which might change in phase with the unknown protein species. In the model, the 12h (0.5 period) phase-shift of the PRC coincides with the maximum of the perturbed variable.

c) We analyzed the dynamics of the oscillator after an initial inhibitor pulse by means of a series of second pulses of the same inhibitor (double pulse treatment). We determined the phase-shifting effects of a series of second perturbing pulses on both the circadian and model oscillator during a 24 h interval after the first perturbing pulse. We were particularly interested whether or not the first pulse might transiently stop the oscillation or whether the oscillation continues after an immediate phase shift [8].

2. The Model

We base our calculations on the following model, originally developed by GOODWIN [3]: synthesis of protein X is inhibited by repressor Z, which formation in turn is regulated by X via an intermediate substance Y.

---

*Dedicated to Prof. Bernhard Rensch on occasion of his 85th birthday.

We choose the following mathematical model (see DRESCHER et al. [5])

$$1.1 \quad \dot{X} = a(1 + Z^m)^{-1} - bX$$
$$(1) \qquad 1.2 \quad \dot{Y} = X \qquad\qquad - bY$$
$$1.3 \quad \dot{Z} = Y \qquad\qquad - bZ$$

It has been shown elsewhere [4] that for this system to show a stable limit cycle it is necessary for $m \geq 9$; thus 1.1 describes the kinetics of an order reaction larger than 9. However, there is experimental evidence [11] that m is as small as 1 or 2.

In order to omit this biochemically unrealistic feature of the model within the same model-class (i.e. negative feedback), one could increase the number of intermediary reactants, which gives rise to a larger system of equations:

$$\dot{X}_1 = a(1 + X_n^m)^{-1} - bX_1$$
$$(2) \qquad \dot{X}_i = X_{i-1} \qquad\qquad - bX_i \qquad , i = 2, .., n$$

The existence of stable oscillations can be proven if $m \geq 2$, $n \geq 8$. Alternatively, one could introduce time-delays into the decay terms or certain V_{max} values into the degradation of Z (see [11] [12]).

We based our simulation of the double-perturbation experiment on model 1, because the PRCs of the two models (1) and (2) did not differ qualitatively.

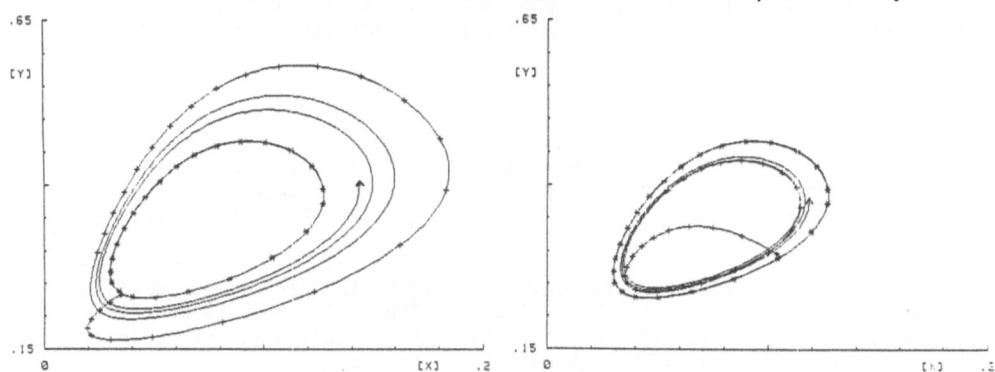

Fig. 1a,b The behaviour of two of the three variables of model 1 (concentrations [X] and [y] after perturbation at different phases a (left cycle) and b (right cycle). The time intervals between points are identical, the velocity of changes are thus different during one cycle.
**** - limit cycle - trajectory before perturbance
++++ - trajectory during and after a perturbance (rate of synthesis a > 1)
——— - trajectory after recovery (a = 1)

3.   Single Perturbing Pulses

We used the data of TAYLOR et al. [7] in order to compare the phase response of the circadian oscillator to single pulses of anisomycin (Fig. 2a) with the responses of the model oscillator to perturbing pulses of similar form. In order to estimate the effect of a 1h or 3h pulse of anisomycin we determined the inhibition of total protein synthesis during and up to 10h after the pulse [8]. The estimated form of the inhibition pulse (inset of Fig. 2a) was then simulated by a similar change of the variable X (protein synthesis in the model (inset of Fig. 2b). The response curves of the two versions of the model (1) and (2) differ only slightly. They show PRCs of approximately similar shape compared to the circadian oscillator.

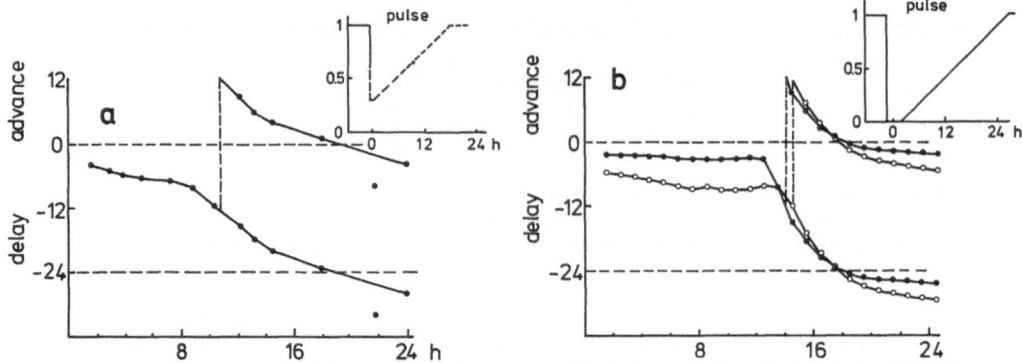

Fig.2a) Phase response of <u>Gonyaulax</u> to single pulses of anisomycin (inhibitor of 80S-ribosome directed protein synthesis) [7]. Ordinate: the phase-shift in h is given either as delays only or as delays and advances, because the direction of the shift is ambiguous. Abscissa: time (h) in constant light (representing the different phases of the oscillation with a period length of ∿ 24h). Inset: pulse form estimated from biochemical determinations of protein synthesis inhibition by a pulse of anisomycin [8]. Ordinate: inhibition; abscissa: time after pulse in hours.
2b) Phase response of the model oscillator to single pulses of an inhibitor (re-setting of the variable X). The two curves plotted in the same way as in a) re-present the two versions of the model 1(O) and 2(●). Ordinate: Phase-shift in h. Abscissa: time in h, representing the phases of the oscillator with a period length of 24 h. Inset: form of pulse applied (other details as in a).

4. Phase Relations

12h phase-shifts with cycloheximide [6] or with anisomycin up to concentrations of 0.5 µM [7] were observed during the "subjective night", i.e. at about 14 - 20h after beginning of constant light. What conclusions can be drawn from this result as far as the phasing of the perturbed variable (possibly the concentration of a protein species) is concerned ?

In the model we find a coincidence between 12 h (1/2 period) phase-shift and the maximum of the perturbed variable (Fig. 3b). Thus, if the model describes the circadian oscillator adequately, one may expect the concentration of a protein in <u>Gonyaulax</u> to be at its maximum between 14 and 20 h.

In order to test this assumption.we measured the synthesis and degradation of proteins [9,10]. In growing cultures the synthesis and degradation of a major-ity of all proteins changes rhythmically, the maximum of which occurring at the expected phase between 14 and 20 h (Fig. 3a). From this result we conclude that in <u>Gonyaulax</u> the protein X, essential for the oscillation, changes rhythmically together with the bulk of other proteins.

5. Double Perturbing Pulses

After an initial pulse of 2 µM anisomycin (3h) we applied a second pulse of 0.2 µM anisomycin at different times during a 24 h interval after the first pulse. We then measured the phase difference of cultures treated with a second pulse with reference to cultures that received the first pulse only [8].

We used this technique of applying 2 subsequent pulses, first introduced by PITTENDRIGH [13], in order to answer the following questions: How does the cir-cadian oscillator react to the first pulse during the following 24 h interval ? Is the reaction to the series of second pulses dependent on the phase at which the first pulse is applied ? Is the oscillator held stationary for some time after the first pulse ? - a possibility that had been discussed previously [7] - and is the

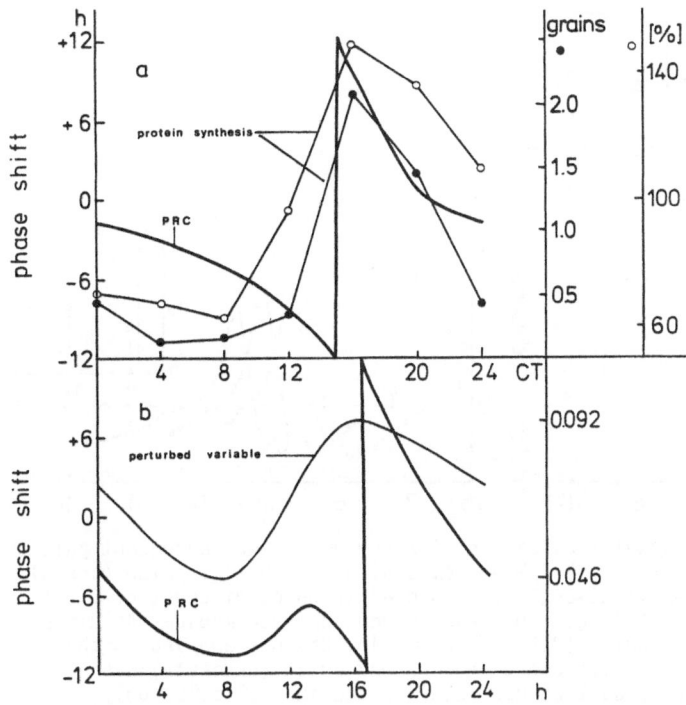

Fig. 3a,b) Phasing of an oscillator variable and phase response to perturbances of that variable in the circadian rhythm of Gonyaulax (a) and the limit cycle model(b). a)Total protein synthesis in growing cultures of Gonyaulax: -o- incorporation of labelled amino acids into total protein (percent of the mean)[9]; -•- labelled proteins in the cytoplasm of EM section (number of grains per 1.7μm²)[10](right ordinates). ——— Phase response curve (PRC) to 1 h, 10 μM cycloheximide pulses (left ordinate: phase-shift in h)[6]. Abscissa: time in h after beginning of constant light. b) Variable X of the model oscillator (right ordinate: values of X) and phase response to a 70% reduction of that variable (left ordinate: phase-shift in h). Abscissa: time in h.

reaction of the model oscillator to a simulated double pulse treatment similar to the reaction observed in the experiments ?

The results showed that the second pulses induced phase-shifts similar to those applied to an unperturbed oscillation (Fig. 4a). The PRCs to the second pulse differ in their phasing, depending on the phase the initial pulse is applied (Fig.4a and 5a) indicating that the oscillation continues, but is immediately phase-shifted to different phases. The different phasing of the circadian oscillation after the first pulse is expressed as different phasing of the PRCs: depending on the phase the first pulse is given (arrow Fig. 4a or abscissa of Fig.5a) the time interval after which the PRCs reach the 12h /1/2 period) phase-shift is different. The longest interval is observed if the pulse is applied at times 16 to 22, i.e., at the time of longest delay shifts (see Fig. 2a).

The model oscillator shows similar reactions: the phasing of the PRCs to second pulses also depends strongly on the phase of the first pulse. This dependency (expressed by the interval between the beginning of the PRC and the crossing of the zeroline from delay to advance phase-shift is strikingly similar in both oscillators (Fig.4b and 5b).

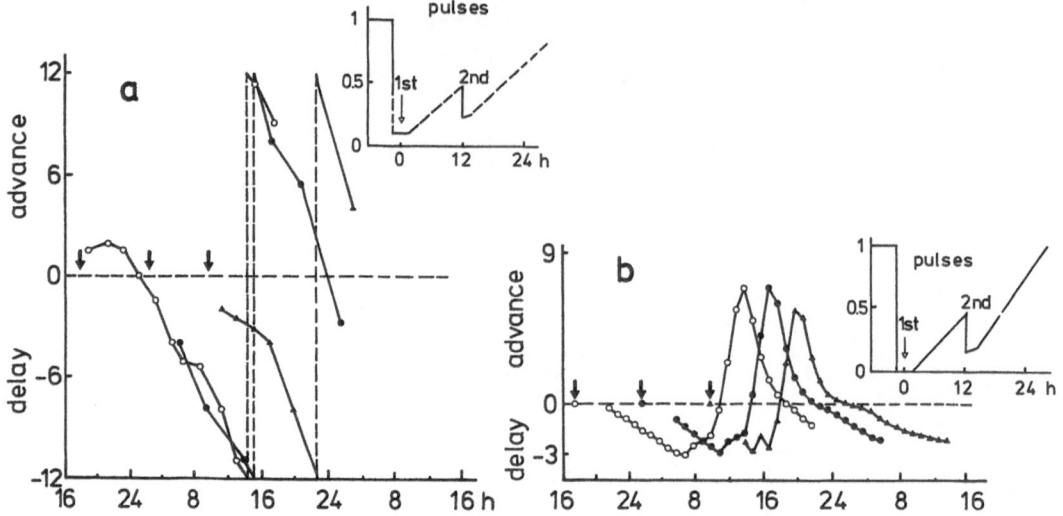

Fig.4a) Phase response of <u>Gonyaulax</u> circadian glow rhythm to two subsequent pulses of anisomycin. Cells were treated with 2µM anisomycin(3h) as initial pulse (arrow) at different phases of the oscillation, given in h after onset of constant light (abscissa). They were then centrifuged and resuspended in fresh medium. At different times thereafter a second pulse (0.2 µM anisomycin, 2h) was applied in the same way. The subsequent phase-shift due to the second pulse (ordinate) was measured in comparison to cultures that had received the initial pulse only (control, base line). Inset: form of a particular pulse sequence as estimated from biochemical determinations of protein synthesis inhibition [8], ordinate: inhibition, abscissa: time after pulse in h. 4b) Phase response of the model oscillator to two subsequent pulses (resetting of the variable X) at different phases of the oscillation. Phase-shifts to the second pulse (ordinate) with respect to a control (first pulse only, base line) depending on the phase of the first pulse (arrow, abscissa). Inset: form of a particular pulse sequence, chosen similar to the experimental pulse forms.

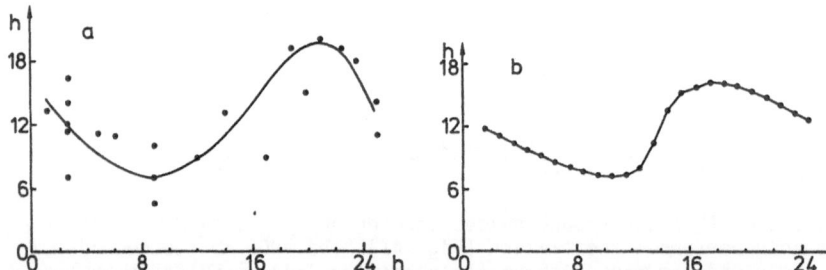

Fig.5) Phasing of the PRC to second pulses as a function of the phase at which the first pulse is applied (abscissa). Phasing of PRCs is expressed as the interval between first pulse (arrow Fig.4) and the delay-advance transition of the PRC. Ordinate: interval duration in h a) <u>Gonyaulax</u> circadian oscillator b) Model oscillator.

6. Concluding Remarks

From the similarity of the responses to perturbing pulses shown by the circadian and the model oscillator, i.e., from the similar dynamic properties, we conclude that the functional structure of the model may represent a possible structure of the circadian oscillator.

References

1 G. Cornelius, L. Rensing: BioSystems 15, 35-47 (1982)
2 J. Feldman, J.C. Dunlap: Photochem. Photobiol. Reviews 7, 319-368 (1983)
3 B.C. Goodwin: Adv. Enzyme Regul. 3425 (1965)
4 J.C. Griffith: J. theor. Biol. 20, 202 (1968)
5 K. Drescher, G. Cornelius, L. Rensing: J. theor. Biol. 94, 345-353 (1982)
6 J.C. Dunlap, W. Taylor, J.W. Hastings: J. Comp. Physiol. 138, 1-8 (1980)
7 W. Taylor, R. Krasnow, J.C. Dunlap, H. Broda, J.W. Hastings:
   J. Comp. Physiol. 148, 11-25 (1982)
8 U. Hobohm, G. Cornelius, W. Taylor, L. Rensing: Comp. Biochem.
   Physiol. in press
9 G. Cornelius, A. Schroeder-Lorenz, L. Rensing: Am. J. Physiol. (B)
   submitted
10 B. Donner, U. Helmboldt-Caesar, L. Rensing: manuscript submitted
11 J.D. Murray: Lectures on Nonlinear-Differential Equation Models in
   Biology, Springer, Heidelberg, 1977
12 P.R. Painter, J.J. Tyson: in "Cell Cycle Clocks", L.N. Edmunds,
   Jr.Ed., Dekker, New York, Basel 1984
13 C.S. Pittendrigh, Z. Pflanzenphysiol. 54, 273-307 (1966).

Effects of Light and Temperature Steps on Circadian Rhythms of *Neurospora* and *Gonyaulax*

Van D. Gooch

Division of Science and Mathematics, University of Minnesota-Morris
Morris, MN 56267, USA

1.   Introduction: Many organisms have physiological activities that fluctuate
on a daily basis; frequently these fluctuations will continue to occur on an
approximately 24-hour basis even in the laboratory under constant conditions. Such
rhythms are termed circadian rhythms. Light, temperature and certain chemicals
have been shown to affect many circadian rhythms. These parameters are usually
applied as a pulse (a step-up followed by a step-down) to a free-running rhythm
under constant conditions. The pulse causes shifts from normal levels, but eventually the cycle returns into its exact normal rhythm with the only permanent
change being a phase-shift relative to the control. Thus, such experiments and
others have led to the development of the limit cycle model for circadian
rhythms [1-4]. In an attempt to answer a variety of questions, there have been
several types of experiments performed using light, temperature and chemicals.
These experiments have names that are jargon to circadian rhythm researchers:
pulse experiments resulting in phase response curves, entrainment experiments,
"holding" experiments, skeleton photoperiod experiments, and step experiments
[5]. It is important to realize, however, that all of these experiments are
variations upon each other, i.e. they all involve a step-up and/or a step-down
of the perturbation in question; the different experiments depend upon when these
step-ups and step-downs are applied relative to each other, and when they are
applied within a cycle. In an attempt to distinguish between the effects of step-
up versus step-down, experiments can be performed on a series of cultures such
that one step is always applied at the same phase while the other step is varied.
Such experiments have been performed in the past [6-11]. Data is presented in
this work for the conidiation rhythms of the bread mold Neurospora crassa and the
luminescent glow rhythm of the single-celled marine alga *Gonyaulax polyedra*.

2.   Methods: *Neurospora crassa,* strain *band a,* was grown on an arginine-
glucose-supplemented agar medium [12] in race tubes with a length of 35 cm and an
internal diameter of 1.3 cm. The progress of the growth front of the *Neurospora*
along the race tube was marked exactly every 24 hours under a red safe light. As
the *Neurospora* grow, they oscillate in circadian behavior in their production of
their reproductive spores called conidia. The peaks of conidiation were visually
determined in three replicate race tubes, and the physical position of this
conidiation peak relative to the 24 hour growth front marks is used to determine
the time of the conidiation peak. The data is entered via a digitizer into an
Apple II microcomputer, and subsequently the necessary calculations are made by
the microcomputer. All of the experiments with *Neurospora* were carried out at
25°C except when specified otherwise.

Spontaneous luminescent glow rhythms of populations of *Gonyaulax polyedra,*
strain GP70 are measured using a photomultiplier tube. A turntable apparatus
holding a maximum of 30 vials of *Gonyaulax* and an Apple II microcomputer system
were used to collect data [13]. The peaks of luminescence were determined using
a curve-fitting technique as previously described [14]. The peaks plotted in the
figures are actually the peaks of the Gaussians determined by the curve-fitting
technique. All experiments with *Gonyaulax* were carried out at 20°C in an f/2
medium using artificial sea water [15-16].

3.   Results: Several cultures of *Neurospora* were first entrained to 12 hours of
light - 12 hours of dark for two days (Fig.1). The average circadian time [3] for
the second entrainment peak of all of the cultures is 20.4 hours. After entrainment, various cultures were exposed to various durations of light as specified.
Finally they are put into constant dark,thus allowing the free running circadian
phases and periods to be determined. The average period length for all of the
cultures under free running conditions is 21.o hours. Thus, the experimental
design is such that the time of the last light onset is different for such culture,
while the last dark onset is the same.

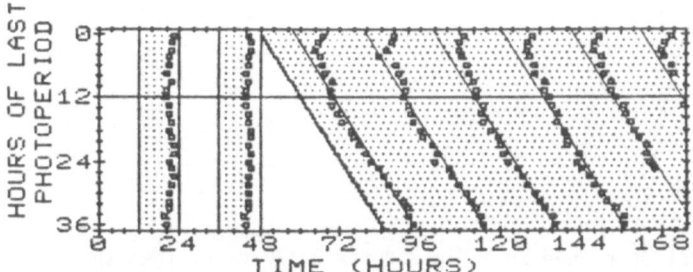

Fig. 1. *Neurospora* was first entrained to 12 hours of light - 12 hours of dark for
two days. Then the various cultures were exposed to various durations of light as
specified. After this treatment they were placed into constant darkness and the
time of the conidial peaks (filled squares) were determined. All shaded areas represent darkness. 25°C. 4000 lux.

        The cultures with a last photoperiod of either 0 hours or 12 hours can be
regarded as controls (considering the 12 hour light-12 hour dark pre-entrainment).
The cultures having a last photoperiod between 0 hours and 8 hours all show phase
advances relative to these controls. The cultures having more than 12 hours of the
last photoperiod tend to show delays. The general tendency is that these cultures
show a first peak of conidiation 9.5 hours after the last dark onset. (If one ex-
trapolates back to the point of the last dark onset then this point of dark onset
corresponds to a circadian time of 9.6 hours [20.4 - (9.5) (24/21.0) = 9.6 where
20.4 is the circadian time the peak is assumed to occur, the 9.5 is the real
hours from the last dark onset to the peak, and the 24/21.0 is the conversion factor needed to convert the 9.5 hours from real time to circadian time]). The
9.5 hour line is drawn diagonally on the graph,and one can see that the data shows
a wiggle about that line with a maximum deflection of 4 hours. This wiggle has
also been previously seen with mosquito and moth data [7,8,11].

        A similar experiment was run with *Gonyaulax*,except the roles of light and
dark are reversed (Fig. 2). Here it is seen that for last dark periods exceeding
6 hours, the glow peaks tend to phase to the last light onset rather than
the last dark onset as was the case for *Neurospora*. Again,this phasing seems to
show a slight wiggle around the average phasing line. Large phase advances are
seen when the last dark period is in the range between 0 and 12 hours. *Neurospora*
can not be used in this type of experiment,since they demonstrate continual
conidiation in the light and thus no rhythms are expressed [17]. Similarly,
Gonyaulax can not be used in the format presented in Fig. 1 because these
photosynthetic organisms quickly die in the constant dark.

        Steps of temperature using *Neurospora* are investigated in Fig.3. In the
experimental configurations shown in the figure,it can be seen that the peaks tend
to phase themselves with the last temperature-change whether it is a step down
(45 → 25°C) or a step up (15 → 25°C. Again it would seem that a slight wiggle could
exist. The 45°C caused large phase advances for data points between 0 and 12 hours
of application. The first and second peak occurred 9.4 hours and 29.3 hours after
the last temperature-transition respectively for 45°C. The situation is quite
different for the 15 and 4°C experiment,where these values are 19.2, 39.5 (4°C).

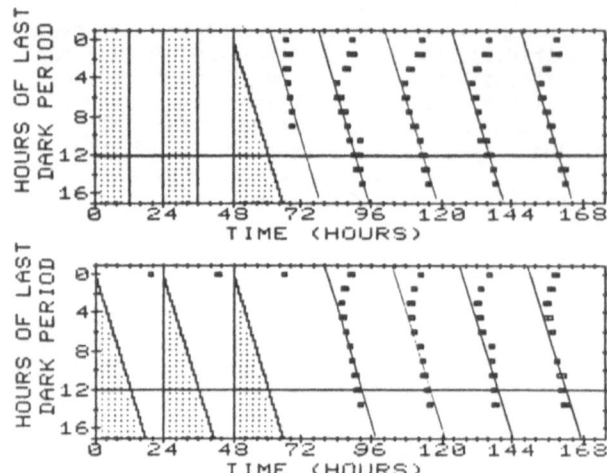

Fig. 2. Gonyaulax were first entrained as indicated for 2 1/2 days. Then the various cultures were exposed to various durations of dark as specified. After this treatment they were placed into constant light and the time of the luminescent glow peaks (filled squares) were determined. All shaded areas represent darkness. 20C. 500 lux. 10,000 cells/ml.

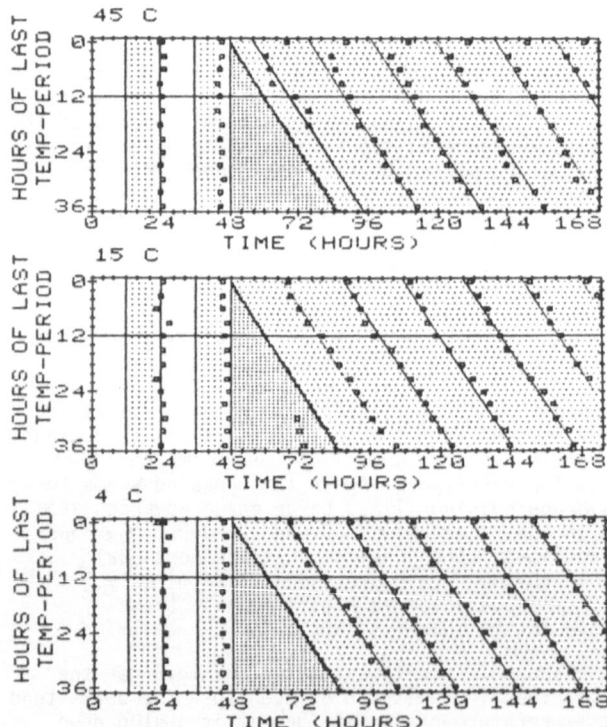

Fig. 3. Effects of different temperatures using Neurospora. The heavily shaded area represent the times when the Neurospora were exposed to the indicated temperatures other than 25°C. All shaded areas, including the heavily shaded area, represent darkness.

4.   Discussion: The type of result seen in Fig. 1 was originally interpreted by assuming a simple clock such that, in bright light,the circadian clock proceeds normally until it reaches a certain phase point,at which point (around CT12) the clock stops [6,18-19]. The clock will hold at that phase point until it is placed back into the dark. Once the light-dark transition has been made,the clock re-starts from that phase point and a peak is subsequently seen a specified number of hours later. In the graphs, lines are drawn where peaks would be expected if perfect holding occurred. The actual results of the experiments are not consistent with this hypothesis for several reasons:

1. There is a distinct "wiggle" which occurs around the predicted holding line in most data. The period length of the wiggle is approximately 24 hours.
2. In many cases,for a last photoperiod of less than 12 hours,it can be seen that phase advances have occurred.
3. In at least one case, when the roles of light and dark have been reversed,the same basic result is still observed. Given the holding hypothesis, one would have to claim that in such a case the circadian clock stops in the dark rather than the light.
4. Temperature has similar effects to light, and the role of high and low temperature can be inverted,yet the same basic effect is still observed.

 PETERSON and SAUNDERS [7,8] interpreted these types of results such that when light levels are different from control levels, the biological clock oscillates within parameter ranges but with approximately the same period length. When the organism is shifted between control levels and experimental levels, new parameter ranges will be reached according to a limit cycle model (Fig. 4). This limit cycle interpretation accounts for the basic result and the anomalies.

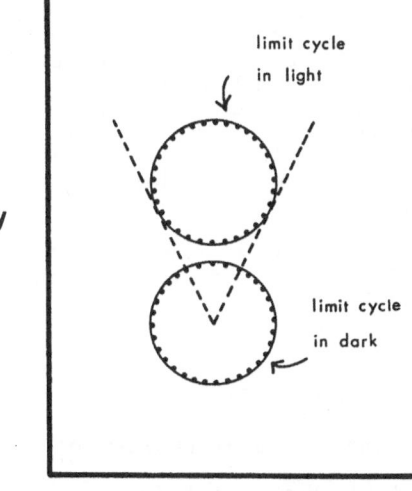

Fig. 4. Theoretical limit cycle diagrams that would yield a hold-ing effect with a four hour am-plitude wiggle. Assumed are perfectly symmetrical, circular limit cycles with perfect 24 hour cycles (each dot along the limit cycle represents one hour of progression). The dotted lines represent isochrons [1] for the dark limit cycle that are separated by four hours. To have a holding effect with a four hour amplitude wiggle, all that would be required is that the light limit cycle be tangent to these isochrons.

In more detail this model assumes the following:

1. The free-running circadian rhythms follow limit cycle behavior.
2. Light or temperature affects the chemical reaction rate constants that generate the circadian rhythm. Then it follows that there will be different singularities for different levels of light and temperature.
3. An altered level of light or temperature allows for a new limit cycle oscilla-tion. The components of the oscillating reactions will then oscillate in a different range. (The experimenter may or may not be able to directly detect this new oscillation depending upon how the oscillating components are coupled to the experimentally observable physiology. For example: Maybe conidiation in

Neurospora always occurs at a cellular pH above 7.5. In the constant dark the pH may oscillate in the range of 7.0 to 7.7, in which case we would see a daily conidiation cycle corresponding to when the pH is above 7.5. In constant bright light, the pH may oscillate in the range of 7.6 to 7.9, in which case the observer would always see conidiation, since the pH is always above 7.5. Thus, the underlying oscillation would be masked from the observer.)

Given these assumptions, when the light or temperature is changed from one level to another the oscillating parameters will immediately begin to seek the corrsponding new limit cycle. If the system is returned to normal levels before the new limit cycle is obtained, the oscillating parameters will return to the old limit cycle in either an advanced or delayed state, depending upon the starting point. This would account for the phase advances seen for short durations of the final photoperiod. Note that this is no different than a classic pulse experiment and thus results in phase response curves that show advances and delays. There is ample literature showing that limit cycle models are excellent in describing this type of phenomenon [1].

When the light or temperature is changed from one level to another, and it is not changed back to its normal level soon thereafter, then the oscillating parameters will attain the new limit cycle. The experimenter may or may not observe this new oscillation, depending upon the amplitude and the coupling of the internal oscillating parameters with the external observable parameters.

When the system is eventually put back into the dark from the light, the amount of time it takes to get back to its free-running dark limit cycle from the free-running light limit cycle only slightly depends upon the phase point from which it started on the light limit cycle. (This is somewhat like saying the amount of time it takes to fly to the earth from the moon only slightly depends upon whether you start on the front of the moon or the back of the moon.) The fact that it is only slightly dependent gives the appearance of holding, but the fact that it is somewhat dependent gives rise to the wiggle about the predicted holding line. The system can actually return more quickly or more slowly to the dark limit cycle depending upon what part of the light limit cycle one started. Thus, the wiggle actually represents the oscillation that was occurring in the light limit cycle and the period length of the wiggle should correspond to the period length of the free-running light limit cycle.

All of the above concepts apply just as well if the roles of light and dark are replaced with high and low temperatures. The concepts also apply if the roles of light and dark (or high and low temperature) are reversed.

5. References:

1. A.T.Winfree: The Geometry of Biological Time (Springer-Verlag Berlin 1980)
2. G. Cornelius and L. Rensing: BioSystems 15, 35-47 (1982)
3. T. Pavlidis: Biological Oscillators: Their Mathematical Analysis (Academic Press, New York 1973)
4. W. Taylor, R. Krasnow, J. Dunlap, H. Broda and J.W. Hastings: J. Comp. Physiol. 148, 11-25 (1982)
5. E. Bunning: The Physiological Clock (Springer-Verlag Berlin 1973)
6. C.S. Pittendrigh: Z. Pflanzenphysiol. Bd. 54, 275-307 (1966)
7. E.L. Peterson and D.S. Saunders: J. Theor. Biol. 86, 265-277 (1980)
8. E.L. Peterson: J. Theor. Biol. 84, 281-310 (1980)
9. D.S. Saunders: J. Comp. Physiol. 110, 111-133 (1976)
10. D. Njus, L. McMurry and J.W. Hastings: J Comp. Physiol. 117, 335-344 (1977)
11. S.K.deF.Roberts: J. Cell Comp. Physiol. 59, 175-186 (1962)
12. M.L. Sargent, W.R. Briggs and D.O. Woodward: Plant Physiol. 41, 1343-1349(1966)
13. V. Gooch, D. Schmidt and H. Broda: "Using the Apple Microcomputer for Long Term Data Collection in Research of Biological Circadian Rhythms", in Proc. of Small College Computing Symposium 15, 142-152 (1982)
14. D. Njus, V.D. Gooch and J.W. Hastings; Cell Biophysics 3, 223-231 (1981)

15. F.M.M. Morel, J.C. Westall, J.G. Rutter and J.P. Chaplick: Water Quality
    Laboratory, Mass. Inst. of Tech., Tech. Note No. 16 (1975)
16. R.R.L. Guillard and J.H. Ryther: Can J. Mibrociol. 8, 229-239 (1962)
17. J.F. Feldman and J.C. Dunlap: Photochem and Photobiol Rev. $\underline{7}$, 319-368 (1983)
18. B.M. Sweeney: Physiologist $\underline{21}$, 118- (1978)
19. H. Nakashima, J. Perlman and J.F. Feldman: Am. J. Physiol. $\underline{241}$, R31-R35 (1981)

Acknowledgements: Most of the *Neurospora* work was done in the lab of Dr. Ludger
Rensing, University of Bremen, Bremen, FRG. The rest of the work was done at
Univ of Minn-Morris with the aid of several students: Dan Zamzow, David Schmidt,
Peter Falkum, Don Patterson, Sue Bludorn, and Matty Benkofske and a Graduate
School Grant-in-aid (0749-5244-72).

Genetic and Physiological Analysis of a Circadian Clock Gene in *Neurospora crassa*

Jerry F. Feldman

Thimann Laboratories, University of California, Santa Cruz
Santa Cruz, CA 95064, USA

1. Introduction

For more than a century developmental biologists have understood the importance of temporal order, or temporal organization, in biological systems. Events during embryogenesis must occur not only in the right position of the embryo but at the correct time and in the correct sequence. Cell biologists, too, have recognized the necessity of specific timing signals in the control of nuclear and cell division in the integrated sequence of events of the cell cycle.

Less well studied and much less well understood are another set of phenomena that represent examples of temporal organization in biological systems - phenomena which couple the biology of the organism to predictable oscillations in the earth's environment. Such oscillations result from specific planetary movements,such as the daily rhythmicity resulting from the earth's rotation upon its axis,or the annual rhythmicity caused by the earth's revolution around the sun. These environmental periodicities produce striking changes in the environment which are so predictable that organisms have been able to evolve sophisticated mechanisms for varying their own physiology and behavior to utilize such periodicities to their own advantage. As pointed out by Pittendrigh [1] a number of years ago, natural selection has effected a temporal organization in living systems particularly suited to their presence on earth, in which the inherited endogenous periodicity of the organism matches some environmental periodicity of the earth.

Among the most thoroughly studied of such endogenous rhythmicities are a particular class of daily rhythms under the control of an endogenous timing device known as a circadian clock. These circadian rhythms exhibit several characteristics which distinguish them from other daily rhythms. They persist in the laboratory under constant environmental conditions with a period length of approximately, but not exactly, 24 hours, and this period length does not vary over a wide range of physiological temperatures. These two characteristics - persistence in constant conditions and a temperature-compensated period length - are features of a self-sustaining, endogenous, time-keeping mechanism referred to as a circadian clock, and are found in organisms representing all taxonomic categories except procaryotes.

2. The Neurospora System

Interest in the cellular or biochemical basis of circadian rhythmicity was stimulated by the discoveries in the 1950's of circadian rhythms in eucaryotic microorganisms such as Euglena [2], Gonyaulax [3], and Neurospora [4], and somewhat later in Acetabularia [5], and a great proportion of the efforts to understand the cellular basis of circadian rhythmicity has focused on these organisms. To a large extent,this is due to the apparent simplicity of such organisms, unencumbered by the complexities of multicellular organization and interactions.

More recently, particular attention has been focused on Neurospora crassa, a filamentous fungus with a long and illustrious history in the area of biochemical genetics. In addition to the obvious advantages of working with a microorganism, Neurospora offers several additional characteristics that make analysis of its circadian clock particularly appealing (for detailed review, see [6]). First, it has a highly developed sophisticated genetic system, with a wide range of mutants and chromosomal aberrations that offer unique opportunities to utilize genetic approaches and strategies. Second, the fundamental biochemistry of the organism has been well studied and analyzed and offers a broad resource of knowledge from which to draw in designing and interpreting biochemical experiments on the clock. Third, molecular studies, using modern techniques of recombinant DNA analysis, have developed to offer a useful and exciting direction of future research.

Fourth, Neurospora has a circadian rhythm of conidiation (asexual spore formation), which is one of the simplest rhythms known to assay [7]. The rhythm is expressed as a culture grows across an agar surface, periodically forming aerial hyphae and conidia at the "correct" time of day at a position on the agar just behind the growing front of the mycelium as it traverses the agar. The conidia remain at the location at which they were formed, and as a result, at the completion of an experiment, the petri dish or "race" tube exhibits bands or patches of conidia alternating with regions of surface mycelia which did not produce conidia at the particular time of day at which they were laid down. In other words, the culture leaves a permanent record of the position (and therefore the time of day) at which the conidia were made during the experiment, and as a result no automated equipment is needed to continuously monitor the culture in order to assay the rhythm. Finally, one can also assay the Neurospora clock in liquid culture by an indirect assay that involves transferring pieces of mycelium from the liquid culture to an agar culture,to determine the phase of the clock in the liquid culture [8]. Such a system has proved extremely valuable for biochemical experiments and manipulations.

3. Biochemical Studies

Efforts to study the biochemical mechanisms of circadian timekeeping have fallen into two general categories. In one approach,various biochemical parameters have been measured to determine which, if any, show circadian rhythmicity. In Neurospora, quite a number of parameters have been so identified, including evolution of CO_2 [9],levels of ATP and energy charge [10,11],synthesis of DNA and RNA [12], the levels of certain unsaturated fatty acids [13], and the activities of such enzymes as glyceraldehyde phosphate dehydrogenase, isocitrate lyase, citrate synthase, NADase, glucose-6-phosphate dehydrogenase and 6 phosphogluconate dehydrogenase [14]. In other organisms,a wide variety of enzymes and substrates have also been shown to oscillate with circadian periodicity. In none of these cases, however, has it been shown that the particular biochemical oscillation is directly or indirectly involved in, or responsible for, the oscillatory mechanism itself. Instead, these oscillations represent processes controlled by the clock - the so-called "hands of the clock" - rather than the clock mechanism itself.

Studies designed to identify biochemical oscillations, even those not directly involved in clock mechanisms, are potentially useful in several ways. In the first place,they define which components of the cell are oscillating,and provide a biochemical map of the circadian cycle. Second, it is hoped that by studying the mechanism by which a particular biochemical component is regulated, one can get clues to the actual timing mechanism,and ultimately identify a rhythmic component that is part of the underlying oscillator. Such studies have been most successful in the study of rhythms of bioluminescence [15,16] and photosynthesis [17,18] in the marine dinoflagellate Gonyaulax,but have not yet identified any of the oscillator components.

A second biochemical approach has been to administer drugs or inhibitors to organisms,and determine their effects on the circadian timekeeping mechanism. Mostly, such experiments involve pulses of the drug to determine if such pulses induce phase-shifts (resetting) of the rhythm. The rationale of such an approach is to identify specific biochemical processes which, when inhibited, alter the clock. Such processes might then be part of, or coupled to, cellular events which are part of the underlying oscillatory mechanism. For some time a great mystique developed in the field because of the apparent lack of effect of many drugs that were tested in different organisms. This led to the notion that the clock was some unique or discrete "block box", insulated from the major metabolic processes of the cell. In recent years, however, the situation has changed dramatically. In Neurospora pulses of drugs affecting mitochondrial respiration [11,19], cytosolic (80S) protein synthesis [20], calmodulin [21], Ca^{2+} transport [21], cyclic AMP metabolism [22], and H^+ transport [23] have all been shown to induce large phase-shifts of the clock.

Such widespread phase-shifting effects have now raised the opposite problem: Do the phase-shifts occur because the primary target of the drug is part of the clock mechanism, or do they occur only as a secondary or tertiary consequence of such action. In several cases, recent results demonstrate that it is clearly the latter interpretation that is correct. For example, in the case of phase-shifting by respiratory inhibitors, Nakashima [19] has shown that the dose of cyanide needed to phase-shift the Neurospora clock is at least an order of magnitude greater than needed to completely inhibit oxygen consumption or to reduce the levels of ATP by more than 10-fold. Similar results have been obtained by Shulz and Rensing [11], who found that cyanide-induced phase-shifting occurred in exactly the same manner, independent of the initial level of ATP in the cell or the extent of reduction of ATP levels by the drug. Taken together, these results clearly indicate that neither ATP itself nor the process of ATP synthesis and degradation are part of the clock mechanism. It seems more reasonable to conclude that respiratory inhibitors phase-shift the clock, because some other process(es) directly or indirectly dependent on the levels of ATP in the cell are part of the clock mechamos. Unfortunately, this includes nearly all of the metabolism of an obligate aerobe, and does not provide a significant clue about clock-related events.

The results with inhibitors of mitochondrial respiration are consistent with observations made in a number of photosynthetic organisms, showing that mitochondrial respiratory inhibitors such as cyanide do not induce phase-shifts in many of these organisms [24,25]. If the crucial process is simply to ensure that the cell has an ample supply of ATP, then inhibition of mitochondrial ATP synthesis in a photosynthetic organism might not likely provide a long-term decrease in ATP levels, since ATP can be synthesized by the alternative photosynthetic electron transport system in the chloroplast.

Drugs which inhibit cytosolic (80S) protein synthesis have also received considerable attention in recent years. In nearly every organism tested, including Neurospora, pulses of drugs such as cycloheximide or anisomycin induce large phase-shifts of the circadian clock. This has led to the suggestion that the synthesis of certain proteins may represent a required step in the circadian cycle. Such suggestions are at least consistent with the finding that there is a strong correlation between the dose response curves for phase-shifting and inhibition of protein synthesis in Neurospora [20]. (Contrast this result with the lack of correlation with cyanide described above.)

On the other hand, a recent result with a Neurospora clock mutant (see below) raises a serious question about this interpretation. This mutant, called frq-7 [26], displays a circadian period length that is considerably longer than that of the normal wild-type strain, but it nevertheless executes the entire circadian cycle in a nearly normal manner [27]. The clock of this mutant has recently been shown to be almost entirely unaffected by cycloheximide - that is, inhibition of protein synthesis in the mutant produces little or no phase-shifting [28]. If the synthesis of a specific protein were a required step in the circadian oscillatory cycle, it is difficult to imagine how such a mutation could arise. Although one could imagine that this step was entirely bypassed in the mutant, it is simpler to hypothesize that the synthesis of a specific protein is not normally a required step in the oscillation, but rather that some protein with a half-life significantly shorter than the circadian cycle must be present at a particular phase of the cycle. Normally, synthesis of that protein must occur at a particular phase of the cycle to restore its level above a critical threshold. However, if a mutation arose which raised the normal level of such a protein or increased its stability, it would result in a clock which was immune to the phase-shifting effects of protein synthesis inhibition, since a short-term inhibition would not lower the level of the protein below the critical level.

Phase-shifting experiments with chemical inhibitors, then, have serious limitations on their ability to identify specific clock components. However, they have recently been used for another purpose - namely, to provide a series of temporal markers throughout the circadian cycle [29]. This temporal map has already served to identify specific phases of the cycle which are altered in clock mutants (see section 5, below).

4. Genetic Approaches

Because of the difficulties of the biochemical approaches described above, a number of laboratories have turned to a genetic approach to study the clock. For many years, genetics has been used as a tool to dissect complex biological processes. The rationale of this approach is that, by inducing a mutation in a single gene, one can alter one component of the system, observe the effects of such alteration in vivo, and then ultimately determine the primary gene product encoded by the mutated gene. The value of this approach for circadian clocks is twofold. In the first place, in contrast to the well-defined chemical systems discussed elsewhere in this symposium in which all of the components of the system are known, biologists have not yet identified even a single component of the circadian pacemaker. Rather than coping with the intricacies of specific molecular models, we are faced with the problem of first identifying which of the many thousands of molecules in a cell are involved in the circadian oscillation. By first altering one component of the system at a time by mutation and thus identifying genes which code for essential clock functions, as evidenced by their effects on the intact system in vivo, one is led immediately to the molecular events required for normal clock function. From such results, one can assign a particular role in the clock to that component, and ultimately the molecular identity of the component can be determined.

To apply such an approach to the study of circadian rhythms, we have isolated approximately 15 mutants of Neurospora crassa in which the period length of the circadian rhythm of conidiation is either shorter or longer than the normal (wild-type) value (for review, see [30]). Genetic analysis has shown that each of these alterations results from a mutation in a single gene and has so far led to the identification of six genetic loci involved in clock function in this organism. Of particular interest is a locus called frq, since more than half of the mutants isolated map to this region of the genome. Frq mutants exhibit a number of genetic and physiological properties, which suggest that this gene or the product(s) encoded in this gene play a key role in the organization and functioning of the Neurospora clock [28]: (1) The mutations are highly specific - only the period length of the circadian oscillation is altered in these mutants; growth rate, development, and morphology are all essentially normal. (2) In heterocaryons in which both mutant and wild-type nuclei are present in the same cytoplasm, the period length of such strains is intermediate between the two parental strains, and the change in period length is proportional to the percent of mutant nuclei present. This "gene dosage" effect means that there is a direct relationship between the number of copies of the frq gene or its product and the quantitative expression of the clock phenotype, as measured by its circadian period length. (3) The period lengths found in frq mutants are highly specific and limited to a small number of values, all of which differ from the wild-type value by 2.5 hours multiplied by some integer (i.e., 2.5, 5.0, or 7.5 hours). This result suggests the presence of some quantum element in clock organization that is controlled by this locus.

Previously, we had found an important difference between frq mutants whose period lengths were shorter than normal and those that were longer than normal. An important and distinguishing property of circadian rhythms in all organisms is that the period length is temperature-compensated - i.e., the clock runs at approximately the same rate over a wide range of physiological temperatures. This temperature compensation feature has been maintained in the short period frq mutants but is altered in the long period mutants [31]. For example, in frq-3, whose period length is about 2.5 hours longer than normal, there is a small but significant increase in the temperature coefficient measuring period length - the clock of frq-3 is slightly less well compensated than the wild type strain. An even greater loss of temperature-compensation was found in frq-7, whose period length is 7.5 hours longer than wild type. Thus, for at least these two mutants, the greater the lengthening of the period, the greater the loss in temperature-compensation. A new mutant, called frq-9, which has entirely lost its temperature compensation feature, is described separately in section 6, below.

5. A Temporal Map of the Cycle

When using the rhythm of conidiation to assay the circadian clock of Neurospora, one can only measure one event per cycle, namely the center of the conidial band. This type of analysis does not give any information about the timing of different parts, or phases, within a single cycle. The existence of mutations which alter the period length of the rhythm raises the question of whether the cycle in such mutants is uniformly altered,or whether the mutation affects only one part of the cycle. It could be, for example, that the cycle consists of a sequence of events, and that the particular mutated gene codes for a product are important for only one of those events.

In order to determine whether the frq mutants are altered throughout the cycle or in only one phase, it was necessary to assay the clock in a manner other than the measurement of the conidiation rhythm. For this analysis,we used the universal property of circadian rhythms that a short pulse of visible light can reset (phase-shift) the clock to a new phase [32]. The amount of the phase-shift and its direction (forward = advance, backward = delay) depends on the phase of the cycle at which the light pulse is administered. A plot of phase-shift vs. phase of the cycle when the light pulse was administered has been called a light phase response curve (PRC), and this PRC gives a continuous assay of the clock throughout its cycle [32]. By comparing the light PRC for the wild type strain to that of a clock mutant, one can compare the progress of the clock throughout its cycle and directly compare the behavior of the mutant clock with that of the wild type strain.

Such analysis has been carried out on all of the frq mutants and has led to the same conclusion with all of them [27]. In all cases,the clock was clearly not affected uniformly through the entire cycle,but rather only for a portion of about one-third of the normal cycle occurring during the middle to late subjective day (the subjective day is defined as that part of the cycle which occurs when, in a 12:12 light-dark cycle, the lights would be on). This portion of the cycle, which normally occupies approximately 7 hours in the wild type strain (period length = 21.5 hrs), has thus been shortened to about 2 hours in frq-1 (period length = 16.5 hrs) and to 4.5 hours in frq-2, frq-4, and frq-6 (period length = 19.0 hrs). In the long period mutants, this 7 hour portion of the cycle has been lengthened to 10.5 hours in frq-3 (period length = 24.0 hrs) and to 14.5 hours in frq-7 and frq-8 (period length = 29 hrs).

This result points up two important issues. First, it does appear as if there are at least two distinct phases of the cycle, one which is affected by the frq mutations and one which is not. It may be that other clock genes will be found which affect different portions of the cycle, and eventually it may be possible to constructs a temporal map to indicate the sequence in which different genes or their gene products function during the circadian cycle. Second, it indicates that the effect of the mutation on clock function may be significantly greater than what one might conclude just by observing differences in overall period length. For example, there is a 7-fold difference in the length of the affected portion of the cycle between frq-1 (2 hours) and frq-7 (14.5 hours). Such differences are much more amenable to any future biochemical analysis than the apparent 1.8-fold difference in overall period length might initially suggest.

The accuracy with which the above results were obtained was somewhat limited by the accuracy with which phase-shifting can be measured in Neurospora. Thus, the estimate of the duration of the altered portion of the cycle could easily have been in error by at least 2 or 3 hours at either end. More recently, however, Nakashima [29] has developed a new, and more accurate, method for carrying out this type of analysis. He has obtained phase response curves for a wide variety of chemical agents, as well as for high-temperature and low-temperature pulses. For each of these agents one can very accurately measure the phase of the cycle at which the agent has a maximum effect on the clock. This results from the fact that just prior to the maximal phase, all of the phase-shifts are phase delays while just after the maximum the phase-shifts are all advances. At the maximum there is a sharp transition from delays to advances, and this "breakpoint" or "crossover" point of the phase response curve can be measured to an accuracy of about + 1 hour in Neurospora. Nakashima [29] has determined the crossover point of the PRC of about 10 different agents,and placed these on a temporal map of the circadian cycle. For example, the crossover point for inhibitors of protein synthesis occurs at CT 5, for antagonists of calmodulin at CT 7, for the calcium ionophore A23187 at CT 12, for low-temperature pulses at CT 14, and for light and high-temperature pulses at CT 20 (CT stands for "circadian time". The cycle of each strain is divided into 24 "circadian hours" with CT 0 representing the beginning of the subjective day.)

Using this information Nakashima [29] then measured the duration of time between each of the various crossover points in the wild type strain and in frq-7 and, in agreement with the previous results, concluded that frq-7 was altered only during about a third of the cycle from about CT 5 to CT 12. This result, determined by a completely different and more precise method than simply using the light PRC, confirms the fact that the frq mutants do not affect the entire circadian cycle but only a portion of it,during the middle to late subjective day.

This conclusion has additional interest because a similar result has been obtained with circadian clock mutants of the fruit by Drosophila melanogaster. Mutations at a locus called per can result in a short period, a long period, or arrhythmia [33]. The short period mutant shows incomplete dominance similar to that of the frq mutants of Neurospora,while the long period and arrhythmic mutants are recessive to wild type (see discussion of frq-9 below). When the short period mutant was analyzed to determine what part of the cycle was altered, using the light PRC method, it was found to affect only the middle to late subjective day phase, just as with frq [34]. Similarities between clock mutants at the frq locus in Neurospora and the per locus in Drosophila thus extend not only to their genetic behavior,but to at least some of their physiological characteristics as well. It may be that the role of these genes in clock function is similar in the two organisms, and molecular analysis of the genes offers an opportunity for direct comparison between them.

6. A Temperature-Dependent Clock Mutant

In section 4 above, the alteration in temperature compensation of the long period frq mutants was described. However, even in the case of frq-7, the clock still shows some degree of temperature compensation. In comparison with the temperature-dependent behavior of growth rate and of most metabolic reactions, whose rate approximately doubles with each 10° C increase in temperature, the clock of frq-7 increases its rate by only about 40% under similar cases. Recently, however, we have isolated a new clock mutant which also maps to the frq locus but which has apparently lost all of its ability to compensate for changes in temperature [35]. Its period length ranges from a very long 32 hours at 18°C to a very short period length of 14 hours at 30°C. Thus, the rate at which the clock runs in frq-9 shows the same temperature-dependence as growth rate.

However, we have demonstrated the clock rate is not simply a function of growth rate in another series of experiments. In the wild type strain and in all of the other clock mutants, the period length of the clock is unaffected by the composition of the medium on which the organism is growing. Thus, the clock is nutritionally compensated as well as temperature compensated. Frq-9, however, has also lost at least some of this nutritional compensation as well. In this strain the period length is dependent on both the nature and concentration of the carbon source in the medium. Strikingly, however, this dependence is in the opposite direction to the dependence of growth-rate on the medium. For example, when growth-rate is increased by increasing the concentration of glucose in the medium, the rate of the clock is decreased (i.e., the period length is longer).

The simultaneous loss of temperature compensation and nutritional compensation in this mutant resembles a phenomenon which we have observed under certain conditions even in the wild type strain. At temperatures above 30°C, even the wild type strain loses some of its temperature compensation and exhibits a Q_{10} of about 1.3. Also in this higher temperature range it loses its ability to compensate for nutritional changes [36]. These results support the suggestion of Pittendrigh and Caldarola [37] that temperature compensation of circadian clocks is only one specific example of a general homeostatic mechanism controlling the period length of the clock. Here in Neurospora we see two situations - one genetic and the other environmental - in which the circadian clock has simultaneously lost its ability to compensate for changes in temperature and nutrients in the environment.

Genetically, frq-9 differs from the other frq mutants as well. In heterocaryons frq-9 is recessive to the wild type frq+ allele [40]. In cases where more than 80% of the nuclei are mutant, the clock of the heterocaryon still behaves exactly like the wild type in terms of period length and temperature compensation. In addition frq-9 is recessive to each of the other frq mutants, in the same manner in which it is recessive to frq+. This result suggests that all of the frq mutants fall in a single functional unit (gene), although we cannot rule out the possibility that frq-9 is a large deletion which overlaps a series of adjacent frq genes or codes for a regulatory element which inactivates several adjacent loci.
Furthermore, the recessive nature of the frq-9 mutation suggests that it is a null mutant - i.e., one in which all expression of the affected locus is eliminated. Since the most striking characteristic of frq-9 is the complete loss of its temperature compensation property, this result suggests that frq or its encoded product plays a key role in the mechanism of temperature compensation in the Neurospora clock.

7. Summary

The analysis of the molecular mechanisms underlying circadian rhythmicity has proceeded slowly during the past decade, and no single component of the oscillator mechanism has yet been identified in any organism. However, recent applications of genetic analysis to the clock problem has led to the identification of specific genes, whose products play an important role in clock function. The availability of recombinant DNA techniques offers the exciting possibility for a molecular analysis of such genes, and the recent cloning of a clock gene in Drosophila [38] and an analysis of the RNA transcripts encoded in this genetic region [39,40] demonstrate the feasibility and power of such studies. It seems likely that the biochemical function of such genes can now be identified and the specific role of such molecules in the organization and functioning of the circadian clock can be identified.

References

1. C.S. Pittendrigh: Cold spr.Harb.Symp.Quant.Biol. 25, 159-184 (1960)
2. V.G. Bruce, C.S. Pittendrigh: Proc. Nat'l Acad. Sci. US 42, 676-682 (1956)
3. J.W. Hastings, B.M. Sweeney: Proc. Nat'l Acad.Sci. US 43, 804-811 (1957)
4. C.S. Pittendrigh, V.G. Bruce, N.S. Rosenzweig, M.L. Rubin: Nature 184, 169-170 (1959)

5. B.M. Sweeney, F.T. Haxo: Science 134, 1361-1363 (1962)
6. J.F. Feldman, J.C. Dunlap: Photochem. Photobiol. Revs. 7, 319-368 (1983)
7. M.L. Sargent, W.R. Briggs, D.O. Woodward: Plant Physiol. 41, 1343-1349 (1966)
8. J. Perlman, H. Nakashima, J.F. Feldman: Plant Physiol. 67, 404-407 (1981)
9. D.O. Woodward, M.L. Sargent: Behaviour of Microorganisms (Plenum, New York 1973) pp. 282-296
10. D.P. Delmer, S. Brody: J. Bacteriol. 121, 548-553 (1975)
11. R. Shulz, L. Rensing: Manuscript in preparation
12. C.L. Martens, M.L. Sargent: J. Bacteriol. 177, 1210-1215 (1973)
13. P.E. Roeder, M.L. Sargent, S. Brody: Biochemistry 21, 4909-4916 (1982)
14. M.L. Hochberg, M.L. Sargent: J. Bacteriol. 120, 1164-1175 (1974)
15. J.C. Dunlap, J.W. Hastings: J. Biol. Chem. 256, 10509-10518 (1981)
16. C.H. Johnson, J.F. Roeber, J.W. Hastings: Science 223, 1428-1430 (1984)
17. B. Prezlin, B.M. Sweeney: Plant Physiol. 60, 388-392 (1977)
18. T. Lonergan, M.L. Sargent: Plant Physiol. 61, 150-153 (1979)
19. H. Nakashima: Plant Physiol., in press (1984)
20. H. Nakashima, J. Perlman, J.F. Feldman: Amer. J. Physiol. 241, R31-35 (1981)
21. H. Nakashima: Plant Physiol. 74, 268-271 (1984)
22. J. Feldman: Science 190, 789-790 (1975)
23. H. Nakashima: Plant Physiol. 70, 892-986 (1982)
24. F. Buhnemann: Biol. Zbl. 74, 691-705 (1955)
25. J.W. Hastings: Cold Spr.Harb.Symp.Quant.Biol.25, 131-140 (1960)
26. G.F. Gardner, J.F. Feldman: Genetics 96, 877-886 (1980)
27. S. Dharmananda: Ph.D. Thesis, Univ. of Calif., Santa Cruz (1980)
28. J.C. Dunlap, J.F. Feldmann: Neurospora Newsletter 29, 12 (1982)
29. H. Nakashima: Sapporo Symposium on Biological Rhythms, in press (1984)
30. J.F. Feldman: Ann.Rev.Plant Physiol. 33, 583-608 (1982)
31. G.F. Gardner, J.F. Feldman: Plant Physiol. 68, 1244-1248 (1981)
32. C.S. Pittendrigh: Proc. Nat'l Acad. Sci. US 58, 1762-1767 (1967)
33. Konopka, S. Benzer: Proc. Nat'l Acad. Sci. US 68, 2112-2116 (1971)
34. R. Konopka; Ph.D. Thesis, Calif.Inst. Tech. (1972)
35. J. Loros, J.F. Feldman: Manuscript in preparation
36. H. Nakashima, J.F. Feldman: Photochem. Photobiol. 32, 247-252 (1980)

37. C.S. Pittendrigh, P.C. Caldarola: Proc. Nat'l Acad. Sci. US 70, 2697-2701 (1973)
38. W.A. Zehring, D.A. Wheeler, P. Reddy, R.J. Konopka, C.P. Kyriacou, M. Rosbash, J.C. Hall: Cell, in press (1984)
39. P. Reddy, W.A. Zehring, D.A. Wheeler, V. Pirrotta, C. Hadfield, J.C. Hall, M. Rosbash: Cell, in press (1984)
40. T.A. Bargiello, M.W. Young: Proc. Nat'l Acad. Sci. US 81, 2142-2146 (1984)

Cyclic Activity of Enzymes of Asparagine-Pyruvate Pathway and Cellular Synchronization. Significance of L-Asparaginase Oscillatory States

Stéphan Jerebzoff

Biorhythm Laboratory, Plant Physiology Center, Toulouse University
118 Route de Narbonne, F-31062 Toulouse Cedex, France

The periodic function of the asparagine-pyruvate pathway and that of the sporulation rhythm are strongly correlated in Leptosphaeria michotii. Indeed,conidiation rhythms are subtended by rhythms of metabolic activities having the same periodicity, which can operate without cell differentiation ; asparagine-pyruvate pathway plays a main part in this periodic system : the activities of asparaginase, aspartate and alanine aminotransferases (AT) and the cellular aspartate pool express a 24 hour periodicity [1]. Sporulation rhythm is regulated by propargylglycine, a substrate analogue, specific for alanine AT in this fungus ; it is,on the other hand, destabilized by cycloheximide. Furthermore, when asparaginase is the only enzyme of the pathway whose activity remains rhythmic, the sporulation rhythm pursues its course, but it is no longer regulated. When both activities of asparaginase and aspartate AT are rhythmic, sporulation rhythm is regulated. Alanine AT activity has also an influence on the function of the sporulation rhythm, but its activity could be still non-rhythmic. So, the synchronized working of the above three components of the asparagine-pyruvate pathway studied is necessary for the normal operation of the sporulation rhythm in L. michotii [2]. In this case,the control of the sporulation rhythm might be carried out through changes in the main metabolic pools driven by asparagine-pyruvate pathway.

However, the mechanisms of the rhythms of enzyme activities have been difficult to understand until now. In L. michotii the asparaginase activity rhythm, which plays a very important part in the periodic system, continues in spite of a dramatic inhibition of protein synthesis ; the mechanisms of this rhythm are studied thoroughly.

Two 150-200 fold purified asparaginases were isolated. Their activities per g dry weight and also their specific activities are cyclic with a period equal to that of crude extracts, but the rhythms of the two enzyme forms are out of phase. The two asparaginase forms differ in their pH_i and K_m for asparagine ; nonMichaelis-Menten kinetics for L-asparagine are observed. As shown by PAGE analyses the form I and form II asparaginases are constituted by a 130-140 kdalton oligomer and its dimer, a high-M_r protein aggregate containing mainly an enzyme polymer, plus a 60 kdalton protein without asparaginase activity [3]. Activation of the form I and form II asparaginases is obtained in vitro by ATP-Mg^{2+} or PO_4^{3-} supply, these effects being modulated according to the phase of the activity rhythm at which enzyme was extracted ; the two enzyme forms are deactivated by dephosphorylation. These facts indicate the presence of a protein kinase and a protein phosphatase in the asparaginase complex. Phosphorylation of the high-M_r protein aggregate and of asparaginase is demonstrated using γ^{32}P-ATP or $^{32}PO_4^{3-}$. By PAGE in SDS of the dephosphorylated aggregate,a radiolabelled protein of 60 kdalton becomes detectable; the 60 kdalton protein is the phosphatase of the complex [4].

The rhythm of asparaginase activity in L. michotii is at least regulated by :
- the balance between the two enzyme forms ; - a reversible phosphorylation process which can be regulated since phosphatase can be phosphorylated.

1. S. Jerebzoff-Quintin and S. Jerebzoff : Physiol. Vég. 18, 147 (1980)

2. S. Jerebzoff and S. Jerebzoff-Quintin : Physiol. Vég. 21, I, 233 and II, 247 (1983)

3. S. Jerebzoff-Quintin and S. Jerebzoff : Physiol. Plant. (in press)

4. S. Jerebzoff and S. Jerebzoff-Quintin : FEBS Letters 171, 67 (1984)

Control and Significance of the Circadian Growth of the Endocuticole in Cockroaches (*Blaberus craniifer*, Blattodea)

Friedrich Weber

Zoological Institute, University of Münster, D-4400 Münster, Fed. Rep. of Germany

In several pterygote insects, the endocuticle which is secreted by the epidermis after molt grows in a daily rhythm[1,2]. Each day, two layers are formed which differ in the orientation of chitin microfibrils. In the one layer they are helicoidally, in the other one unidirectionally arranged. In some cases it has been proved that this growth rhythm is controlled by an endogenous circadian clock[2]. In cockroaches this clock exhibits two characteristic features of circadian oscillators - frequency (~ 1 d^{-1}) and temperature compensation -, but it lacks the third main feature: phase-dependent light sensitivity[3,4]. - The circadian growth of the endocuticle raises two problems: (1) What is its biological significance? (2) Where is the clock localized which controls this developmental rhythm?

Ad(1). Two hypothesis can be propounded: - only the temperature compensation of the growth rhythm (that means, the frequency control by the clock) is significant; - the phase control by the clock is significant because it enables an internal adaptation of the growth rhythm to the circadian multioscillator system which rules behaviour and metabolism. - The 2nd hypothesis requires a characteristic internal phase relationship between the growth rhythm and other circadian endogenous rhythms. Such a relationship is established if the molt of the animals is gated by combined light-dark/cold-warm cycles[5]. Then, the helicoidally structured layers are deposited during the light/warm phases in which cockroaches rest. The same relationship between endocuticle growth and activity distribution has been found in light-active locusts[6]and bugs[7]. According to Neville, the deposition of helicoidally structured chitin-layers consumes much less energy than the deposition of unidirectionally structured material[1]. Thus, the hypothesis is confirmed that the control of the endocuticle growth by a circadian clock is advantageous,because in this way an internal temporal adaptation becomes possible.

Ad(2). The circadian growth rhythm continues some days after the newly molted cockroaches have been decapitated or parts of their ventral nervous cord have been removed[3,8]. Therefore, it has been concluded that the clock is not localized centrally in the nervous system or in a hormone gland of the head, but peripherically, perhaps in the epidermis cells themselves. This hypothesis includes the assumption that the clock inside each cell is started by one of the numerous events which occur around the molting process. - In decapitated animals the rhythmic deposition does not exhibit conspicuous desynchronisations between different legs. This could mean that the hypothetical clock, once started, is running very precisely so that no desynchronisation becomes visible. On the other hand, such stable synchronisation could be due to an internal (peripherical) entrainment mechanism,which synchronizes the clocks inside the epidermis cells throughout the days in which the endocuticle grows. - Recently, the hypothesis of the peripheral localisation of the circadian clock which controls endocuticle growth has been confirmed by in vitro experiments[9]. In cultivated leg pieces taken from newly molted imaginal cockroaches,the deposition of circadian-like endocuticle has been observed. However, such cuticle was formed only in small areas,which are mostly localized at the basis of the spines and which cover some dozen cells at most. Mainly, multilamellate (exocuticle-like) material was deposited in vitro.

Unresolved problems. (1) It is unknown by what mechanism the change is triggered between the deposition of exocuticle and endocuticle. During this change the clock is started which controls endocuticle deposition. - (2) It is probable, but not yet proved,that the deposition of circadian-like endocuticle in vitro is really controlled by a circadian clock. The temperature-dependence of the deposition of circadian-like endocuticle is not yet precisely known. - (3) It is unknown by what mechanism the switch is triggered between the secretion of exocuticle-like and circadian-like material. To solve this problem, the conditions have to be defined under which the deposition of circadian-like structured material is promoted. It can be hypothesized that a circadian rhythm of deposition depends on some interaction between neighbouring epidermis cells. Epidermis cells are electrically coupled by gap junctions. Recently, it has been reported that the degree of coupling is influenced by ecdysone [10]. The stable synchronisation of endocuticle formation in vivo may also be caused by electrical coupling of the epidermis cells. - (4) As well as other multicellular organisms,cockroaches seem to be hierarchically-structured multioscillator systems. Until now the advantages of multioscillator systems in contrast to imaginable unioscillator systems are insufficiently known.

Conclusion. Endocuticle deposition in cockroaches is controlled by an endogenous circadian clock. This clock seems to reside in the periphery, perhaps in the epidermis itself. The existence of central and peripheral circadian clocks demonstrated in other vertebrates as well as in insects provokes questions as to the significance of multioscillator systems and   the mechanisms for   internal synchronization. The methodical advantage of the epidermis system is the fact that the circadian rhythm is recorded by the system itself.

References

1. A.C.Neville: Biology of the arthropod cuticle. Springer Berlin 1975.
2. A.C.Neville: J.Insect Physiol. 29, 211-219 (1983).
3. R.Lukat: Experientia 34, 477 (1978).
4. G.Wiedenmann: Dissertation. Fachbereich Biologie, Tübingen 1978.
5. T. Runte a. F.Weber: Experientia 38, 1049-1051 (1982).
6. A.C.Neville: Q.J. microsc. Sci. 106, 315-325 (1965).
7. H.Dingle, R.L. Caldwell, I.B., Halskell,J.: Insect Physiol. 15, 373-378 (1969).
8. A.C.Neville: J. Insect Physiol. 13, 933-939 (1967)
9. F.Weber: Postmolt cuticle growth in a cockroach: in vitro deposition of multilamellate and circadian-like layered endocuticle: Experientia (in press).
10. S.Caveney, M.G. Blennerhasset: J. Insect Physiol. 26, 13-25 (1980).

The Hypothalamic Control of Eating and Circadian Rhythms: Opponent Processes and Their Chemical Modulators

Stephen Grossberg*

Center for Adaptive Systems, Boston University, Department of Mathematics
111 Cummington Street, Boston, MA 02215, USA

1. Homeostatic and Nonhomeostatic Factors in the Control of Eating and Circadian Rhythms

Many types of behavior are difficult to understand due to the complex interplay of external and internal factors in their regulation. For example, eating behavior is modulated by internal homeostatic influences, such as satiety signals, in order to prevent undue gastric distention and metabolic overload from occurring due to unrestricted eating. Eating behavior is also strongly influenced by non-homeostatic influences, such as oropharyngeal factors and reinforcing events, that endow food-predictive cues with their gustatory appeal. A perfect internal homeostatic control would badly serve an animal's survival if it could not be modulated by external signals of food availability and quality. In a similar fashion, a perfectly clock-like circadian oscillator would be of limited value if it could not be modulated by seasonal changes in the duration of day and night. A nocturnal mammal on a summer night has less time available to carry out vital consummatory activities than it does on a winter night.

This article outlines a neural network design that has been used to explain and predict the interplay of external and internal signals on operant eating behavior [1, 2, 3] and circadian activity-rest cycles [4, 5, 6]. I speak of a "neural network design" because every process in the eating model is homologous to a process in the circadian model. The models differ in terms of their specialized internal circuit diagrams,and in terms of the input and output factors with which they interact, but not in terms of their component processes. These model components are assumed to represent processes taking place in the mammalian hypothalamus,and in neural tissues that interact closely with the hypothalamus.

The present article outlines some of the main processes that are used in the eating model, describes the mechanistic homolog that exists between the eating model and the circadian model, and indicates how the circadian model influences the operation of the eating model. The article ends by noting how breakdowns in the interactions between the circadian model and other gated dipole circuits can lead to formal syndromes that resemble several mental disorders.

2. Antagonistic Rebounds within Opponent Processes

At the core of our theory is a neural network model of opponent processes. The type of opponent process that is needed can be understood

---
* Supported in part by the Air Force Office of Scientific Research (AFOSR 82-0148) and the Office of Naval Research (ONR-N00014-83-K0337).

by considering the property of *antagonistic rebound*. By antagonistic rebound, I mean the type of reaction whereby offset of a reinforcing cue of negative (positive) motivational sign can trigger a motivational reaction of positive (negative) sign [7, 8]. For example, while a sustained shock remains on, it can elicit fear, or a negative reaction [9]. Offset of the shock can elicit a wave of relief, or a positive reaction [10]. While a sustained food cue remains on, it can elicit autonomic and motor preparations for eating, or a positive reaction. Offset of the food cue can elicit a wave of frustration, or a negative reaction [11, 12, 13, 14].

To model such an opponent process, one needs the following ingredients: (1) <u>Opponent Channels</u>: An on-channel is turned on by the eliciting cue, just as shock elicits fear. An off-channel is turned on transiently by offset of the eliciting cue, just as offset of shock elicits relief. The on-channel and the off-channel compete so that only one channel can elicit an output signal at any time. (2) <u>Tonic Arousal</u>: In order to generate an antagonistic rebound after the eliciting cue turns off, there must exist another input source that is capable of energizing the rebound. The simplest hypothesis is that an internally generated, tonically active input equally energizes, or arouses, both channels, and that the phasic cue causes its on-reaction and off-rebound by altering the relative input sizes to the two channels. (3) <u>A Slowly Varying Process</u>: After the eliciting cue shuts off, how does the network calibrate an off-rebound? Some measure of the prior occurrence of the eliciting cue must persist after its offset. Somehow, this slowly varying process helps to cause an off-rebound before it can re-equilibrate to the absence of the eliciting cue.

What type of slowly varying process has these properties? I now explain the following remarkable fact: A chemical transmitter process that attempts to generate *unbiased* signals within the on-channel and the off-channel instantiates the desired properties if it reacts slowly to the eliciting cue and its offset.

3. Chemical Gates: Unbiased Transmitter-Modulated Signalling

The simplest rule whereby one nerve cell site can send unbiased signals to another nerve cell site is the following rule. If $S(t)$ is the input signal to one cell site and $T(t)$ is the output signal to the next nerve site, then the linear relationship

$$T = SB, \tag{1}$$

where B is a positive constant, is the simplest law of unbiased transmission. By (1), the outgoing signal is proportional to the incoming signal, and the signal is relayed perfectly.

When the output signal $T(t)$ is due to the release of a chemical transmitter $z(t)$ in response to the input signal $S(t)$, further consideration is necessary. How is a large and sustained input $S(t)$ prevented from depleting $z(t)$ and thereby causing a progressively smaller signal $T(t)$? In other words, when $T(t)$ is due to the release of a transmitter, the term B in (1) may not be constant. It may decrease through time as $z(t)$ is depleted, thereby reducing the sensitivity of $T(t)$ to $S(t)$. In this situation, (1) is replaced by the equation

$$T = Sz. \tag{2}$$

Our task is to analyse how $z(t)$ approximates a constant B,

$$z \cong B, \tag{3}$$

despite its depletion due to input S.

Equation (2) says that transmitter z is released at a rate (proportional to) T in response to input S. In other words, z *gates* S to generate T, or T is caused by a *mass action* interaction between S and z. By (2), either an increase in S or in z can increase T, and no output signal T can be released if either no input signal occurs (S = 0) or if no transmitter is available (z = 0).

Equation (3) requires that the *sensitivity* of T to S be maintained through time. If both (2) and (3) are simultaneously implemented, as in (1), then unbiased transmission by a depletable chemical is achieved. Equation (1) means that z(t) is replenished instantaneously, or at least at a rate that is rapid relative to the rate of gated release. The property of antagonistic rebound occurs when the rate of accumulation is slow relative to the rate of gated release. In order to represent this type of process, an algebraic equation is insufficient. A differential equation is needed. I will use the simplest differential equation that is capable of reconciling (2) and (3) when both the accumulation and gated release processes take place at a finite rate relative to the rate with which the signal S can fluctuate. In this situation, (2) and (3) are not both exactly satisfied at any one time. The process attempts to achieve unbiased transmission, but can do so only approximately due to its finite reaction rates.

The simplest differential equation capable of simultaneously implementing (2) and (3) is the following one [8, 15, 16]:

$$\frac{d}{dt}z = A(B - z) - CSz, \tag{4}$$

where A, B, and C are positive. In (4), the notation $(d/dt)z$ denotes the net production rate of z. Term $A(B - z)$ says that z accumulates at a rate A until it reaches the target level B, as required by (3). Term $-CSz$ says that the loss of transmitter per unit time due to gated release is proportional to Sz, as required by (2). Henceforth I choose C = 1 for notational simplicity. This amounts to rescaling the size of S.

Term $A(B - z)$ may be physically instantiated in more than one way. For example, a passive accumulation of z may occur onto unoccupied sites whose total number is B. Alternatively, transmitter precursors may actively be produced at a rate AB, but feedback inhibition via term $-Az$ of transmitter z onto an intermediate stage of production may reduce the net production level to $A(B - z)$. Without such feedback inhibition, transmitter production would continue unabated until the cell ruptured.

In response to a constant signal of size S, (4) implies that the transmitter z approaches the equilibrium value

$$z = \frac{AB}{A + S}. \tag{5}$$

In other words, larger signals S deplete more transmitter. On the other hand, the output signal that is generated by an input S does not equal z. The output signal is equal to T = Sz, due to (2).

Figure 1 describes how the output T reacts to changes in the size of the input S. A rapid increase in S from S_0 to S_1 elicits a slow decrease in z. Multiplication of the graphs of S(t) and z(t) shows that a rapid increase in S generates a rapid increase in T followed by a slow decrease, or habituation, of T to an intermediate level. In a similar way, a rapid decrease in S from S_1 to S_0 generates a rapid de-

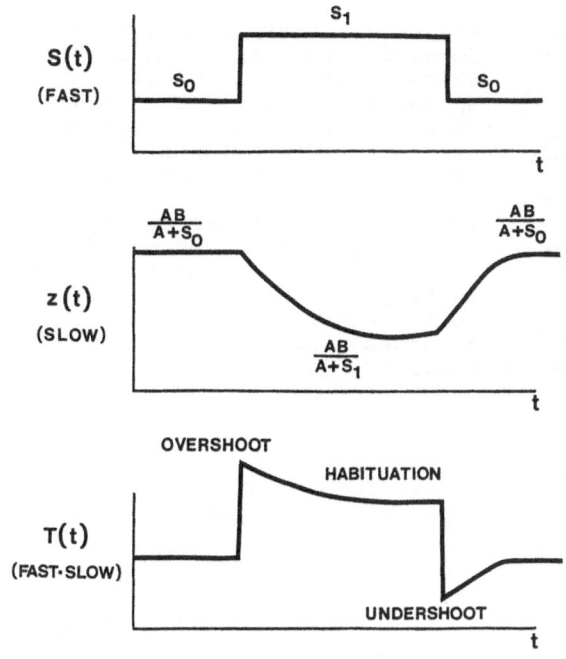

$S(t)$
(FAST)

S_0 S_1 S_0

t

$z(t)$
(SLOW)

$\frac{AB}{A+S_0}$ $\frac{AB}{A+S_0}$

$\frac{AB}{A+S_1}$

t

$T(t)$
(FAST·SLOW)

OVERSHOOT

HABITUATION

UNDERSHOOT

t

Figure 1: Reaction of output signal T and transmitter gate z to changes in input S. The output T is the product of a fast process S and a slow process z. Overshoots and undershoots in T are caused by z's slow habituation to fast changes in S.

crease in T followed by a slow increase, or habituation, to an intermediate level. In all, rapid increases and decreases in the input S generate overshoots and undershoots in the output T due to the slow rate of reaction, or habituation, of the transmitter as it seeks to generate unbiased signals.

4. Gated Dipoles: Tonically Aroused Chemical Modulators in Opponent Processes

Figure 2 describes how a phasic input (J) can elicit a sustained on-response and a transient off-rebound from opponent channels whose inputs are gated by slowly varying chemical transmitters. This type of opponent process model is called a *gated dipole* [8]. I wish to emphasize that the opponent channels of a gated dipole network can be instantiated just as well by an intracellular network of reactions [17] as they can by an intercellular network of whole cells. An intercellular interpretation will be given here for definiteness.

The left-hand series of stages in Fig. 2 represents the on-channel, and the right-hand series of stages represents the off-channel. Both channels receive an equal arousal input, denoted by I, that is constant through time. The arousal input energizes the antagonistic rebound after the on-input shuts off. The on-input, denoted by J, is delivered only to the on-channel. Input J is switched from zero to a positive level and held at that level long enough for gate equilibration to occur. Then J is shut off.

Inputs I and J are added by the activity (or potential) $x_1(t)$. Activity $x_1(t)$ responds quickly to input fluctuations, relative to the reaction rate of the network's slow gates. The graph of $x_1(t)$ has the

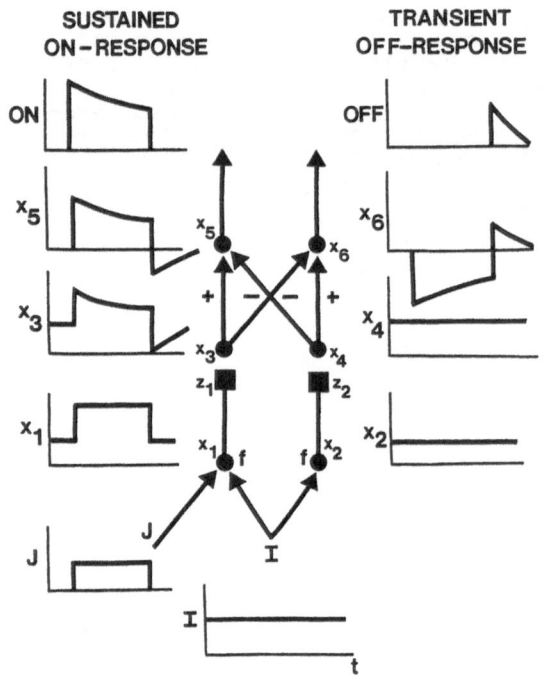

SUSTAINED ON-RESPONSE

TRANSIENT OFF-RESPONSE

Figure 2: Example of a feed-forward gated dipole. A sustained habituating on-response (top left) and a transient off-rebound (top right) are elicited in response to onset and offset, respectively, of a phasic input J (bottom left) when tonic arousal I (bottom center) and opponent processing (diagonal pathways) supplement the slow gating actions (square synapses). See text for details.

same form as the top graph in Fig. 1: a rapid switch from a lower positive activity to a higher positive activity, followed by a rapid return to the lower level. The activity $x_1(t)$ generates an output signal $f(x_1(t))$ in its pathway that again has the form of a double switch between two positive values. The output signal $f(x_1(t))$ is gated by a slow transmitter $z_1(t)$ that accumulates and is released from the square synapses in the on-channel. Figure 1 describes the effect of this slow gate on the input to the next stage. Consequently, activity $x_3(t)$ follows an overshoot-habituation-undershoot-habituation sequence through time. Then $x_3(t)$ relays an output signal of the same form to $x_5(t)$. Activity $x_5(t)$ also receives an inhibitory signal from $x_4(t)$. To determine what happens next, we consider the dynamics of the off-channel.

The off-channel receives only the constant tonic input I. Hence $x_2(t)$ and the slow gate $z_2(t)$ in the off-channel square synapses are constant through time. The activity $x_4(t)$ is therefore also constant through time. For definiteness, I make the simplest assumption that corresponding stages in the on-channel and the off-channel possess the same parameters. Since the arousal input I to both channels is also equal, the size of x_4 equals the baseline activity level of $x_3(t)$. This is not always true, but its violation is easy to analyse after the symmetric case is understood.

We can now determine the reactions of activity $x_5(t)$ through time. Since the signals from $x_3(t)$ and $x_4(t)$ subtract before perturbing $x_5(t)$, and their baseline activities are the same, the baseline activity of $x_5(t)$ equals zero. Activity $x_5(t)$ thus overshoots and undershoots a zero baseline when the input J is turned on and off. By contrast, activity $x_6(t)$ responds in an opposite way from $x_5(t)$ because x_3 excites x_5 and inhibits x_6, whereas x_4 inhibits x_5 and excites x_6.

The final assumption is that the output signals caused by activities $x_5(t)$ and $x_6(t)$ are rectified: Outputs are generated only if these activities exceed a nonnegative signal threshold. As a result, the on-channel generates a sustained output signal while the input J is on. This output signal habituates as the gate $z_1(t)$ slowly equilibrates to the input. By contrast, the off-channel generates a transient off-response, or antagonistic rebound, after the input J shuts off.

5. From Phasic Off-Rebounds to Circadian Oscillations

Why can a suitably designed gated dipole generate an endogenously active circadian rhythm? The answer is that the rhythm is due to an unending succession of self-generated antagonistic rebounds. Such a rhythm does not occur in the network of Fig. 2 because this network does not possess positive feedback loops within its on-channel and its off-channel. This type of anatomical change, from a feedforward anatomy to a feedback anatomy, can transform the phasically reactive network of Fig. 2 into an endogenous oscillator.

When a phasic input to the on-channel of any gated dipole turns off, an antagonistic rebound can occur in its off-channel. If, moreover, the on-channel and the off-channel excite themselves via positive feedback, then offset of the rebound in the off-channel may act like offset of a phasic cue to the off-channel. A rebound in the on-channel is hereby caused. Then offset of the rebound in the on-channel triggers another rebound in the off-channel, and the process continues indefinitely.

The period of this oscillator varies inversely with the transmitter accumulation rate over a wide numerical range [4]. Any genetic or prenatal hormonal factor that is capable of tuning the transmitter accumulation rate determines the free-running period of such a clock in the dark.

6. Inverted U in Sensitivity: Circadian Modulation of Appetitive

Circuitry

The simple circuit in Fig. 2 possesses a number of mathematical properties, as it stands, that made our present theory possible. We assume, for example, that the output of the hypothalamic circadian pacemaker modulates the arousal level (parameter I in Fig. 2) that energizes appetitive hypothalamic circuits, such as the eating circuit [6]. This hypothesis was suggested by the following basic property of gated dipole circuits [3, 8]. When a gated dipole's arousal level is parametrically increased, the outputs of the dipole are altered according to an *inverted U* law: Dipole outputs are depressed in response to input signals if the arousal level is chosen either too low or too high. The dipole optimally responds to phasic inputs at intermediate levels.

We hypothesize that the pacemaker causes a circadian oscillation to occur in the arousal level of appetitive circuits. The inverted U property of the gated dipoles in appetitive circuits prevents these dipoles from reacting when their arousal level is too high or too low.

Behaviors that require appetitively activated motivational signals for
their execution will thus not be emitted if the circadian output is
too high or too low. We hypothesize that, under normal conditions, the
arousal level during the wakeful state brings appetitive dipoles into
the middle range of their inverted U.

It remains to discuss whether the arousal level becomes smaller
("underarousal") or larger ("overarousal") during sleep than it is
during the wakeful state. We hypothesize that underarousal occurs during sleep. This hypothesis is based upon the following property of
gated dipoles in the underaroused state. During underarousal, the
threshold intensity that an input must exceed to generate an output
signal is significantly raised. The network will not react to many inputs that would cause a reaction during the wakeful state. If, however,
an input is sufficiently intense to exceed this elevated threshold,
then the dipole output is hypersensitive to the input; that is, supra-
threshold increments in the input cause larger than normal increments
in dipole output. Thus a dipole that is underaroused during sleep can
react to intense inputs. An overaroused dipole, by contrast, does not
react to inputs of any intensity.

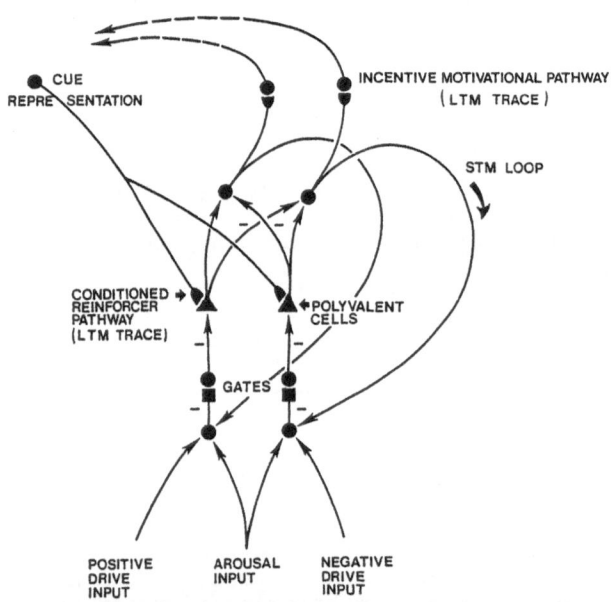

Figure 3: A motivational network: A positive drive input and a negative drive input homeostatically modulate the sensitivity of the gated
dipole to cue inputs on an ultradian time scale. The arousal input modulates dipole sensitivity on a circadian time scale. Each cue representation sends a pair of conditionable pathways to the dipole's on-
channel and off-channel. A cue (CS) becomes a conditioned reinforcer
when one of its long-term memory (LTM) traces grows larger than the
other due to pairing with an unconditioned stimulus (US). The output
signals from the gated dipole are called incentive motivation signals.
These signals modulate attention to the cue and the release of overt
behavior by the cue. The short-term memory (STM) loop enables the dipole to maintain a steady motivational baseline and to provide enough
hysteresis to prevent reset of its motivational decision in response
to small fluctuations in cue inputs.

7. Homeostatic and Nonhomeostatic Influences on Eating: Satiety and Reinforcement

The manner in which homeostatic and nonhomeostatic inputs influence appetitive gated dipole circuits is illustrated in Fig. 3. In Fig. 3, two additional sources of inputs reach the gated dipole: homeostatic internal drive inputs and nonhomeostatic external cue inputs. In the case of eating, the positive drive input increases with hunger, and the negative drive input increases with satiety, owing either to gastric distention or to slower metabolic factors [18, 19, 20, 21]. Thus the total internally activated input to the on-channel (off-channel) is a sum of a positive (negative) drive input and an arousal input. The total internal inputs to the on-channel and the off-channel are no longer equal.

These inputs are gated by slowly varying transmitters in the square synapses, just as in Fig. 2. Figure 3 depicts a variation on the gated dipole design, in which the transmitters are inhibitory and are part of a 2-stage disinhibitory pathway, instead of being excitatory transmitters in a 1-stage excitatory pathway.

Each external cue is assumed to activate a pair of conditionable pathways that lead to the on-channel and the off-channel of the dipole. Each pathway is conditionable because it contains a long-term memory (LTM) trace within its synaptic knob (hemispherical end of pathway). Each LTM trace $z(t)$ obeys a classical conditioning law of the following form [15, 16]:

$$\frac{dz}{dt} = -Uz + Vf(x).\tag{6}$$

In (6), the term V is a nonnegative learning signal that is positive only when the synaptic knob receives an input from its cue representation. Function $x(t)$ is the activity, or potential, of the postsynaptic gated dipole cell that abuts the synaptic knob. Function $f(x)$ is a nonnegative and increasing function of x that is positive only when x is positive. Thus term $Vf(x)$ in (6) is positive only when both the cue-activated input to the synaptic knob and the postsynaptic potential are positive. Term $-Uz$ in (6) says that z time-averages the product $Vf(x)$ at the rate U. I have written U and V in a general form because they can be realized in different ways in specialized learning networks [2, 6].

The cells at which external cue, internal drive, and arousal inputs converge are assumed to be *polyvalent*: These cells vigorously fire when both their external cue and their internal drive inputs are sufficiently large. Then the outputs from the polyvalent cells compete before generating a net dipole output in either the on-cell or the off-cell channel, but not both. These outputs carry *incentive motivation* signals that modulate the sensory-motor circuits that control observable behavior. Due to the polyvalent cells, incentive motivation is not released even if internal drive is high in the absence of compatible cues, and is not released in response to compatible cues if internal drive is too low.

Cues that are not innate reinforcers become "compatible cues," or conditioned reinforcers, in the following way. Suppose that an indifferent cue, or conditioned stimulus (CS), occurs just before an unconditioned stimulus (US), such as the taste of food, turns on the on-channel. A US can elicit positive incentive output from the on-channel even before learning occurs. By associating the CS with the on-channel activation that is caused by the US, the CS's LTM trace to the on-channel can grow larger than its LTM trace to the off-channel. When the CS occurs later on, it generates equal signals in both of its pathways. The LTM traces in these pathways multiplicatively gate the signals before they can reach their respective dipole channels. Since the LTM trace abutting the on-channel is larger than the LTM trace abutting the off-channel, the on-channel receives a larger LTM-gated input than does the off-channel. After the polyvalent cells compete, a positive incentive signal is emitted from the on-channel. The indifferent CS has become a positive conditioned reinforcer.

Any mechanism that causes the LTM trace in the off-channel to grow as large or larger than the LTM trace in the on-channel can extinguish the CS's conditioned reinforcer potency. Antagonistic rebounds play an important role in this type of extinction-through-counterconditioning.

8. Secondary Conditioning

Secondary conditioning can occur in the circuit of Figure 3 due to the action of its feedback pathways (STM loops). Suppose, for example, that the LTM trace within a pathway from a prescribed cue representation to a positive drive representation has grown due to prior conditioning. The corresponding cue hereby becomes a positive conditioned reinforcer. Signals from this cue representation can thus activate the positive drive representation, and thereby habituate the transmitter gate in its STM loop. Suppose that, after a sufficiently long time interval of cue representational signalling, the cue representation is suddenly shut off due to the occurrence of an unexpected event. An antagonistic rebound can hereby be caused that activates the negative drive representation. Any cue representation that is activated while the negative rebound is active can become a negative conditioned reinforcer due to growth of the LTM trace in the pathway between this cue representation and the negative drive representation. The cue hereby acquires negative conditioned reinforcer properties despite the fact that no negative reinforcer has been externally presented. In this way, offset of a positive conditioned reinforcer can support secondary emotional conditioning of a negative conditioned reinforcer. See GROSSBERG [1, 3] for details.

9. Normal and Hyperphagic Feeding Cycles

This section indicates how the inverted U property of gated dipole motivational circuits enables internal homeostatic signals to modulate the responsiveness of these circuits to external nonhomeostatic signals. An analysis of normal and hyperphagic feeding cycles can be given in terms of these properties [3].

Figure 4 summarizes the idealized changes through time in homeostatic hunger input (K_1) and satiety input (K_2) to a motivational dipole which receives a normal level of arousal (L). In the full theory, a hunger dipole competes with other motivational dipoles (thirst, sex) to decide which dipole will win the motivational competition. These additional complexities are discussed in GROSSBERG [1, 2, 3].

Figure 4a depicts typical sizes of K_1, K_2, and L just before a meal begins. Because the nonspecific arousal level L changes on a circadian time scale, it is assumed to be approximately constant during the times under consideration. Because the hunger input K_1 exceeds the satiety input K_2 just before a meal begins, K_2 + L equals the net arousal level

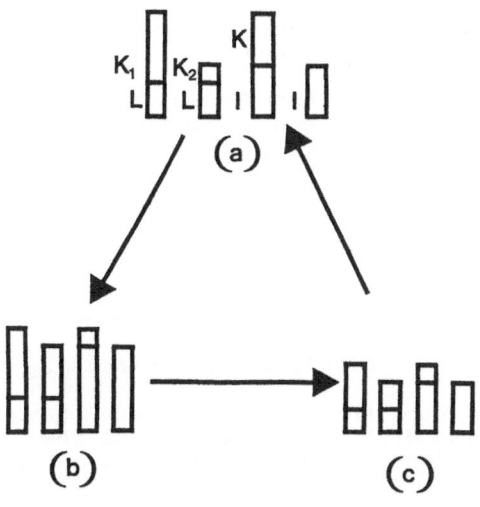

Figure 4: A normal gated dipole feeding cycle: (a) A large hunger input K_1 and small satiety input K_2 trigger eating by keeping the net arousal level I moderate and the net drive input K large; (b) fast growth of the satiety input shuts off eating by increasing I and decreasing K; (c) digestion decreases I while keeping K small. Then K_1 increases and the cycle begins again.

I at this time (Figure 4a). The net homeostatic drive input K then equals the difference $K_1 - K_2$ of the two drive inputs (Figure 4a). This net drive input should not be confused with phasic nonhomeostatic inputs that are rapidly elicited by presentation of external cues. The homeostatic quantities I and K vary on ultradian and circadian time scales to determine the dipole's responsiveness to external cues.

Just before a typical meal begins (Figure 4a), the satiety input K_2 is small. Thus the net arousal level I falls in the range of normal dipole sensitivity. The net drive input K is large since it is the difference of a large hunger input K_1 and a small satiety input K_2. As K increases due to an increase in hunger, it biases the dipole's on-channel to respond vigorously to food-related conditioned reinforcers. Eventually K becomes so large that the eating circuit wins the motivational competition, and eating begins.

As eating proceeds, a fast satiety signal due to gastric distention causes an increase in K_2 before K_1 can significantly change (Figure 4b). As a result of this increase in K_2, the net arousal I increases and the net drive input K decreases to keep the sum I + K constant. The increase in I can overarouse the dipole, thereby rendering it insensitive due to the dipole's inverted U property. The decrease in K reduces the hunger representation's bias to fire in response to conditioned reinforcer cues. Satiety can hereby cause a coordinated reduction in overall dipole sensitivity and in net appetitive drive. In the satiety-induced overaroused state, a fixed net drive input K sensitizes the dipole less than it does in a normally aroused state.

After eating ceases and digestion proceeds, both the hunger input K_1 and the satiety input K_2 slowly decrease (Figure 4c). As a result, the net arousal level I decreases towards the normal range, while the net drive input K remains small. The dipole's sensitivity hereby increases, but it cannot easily win the motivational competition because the drive-mediated priming of food-related conditioned reinforcers remains weak. Eventually, the hunger input begins to grow again (Figure 4a). Net arousal I remains normal and the net homeostatic drive input K grows until eating is triggered.

The critical observations in this summary concern the way that a homeostatic input can alter *both* net arousal and net homeostatic drive;

259

in particular, that an *increase* in satiety input can both *desensitize* the dipole and cause a *decrease* in the net hunger input.

GROSSBERG [3] describes how an anatomical or pharmacological lesion in the arousal pathway and/or the satiety input pathway to the eating dipole's off-channel can cause the dipole to become functionally under-aroused. In this underaroused state, the dipole exhibits hyperexcitable properties that closely resemble the symptoms of hypothalamic hyperphagia. Thus,the particular inverted U properties that obtain in a gated dipole opponent process help to explain both the normal course of an appetitive cycle and its breakdown due to lesions which upset the inverted U balance.

10. From Modulators of Eating to Modulators of Circadian Rhythmicity

The homeostatic and nonhomeostatic processes that Carpenter and I have used to quantitatively simulate circadian data are homologous to the satiety and conditioned reinforcer processes, respectively, that I have just outlined.

The homolog of the conditioned reinforcement process is a slow gain control process, which also obeys an LTM equation of the form (6). This slow gain control process buffers the circadian clock against adventitious lighting changes, such as a cloudy day, and enables the clock to adapt to pervasive lighting changes, such as seasonal changes in the number of hours of sunshine per day. This ecologically plausible slow gain control process enables us to explain paradoxical parametric data about long-term after-effects and split rhythms [6, 22, 23, 24]. The homolog of the homeostatic satiety signal is a homeostatic fatigue signal that prevents the circadian clock from generating intense bouts of activity during periods of severe metabolic debt. This ecologically plausible process enables us to explain paradoxical data about Aschoff's rule and its violations in diurnal mammals, the circadian rule, and split rhythms [5, 6, 22, 25, 26].

11. A Biochemical Homolog?

The coexistence of habituative dipole gates and LTM gates in the eating circuit suggests that several distinct transmitter systems cooperate to regulate eating behavior. The hypothesis in GROSSBERG [8] that the habituative arousal transmitter system is catecholaminergic and the LTM transmitter system is cholinergic is so far compatible with a large body of psychopharmacological data [3, 27, 28, 29, 30].

It remains to be seen whether the hypothesized homolog between the processes within the hypothalamic eating circuit and the circadian circuit holds on the biochemical level. In particular, is the slow gain control process of the pacemaker cholinergic and its habituative gating process catecholaminergic? The elaboration of these biochemical mechanisms may be aided by the detailed analysis of processes U and V in (6) that CARPENTER AND GROSSBERG [6] provide. This analysis links the choice of terms U and V in (6) to properties of behavioral data about after-effects and split rhythms. This analysis may help to discover the putative gain control transmitter system by predicting how transmitter agonists and antagonists should alter after-effects or split rhythms.

12. Eating Rhythm and Internal Zeitgeber

The discussion in Section 9 illustrates how a motivated behavior can occur rhythmically and can be phase-shifted, even if it is not controlled by a circuit that is endogenously oscillatory in the absence of homeostatic feedback. A behavioral rhythm can be maintained by homeostatic feedback signals, such as the satiety signal, which build up and dissipate on an ultradian time scale. As Section 9 showed, if food and food-related reinforcers remain constantly available, then after a meal is eaten, the increase in the satiety input and the subsequent decreases in both the hunger input and the satiety input can delay food intake until these processes recover from the meal. Then the eating circuit becomes sensitive to food-related external cues once again and the next meal can begin.

Phase-shifting of the eating cycle can occur in response to a food cue that is an unusually strong positive conditioned reinforcer. A large conditioned reinforcing signal can offset a relatively weak drive input to enable the corresponding polyvalent cells (Figure 3) to trigger an eating bout earlier than it would otherwise have occurred.

Internal homeostatic signals, such as the satiety signal, also help to synchronize the many off-cell sites that constitute the off-population of a gated dipole circuit. In the circadian model, the fatigue signal also has a synchronizing effect. It acts like an internal Zeitgeber that helps to entrain the on-population and off-population of the circadian pacemaker. CARPENTER AND GROSSBERG [6] analyse how an abnormally small fatigue signal can help to cause split rhythms and to permit individual dipoles in the pacemaker population to drift out of phase.

13. Mental Disorders and the Chemical Calibration of an Optimal Arousal Level

From the perspective of the circadian wake-sleep cycle, the inverted U property of a gated dipole circuit is a valuable one. This property enables the pacemaker to desensitize appetitive circuits during the sleep state and to resensitize them during the wakeful state. Drive inputs more finely modulate the wakeful sensitivity level on an ultradian time scale. Once it is agreed that the arousal level of appetitive circuits is modulated for this purpose on circadian and ultradian time scales, it becomes natural to ask how an optimal level of arousal is chosen during the wakeful state. The urgency of this problem becomes clear when one realizes that symptoms of several mental disorders have been interpreted, and indeed predicted, using the formal syndromes of behavioral properties that obtain in underaroused and overaroused gated dipoles. These mental disorders include hypothalamic hyperphagia, juvenile hyperactivity, Parkinsonism, and simple schizophrenia [3, 8, 31].

The gated dipole theory hereby sheds new light on the issue of how cholinergic-catecholaminergic imbalances can contribute to mental disorders. It also emphasizes the fundamental importance of the following unsolved problem:
Self-Tuning of the Inverted U: What auxiliary feedback mechanisms among the circadian circuit, the ultradian drive inputs, and their target gated dipole appetitive circuits enable the circadian circuit to deliver an arousal level in the normal range during the wakeful state?

REFERENCES

1. S. Grossberg: Psych. Rev. 89, 529-572 (1982).
2. S. Grossberg: J. Theor. Neurobiol. 1, 286-369 (1982).
3. S. Grossberg: "Some psychophysiological and pharmacological corre-

lates of a development, cognitive, and motivational theory",
in Brain and Information: Event Related Potentials, P. Karrer,
J. Cohen, and P. Tueting (Eds.) (New York Academy of Sciences,
New York, 1984).

4. G.A. Carpenter and S. Grossberg: Biol. Cyber. 48, 35-59 (1983).
5. G.A. Carpenter and S. Grossberg: Am. J. Physiol., in press (1984).
6. G.A. Carpenter and S. Grossberg: J. Theor. Biol., in press (1985).
7. S. Grossberg: Math. Biosci. 15, 39-67 (1972).
8. S. Grossberg: Math. Biosci. 15, 253-285 (1972).
9. W.K. Estes and B.F. Skinner: J. Exp. Psych. 29, 390-400 (1941).
10. M.R. Denny: "Relaxation theory and experiments", in Aversive Con-
    ditioning and Learning, F.P. Brush (Ed.) (Academic Press, New
    York, 1971).
11. A. Amsel: Psych. Bull. 55, 102-119 (1958).
12. A. Amsel: Psych. Rev. 69, 309-328 (1962).
13. W.R. McAllister and D.E. McAllister: "Behavioral measurement of
    conditioned fear", in Aversive Conditioning and Learning,
    F.R. Brush (Ed.) (Academic Press, New York, 1971).
14. A.R. Wagner: "A variety of punishment", in Punishment and Aversive
    Behavior, B.A. Campbell and R.M. Church (Eds.) (Appleton-Cen-
    tury-Crofts, New York, 1969).
15. S. Grossberg: Proc. Natl. Acad. Sci. 60, 758-765 (1968).
16. S. Grossberg: J. Theor. Biol. 22, 325-364 (1969).
17. G.A. Carpenter and S. Grossberg: J. Theor. Neurobiol. 1, 1-42
    (1981).
18. B.K. Anand and R.V. Pillai: J. Physiol. (Lon.) 192, 63-77 (1967).
19. H.D. Janowitz, M.E. Hanson, and M.I. Grossmann: Am. J. Physiol.
    156, 87-91 (1949).
20. J. LeMagnen: "Regulation of food intake", in Hunger and Satiety in
    Health and Disease, F. Reichsman (Ed.) (S. Karger, Switzer-
    land, 1972).
21. K.N. Sharma, B.K. Anand, S. Dua, and B. Singh: Am. J. Physiol.
    201, 593-598 (1961).
22. G.E. Pickard and F.W. Turek: Science 215, 1119-1121 (1982).
23. C.S. Pittendrigh and S. Daan: J. Comp. Physiol. 106, 223-252 (1976).
24. C.S. Pittendrigh: "Circadian oscillations in cells and the circa-
    dian organization of multicellular systems", in Circadian
    Oscillations and Organization in Nervous Systems, C.S. Pit-
    tendrigh (Ed.) (MIT Press, Cambridge, MA, 1974).
25. J. Aschoff: Cold Spring Harbor Symp. Quant. Biol. 25, 11-28 (1960).
26. J. Aschoff: Z. Tierpsychol. 49, 225-249 (1979).
27. A.E. Fisher: "Relations between cholinergic and other dipsogens in
    the central mediation of thirst", in The Neuropsychology of
    Thirst: New Findings and Advances in Concepts, A.N. Epstein,
    H.R. Kissileff, and E. Stellar (Eds.) (V.H. Winston, Washing-
    ton, D.C., 1973).
28. A.J. Friedhoff (Ed.): Catecholamines and Behavior, Vol. I: Basic
    Neurobiology (Plenum Press, New York, 1975).
29. A.J. Friedhoff (Ed.): Catecholamines and Behavior, Vol. II: Neuro-
    psychopharmacology (Plenum Press, New York, 1975).
30. E.M. Stricker and M.J. Zigmond: "Brain catecholamines and the la-
    teral hypothalamic syndrome", in Hunger: Basic Mechanisms and
    Clinical Implications, D. Novin, W. Wyrwicka, and G. Bray
    (Eds.) (Raven Press, New York, 1976).
31. S. Grossberg: Biol. Psychiat. 19, 1075-1118 (1984).

The Circadian Activity Rhythm of Mammals: A Comparison of Models and Experiments

Gail A. Carpenter*

Center for Adaptive Systems, Boston University, Boston, MA 02215, USA, and

Department of Mathematics, Northeastern University, Boston, MA 02115, USA

A classic volume on the mechanisms, models, and significance of oscillations in biological systems is "Biological Clocks" [1], the proceedings of a 1960 Cold Spring Harbor Symposium on Quantitative Biology. Articles in that collection study important aspects of circadian rhythms at levels ranging from the biochemical to the ecological. In particular, Pittendrigh [2] and Aschoff [3], considering the activity patterns of free-running animals, describe split rhythms [2], after-effects [2], Aschoff's rule[1] [3], and the circadian rule[1] [3]. The rich parametric data base which has grown out of the study of these behavioral phenomena now provides critical tests with which to access models of circadian activity rhythms. Two classes of models of mammalian circadian rhythms, and the relationship between these models and experimentally observed free-running activity patterns, will be considered in this article.

Split rhythm [2] experiments have played a fundamental role in the development of *two-oscillator* and *multi-oscillator* models of circadian activity rhythms [2, 5-11]. In a typical split rhythm experiment, the consolidated daily activity cycle of a (nocturnal) hamster, placed in constant light for several weeks, splits into two components [9]. Simulating this pattern, two coupled oscillators, or oscillating populations, which are initially in-phase can "split," or oscillate out-of-phase, in certain parameter ranges. To simulate after-effect experiments, two-oscillator models require elaboration. In 1960, Pittendrigh [2] noted that exposure to a particular light regime, such as alternating light-dark cycles or constant light, could induce long-term changes in an animal's free-running activity in the dark. In particular, "the" circadian period can depend significantly upon an animal's environment many weeks prior to the time at which the period is being measured. While it is true generally that the period of a weakly coupled two- or multi-oscillator system can depend upon initial phase-differences, and thus reflect its prior input environment, two-oscillator simulations of certain paradoxical after-effect results use a special mechanism for the processing of light inputs [5, 9, 10, 12]. The paradoxical results show that, in some cases, an *increase* in the number of hours of light per day causes a *decrease* in the subsequent period in the dark; while in other cases, an *increase* in the number of hours of light per day causes an *increase* in the subsequent period in the dark [12]. Specifically, nocturnal mammals exposed for several weeks to a "long photoperiod" light regime, with 18 hours of light and 6 hours of darkness per day, tend to have unusually short subsequent activity periods in the dark; while nocturnal mammals exposed for several weeks to constant light, have unusually long subsequent activity periods in the dark. To account for these findings, Pittendrigh and his colleagues [5, 9, 10] assume that an external on-off, or *non-parametric*, light source acts by differentially phase-resetting the constituent oscillators when lights are turned on or off; while an external constant, or *parametric*, light source acts by altering a constant pacemaker parameter.

---

* Supported in part by the National Science Foundation (NSF MCS-82-07778) and the Office of Naval Research (ONR-N00014-83-K0337).

1. Term coined by Pittendrigh [2] and Hoffmann [4].

The assumption of distinct modes of action for parametric and non-parametric
light sources accounts for the results of parametric and non-parametric after-effect
experiments, but at the same time raises some new questions. How, for example, could
an organism receiving a light input at a given moment know whether to process that
input as parametric or non-parametric? How can 18 hours of light per day have no
effect upon an internal parameter when 24 hours of light per day effects important
changes in that parameter? Would 19, 20, or 23.9 hours of light per day be proces-
sed parametrically or non-parametrically? Moreover, a parametric light source is
not necessarily registered as a parametric input by the nervous system. During sleep,
eye-closure, retreat to a dark nest, internal feedback, or self-selection of the
light-dark cycle can attenuate or eliminate the light signal to the nervous system,
which thus registers a parametric external light source as a non-parametric internal
light input. Since these various difficulties arise when attempting to simulate both
split rhythm and after-effect results with a single, two-oscillator model, it is
worth noting that the hamster, the principal subject for split rhythm experiments
on nocturnal mammals, exhibits inconsistent after-effects on period [12].

An alternative model of circadian rhythms [13, 14, 15] has been used to simulate important properties of mammalian activity rhythms, including split rhythm and
after-effect experiments. Equations defining this model are outlined at the end of
this article. The model describes a basic *gated pacemaker* circuit consisting of elementary neural components. An essential part of this circuit is a slowly accumulating transmitter, whose accumulation rate determines the approximate period of the
basic gated pacemaker. The model thus suggests that a transmitter accumulation rate
plays an important role in establishing a circadian time scale. In the context of a
mammalian circadian activity system, these gated pacemaker circuits can be interpreted either as intracellular entities or as small neural networks distributed
throughout each suprachiasmatic nucleus (SCN). The net output of these distributed
gated pacemaker circuits is assumed to drive observable activity. Analysis of the
gated pacemaker system includes consideration of how various ecological factors
alter the oscillatory structure of the basic pacemaker. These ecological factors
include the attenuation of light input during sleep; metabolic feedback, or "fatigue", which builds up with the net pacemaker output and decays on an ultradian
(e.g., 4-hour) time scale, and which can act as an "internal Zeitgeber" to help
synchronize the distributed small circuits; and a slow gain control process, which
averages pacemaker output on a time scale of weeks. Light is registered as a real-
time input, with no *a priori* assumptions about distinct mechanisms for processing
parametric and non-parametric light sources. Indeed, whenever light attenuation
occurs during sleep, all light input functions are non-parametric. Nevertheless,
both parametric and non-parametric after-effect experiments [12] are simulated
[15]. This is accomplished by taking into account the nature of a model or experimental animal's activity (or pacemaker output) during the various light regimes.
With 18 hours of light per day, a nocturnal mammal is vigorously active during the
remaining 6 hours of darkness. In contrast, a nocturnal mammal kept in constant,
and therefore unavoidable, light is only weakly active. It is these different levels
of activity which account for the opposite after-effects on period in the gated
pacemaker analysis. Moreover, versions of the slow gain control process that bring
about the slow onset of split rhythms incidentally produce inconsistent after-effects on period. Thus, both in the hamster and in the gated pacemaker model, split
rhythms and inconsistent after-effects on period occur together.

Attenuation of light during sleep plays an important role in the gated pacemaker analysis of other circadian phenomena. For example, as *in vivo*, a phase response curve (PRC) of a nocturnal model is similar in shape to a PRC of a diurnal
model, with phase delays during early subjective night and phase advances during
late subjective night. In addition, when light attenuation during sleep is added
to the model, the typical differences between nocturnal and diurnal PRC's appear
[13]. That is, the subjective day "dead zone" of a nocturnal PRC and the comparatively even day/night portions of a diurnal PRC occur when light input is partially attenuated during sleep.

Light attenuation during sleep also plays a key role in the gated pacemaker
analysis of Aschoff's rule [14]. In 1960 [3], Aschoff noted that, in experiments

performed on a variety of species up to that date, the circadian activity period of a nocturnal animal increased as the intensity of a constant light source increased. In contrast, the circadian activity period of a diurnal animal decreased as the intensity of a constant light source increased. In 1979 [16], Aschoff reviewed the greatly expanded data base and found that, while the original rule still holds for many groups, including the nocturnal mammals, only a minority of diurnal mammals obey the original rule. The circadian activity period of most diurnal mammals either increases or first decreases and then increases as light intensity increases. For the basic gated pacemaker, nocturnal and diurnal circuits are symmetric, and solutions therefore have identical periods at a given light level. Thus, the differences predicted by Aschoff's rule could never appear. However, when ecological factors are considered one at a time, differences emerge. First, with negative feedback, or "fatigue", but no light attenuation during sleep, both nocturnal and diurnal models obey the 1960 version of Aschoff's rule. With both fatigue and light attenuation during sleep, the nocturnal model continues to show an increase in period as light intensity increases, but the period of the diurnal model first decreases then increases as light intensity increases. Thus a combination of both negative metabolic feedback and light attenuation during sleep produces the 1979 version of Aschoff's rule, with a consistent nocturnal pattern and an inconsistent diurnal pattern. Analysis of model solutions reveals that the consistent nocturnal increase of period with light intensity is due to the fact that the external Zeitgeber, light, and the internal Zeitgeber, fatigue, both act to inhibit pacemaker output; but that the inconsistent diurnal pattern is due to the fact that light, which enhances pacemaker output, and fatigue, which inhibits pacemaker output, trade off against one another. In contrast to the complex situation with respect to Aschoff's rule, the circadian rule still holds for both diurnal and nocturnal models, with light-stimulating diurnal model output and inhibiting nocturnal model output.

In two-oscillator model analysis, parameters are typically chosen so that frequency and amplitude covary [5, 17], as suggested by the 1960 version of Aschoff's rule and the circadian rule. In a two-oscillator nocturnal model, then, amplitude decreases when period increases, and this occurs when a parameter is altered by a model parametric light source. This analysis does not account for Aschoff's 1979 finding that frequency and amplitude usually do not covary in the activity rhythms of diurnal mammals. Enright's analysis of his multi-oscillator coupled stochastic system [6] shows how exceptions to Aschoff's rule can occur when parameters are chosen so that some of the constituent oscillators fire twice in one day. However, this occurs at extreme parameter values, and Enright regards exceptions to Aschoff's rule as rare.

The Gated Pacemaker Model

The basic gated pacemaker circuit describes the dynamics of on-cell/off-cell[1] pairs, called *gated dipoles*, in which the on-cells and the off-cells mutually inhibit one another. Processes (A)-(E) define the basic gated pacemaker dynamics.

(A) Slowly accumulating transmitter substances are depleted, or habituated, by gating the release of feedback signals.

(B) The feedback signals are organized as an on-center off-surround, or competitive, anatomy.

(C) Both on-cells and off-cells are tonically aroused.

(D) Light excites the on-cells of a diurnal model and the off-cells of a nocturnal model.

(E) The on-cells drive observable activity, such as wheel-turning, in both the diurnal model and the nocturnal model.

Processes (F) and (G) define the additional fatigue and slow gain control dynamics.

---

1. If the gated pacemaker is interpreted as an intracellular entity, "on-cells" and "off-cells" would correspond to subcellular functional units.

(F) On-cell activity gives rise to a fatigue signal that excites the off-cells in both the diurnal model and the nocturnal model. The fatigue signal is a time-average of the on-cell output signal on an ultradian time scale.

(G) On-cell activity gives rise to a slowly varying gain control signal that excites the on-cells in both the diurnal model and the nocturnal model. The gain control signal is a time-average of the output signal on a time scale of months.

The general model for a nocturnal gated pacemaker is defined in equations (1)-(6) and illustrated in Figure 1.

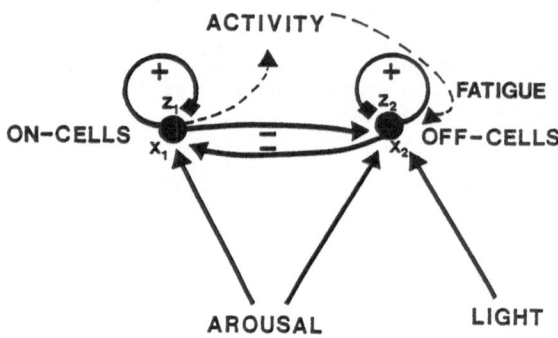

Figure 1: Gated pacemaker circuit of the nocturnal model with fatigue feedback.

NOCTURNAL MODEL

ON-POTENTIAL

$$\frac{dx_1}{dt} = -Ax_1 + (B-x_1)[I + f(x_1)z_1 + Sy] - (x_1+C)g(x_2),$$ (1)

OFF-POTENTIAL

$$\frac{dx_2}{dt} = -Ax_2 + (B-x_2)[I + f(x_2)z_2 + F + J(t)] - (x_2+C)g(x_1),$$ (2)

ON-GATE

$$\frac{dz_1}{dt} = D(E - z_1) - Hf(x_1)z_1,$$ (3)

OFF-GATE

$$\frac{dz_2}{dt} = D(E - z_2) - Hf(x_2)z_2,$$ (4)

FATIGUE

$$\frac{dF}{dt} = -KF + h(x_1),$$ (5)

GAIN CONTROL

$$\frac{dy}{dt} = -Uy + Vf(x_1).$$ (6)

266

Variable x_1 in equation (1) is the potential of an on-cell (or cell population, or subcellular unit) v_1. Variable x_2 in equation (2) is the potential of an off-cell v_2. Both x_1 and x_2 obey membrane equations. In (1) and (2), the parameter $-A$ in the terms $-Ax_1$ and $-Ax_2$ determines the fast decay rate of the potentials x_1 and x_2. Also in (1) and (2), term I represents the constant arousal level that equally excites v_1 and v_2. In (1), the transmitter substance z_1 gates the nonnegative feedback signal $f(x_1)$ from v_1 to itself. Term $f(x_1)z_1$ is proportional to the rate at which transmitter is released from the feedback pathway from v_1 to itself, thereby re-exciting x_1. Term Sy describes the effect of the gain control process y on v_1. Term S is a signal that is gated by y, thereby generating a net excitatory input Sy at the on-cells v_1. The choice of signal S is described below. The off-cells v_2 inhibit the on-cells v_1 via the nonnegative signal $g(x_2)$ in term $-(x_1+C)g(x_2)$. Equation (2) is the same as equation (1), except that the indices 1 and 2 are interchanged; both the light input $J(t)$ and the fatigue signal F excite v_2 but not v_1; and the slow gain control process excites v_1 but not v_2.

Equations (3) and (4) define the transmitter processes z_1 and z_2. In (3), the transmitter z_1 accumulates to its maximal level E at a slow constant rate D via the term $D(E - z_1)$. This slow accumulation process is balanced by the release of z_1 at rate $Hf(x_1)z_1$, leading to the excitation of x_1 in equation (1). A similar combination of slow accumulation and gated release defines the dynamics of transmitter z_2 in (4).

The remaining processes F and y modulate the behavioral patterns that are generated by the pacemaker, as during split rhythms and long-term after-effects, but are not the source of the pacemaker's clock-like properties. Both F and y average indices of pacemaker activity, but are not independent oscillators.

The fatigue signal F in (5) is a time-average of $h(x_1)$, which increases with on-cell activity x_1. Speaking intuitively, an increase in x_1 and a decrease in x_2 arouse neural circuits that support the awake state. Fatigue builds up as a function of increased metabolic activity during the awake state, including but not restricted to overt action. Fatigue, in this sense, can thus build up in an alert but physically restrained animal. Since F excites the off-cells v_2 in (2), it tends to inhibit the arousal generated by the pacemaker. The decay rate K of the fatigue signal F is assumed to be ultradian. In particular, $A > K > D$ so that the potentials x_1 and x_2 react faster than the fatigue signal F, which in turn reacts faster than the pacemaker gates z_1 and z_2.

The slow gain control process y in (6) is also a time-average, but on a time scale that is much slower than the circadian time scale. Process y averages term $Vf(x_1)$ at an averaging rate U. Then Sy in (1) acts as an excitatory input to the on-cells v_1. Term Sy in equation (1) combined with equation (6) formally define a long-term memory trace y [18]. In all cases, terms S, U, and V are chosen to be constant, or to vary as a function of light or on-cell activity.

The diurnal model differs from the nocturnal model only in the equations (1) and (2) that define its on-cell and off-cell potentials. In particular, light input $J(t)$ excites the on-cells but not the off-cells of the diurnal model. By contrast, the fatigue input F excites off-cells in both the diurnal and the nocturnal models, and the slow gain input y excites on-cells in both the diurnal and the nocturnal models.

The models in equations (1)-(6) are completely defined by a choice of the signal functions f, g, and h; the light input J(t); the signals S, U, and V; and the parameters. For simplicity, the signal functions f(w) and g(w) in (1)-(6) are chosen to be threshold-linear functions of activity w:

$$f(w) = g(w) = \max(w,0).\qquad(7)$$

The signal function h(w) in (5) is defined by

$$h(w) = M \max[f(w) - N,0].\qquad(8)$$

The definition of h(w) can be interpreted as follows. We assume that $f(x_1(t))$ is the output signal of the pacemaker. Behavioral activity is triggered when $f(x_1(t))$ exceeds the positive threshold N. We assume that the function $h(x_1(t))$ defined by (8) provides an index of behavioral activity. Since by (7), $f(w) = w$ when $w \geq 0$, we can simplify the definition of h(w) in (8) to

$$h(w) = M \max(w-N,0).\qquad(9)$$

Equation (5) says that, in this simple version of the model, the fatigue signal F builds up at a rate proportional to behavioral activity $h(x_1(t))$. Activity ceases when $x_1(t) \leq N$. During such a time interval, fatigue decays at the ultradian rate K.

We operationally define sleep in terms of its effects on the pacemaker. The main effect is that eye-closure (or entering a dark nest) can attenuate the light input to the pacemaker. Letting L(t) be the light input that reaches the pacemaker when its "eyes" are open, we define the net light input in equations (1) and (2) to be

$$J(t) = \begin{cases} L(t) & \text{if } x_1(t) > P \\ \\ \theta L(t) & \text{if } x_1(t) \leq P \end{cases}.\qquad(10)$$

Parameter θ is a light attenuation factor due to eye-closure. A value of $\theta = 1$ means that no light attenuation occurs during sleep. A value of $\theta = 0$ means that complete light attenuation occurs during sleep.

Different choices of the slow gain control process can produce different long-term properties of solutions. A variety of after-effect experiments on nocturnal mammals can be simulated [15] using any one of a small class of slow gain control processes, described in paragraphs (I)-(III) below. A gain control process is chosen by defining the performance signal S in equation (1) and the averaging rates U and V in equation (6).

A distinct but closely related class of slow gain control processes, described in paragraphs (IV)-(VI) below, generates not only split rhythms but also inconsistent after-effects on period. The precise definitions of the two classes, given in paragraphs (I)-(III) and (IV)-(VI), predict dynamic properties which can be sought in the physiological and biochemical processes of the two corresponding classes of nocturnal mammals.

Slow Gain Processes which Generate Consistent After-Effects

(I) The long-term memory (LTM) trace y gates a tonically active pathway. Hence the performance signal is given by

$$S(t) = S_{tonic}(t) \equiv Q\qquad(11)$$

in (1), where Q is a positive constant. The terms U and V in (6) are both switched on when the model wakes up, and are switched off when the model goes to sleep. Thus

$$U(t) = RS_{awake}(t),$$ (12)

where R is a positive constant, and

$$V(t) = S_{awake}(t),$$ (13)

where

$$S_{awake}(t) = \begin{cases} 1 \text{ if } x_1(t) > P \\ \\ 0 \text{ if } x_1(t) \leq P \end{cases}.$$ (14)

Parameter P in (14) is the sleep threshold that was defined in equation (1). By (6), (12) and (13), equation (6) for the LTM trace is

$$\frac{d}{dt} y = S_{awake}[-Ry + f(x_1)].$$ (15)

By (15), the sampling signal S_{awake} gates the sensitivity of y to the postsynaptic potential x_1. Process y is therefore located in Figure 2 abutting synapses where it can both sample the postsynaptic potential x_1 and receive the presynaptic sampling signal S_{awake}.

Light plays no direct role in the slow gain process (I). In contrast, light plays a direct, inhibitory role in the slow gain processes (II) and (III).

(II) As in (I),

$$S(t) = S_{tonic}(t) \equiv Q.$$ (11)

In the averaging equation (6), light reduces the rates U and V:

$$U(t) = RS_{light}(t) \quad \text{and}$$ (16)

$$V(t) = S_{light}(t), \quad \text{where}$$ (17)

$$S_{light}(t) = \frac{1}{1 + WJ(t)}$$ (18)

and J(t) is the light intensity defined by (10). By (6), (16), and (17), the equation for the LTM trace is

$$\frac{d}{dt} y = S_{light}[-Ry + f(x_1)].$$ (19)

As in Figure 2, the LTM trace gates a tonically active pathway. However, the plasticity of the LTM trace is presynaptically gated by a pathway that is tonically on when light is off, but is inhibited when light turns on.

(III) As in (II),

$$\frac{d}{dt} y = S_{light}[-Ry + f(x_1)].$$ (19)

The performance signal S is, in this case, no longer constant. Instead,

$$S(t) = QS_{light}(t).$$ (20)

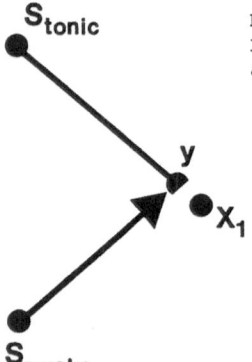

S_{tonic}

y

x_1

S_{awake}

Figure 2: The slow gain control process described in (I). The LTM trace y time-averages x_1 when the model is awake and gates a tonically active performance signal.

Thus $S_{light}(t)$ is both the performance signal and the averaging signal in this process.

Slow Gain Processes which Generate Split Rhythms and Inconsistent After-Effects

In every case (IV)-(VI), the performance signal is the same; namely,

$$S(t) = QS_{awake}(t), \tag{21}$$

where

$$S_{awake}(t) = \begin{cases} 1 \text{ if } x_1(t) > P \\ \\ 0 \text{ if } x_1(t) \leq P \end{cases}. \tag{14}$$

It is this choice of performance signal which determines the primary difference between class (I)-(III) and class (IV)-(VI). The functional property that all the LTM traces guarantee is that y becomes smaller as light intensity increases.

(IV) $\frac{d}{dt} y = -Ry + S_{awake} f(x_1);$ $\hspace{2cm}$ (22)

that is, $U = R$ and $V = S_{awake}$.

(V) $\frac{d}{dt} y = -Ry + f(x_1);$ $\hspace{3cm}$ (23)

that is, $U = R$ and $V = 1$.

(VI) As in (II) and (III),

$$\frac{d}{dt} y = S_{light}[-Ry + f(x_1)]; \tag{19}$$

that is, $U = RS_{light}$ and $V = S_{light}$, with S_{light} given in equation (18).

Mathematical analysis reveals why cases (I)-(III) have consistent after- effects on period and cases (IV)-(VI) have both inconsistent after-effects on period and split rhythms [15]. The following sketch is part of that analysis, in which a connection is made between the diurnal mammals, which inconsistently obey Aschoff's rule, and the nocturnal mammals which have inconsistent after-effects. The formal connection in the equations links the case when $\theta = 1$ in equation (10) with the case when $S = S_{tonic} \equiv Q$ in equation (11); and the case when $\theta = 0$ in equation (10) with

270

the case when $S = QS_{awake}$ in equation (21). Consistent after-effects on period (τ) occur when τ is a monotonic (in fact, a decreasing) function of y. Inconsistent after-effects on period occur when τ is a non-monotonic function of y: even if a given experimental paradigm consistently decreased y, the effect on τ would be indeterminant. The formal connection between after-effect results for the nocturnal model and Aschoff's rule for the diurnal model can be made by inspection of the on-potential equation corresponding to (1) of the diurnal model:

$$\frac{dx_1}{dt} = -Ax_1 + (B-x_1)[I + f(x_1)z_1 + Sy + J(t)] - (x_1+C)g(x_2). \qquad (24)$$

When $L(t) \equiv L$ is a constant external light source, the internal light input function $J(t)$ from equation (10) is given by

$$J(t) = \begin{cases} L & \text{if } x_1(t) > P \\ \\ \theta L & \text{if } x_1(t) \leq P \end{cases} . \qquad (25)$$

Note that Sy and $J(t)$ appear in the same term of equation (24), and appear nowhere else in the diurnal model equations (1)-(5). Note, too, that when $\theta = 1$ (that is, there is no light attenuation during sleep), $J(t) \equiv L$, a constant. On a time scale in which y is nearly constant, the effect on τ of $Sy \cong Qy$, as y increases, is the same as the effect on τ of $J(t) \equiv L$ as L increases. Both Sy and $J(t)$ act as constant input functions on a short time scale. The case $J(t) \equiv L$ (i.e., $\theta = 1$) is the case in which the diurnal model obeys Aschoff's rule, and τ depends monotonically on L. Thus, when $S = S_{tonic} = Q$, consistent after-effects on τ occur.

   Similarly, the cases with inconsistent after-effects on τ are formally equivalent to the diurnal model case in which $\theta = 0$, that is, when total attenuation of the light input occurs during sleep. In this case, $J(t) = S_{awake}(t) \cdot L$, by equations (14) and (25). On a time scale in which y is nearly constant, the effect on τ of $Sy \cong S_{awake} \cdot Qy$, as y increases, is the same as the effect on τ of $J(t) = S_{awake}(t) \cdot L$ as L increases. Both Sy and $J(t)$ act as constants while $x_1(t) > P$ but equal 0 when $x_1(t) \leq P$. The case $J(t) \equiv S_{awake}(t) \cdot L$ (i.e., $\theta = 0$) is a case in which the diurnal model inconsistently obeys Aschoff's rule, and τ depends non-monotonically on L. Thus when $S = QS_{awake}$, inconsistent after-effects on τ occur.

Acknowledgements: Thanks to Cynthia Suchta for her valuable assistance in the preparation of the manuscript and illustrations.

REFERENCES

1. "Biological Clocks", Cold Spring Harbor Symp. Quant. Biol. 25, (1960).
2. C.S. Pittendrigh: Cold Spring Harbor Symp. Quant. Biol. 25, 159-185 (1960).
3. J. Aschoff: Cold Spring Harbor Symp. Quant. Biol. 25, 11-28 (1960).
4. K. Hoffmann: Overt circadian frequencies and circadian rule, in Circadian Clocks, J. Aschoff, (Ed.) (Amsterdam: North-Holland 1965).
5. S. Daan and C. Berde: J. Theor. Biol. 70, 297-313 (1978).
6. J.T. Enright: The Timing of Sleep and Wakefulness (Berlin, Heidelberg, New York: Springer-Verlag 1980).
7. M. Kawato and R. Suzuki: J. Theor. Biol. 86, 547-575 (1980).
8. T. Pavlidis: Bull. Math. Biol. 40, 675-692 (1978).
9. C.S. Pittendrigh: Circadian oscillations in cells and the circadian organization of multicellular systems, in Circadian Oscillations and Organization in Nervous Systems, C.S. Pittendrigh (Ed.) (Cambridge, MA: MIT Press 1974).

10. C.S. Pittendrigh and S. Daan: J. Comp. Physiol. $\underline{106}$, 333-355 (1976).
11. A.T. Winfree: J. Theor. Biol. $\underline{16}$, 15-42 (1967).
12. C.S. Pittendrigh and S. Daan: J. Comp. Physiol. $\underline{106}$, 223-252 (1976).
13. G.A. Carpenter and S. Grossberg: Biol. Cybernetics $\underline{48}$, 35-59 (1983).
14. G.A. Carpenter and S. Grossberg: Am. J. Physiol. (1984).
15. G.A. Carpenter and S. Grossberg: A neural theory of circadian rhythms: Split rhythms, after-effects, and motivational interactions. Submitted for publication (1984).
16. J. Aschoff: Z. Tierpsychol. $\underline{49}$, 225-249 (1979).
17. R. Wever: Kybernetik $\underline{1}$, 139-154 (1962).
18. S. Grossberg: The hypothalamic control of eating and circadian rhythms: Opponent processes and their chemical modulators. This volume.

Weak Coupling Between the Two Pacemakers of the Bilateral Circadian Mechanism in Crickets

Gottfried Wiedenmann

Arbeitskreis N.c.R. Zoologisches Institut, J.W. Goethe Universität
D-6000 Frankfurt/Main, Fed. Rep. of Germany

The circadian calling song rhythm of the cricket Teleogryllus commodus is driven by a dual control mechanism [1]. This only becomes apparent when the subunits reveal individual oscillatory features [2], which in unimodal patterns are expressed in the different slopes of onset and end of activity (Fig. 1a above arrows). In split rhythms, the activity components are quite distinct (Figs. 1b,2a). The two pacemakers either function independently or are mutually coupled and have the same period. Transitions from one state to the other can occur spontaneously, indicating that the coupling is weak. Interactions become evident when two different periods change to a common one (Figs. 1a, b 2b), which in most cases corresponds closely to one of the original periods. After splitting, an independent continuation (Fig. 2a), as well as obvious interactions and a reunion of the two components (Fig. 2b) are found. When interactions occur, certain phase

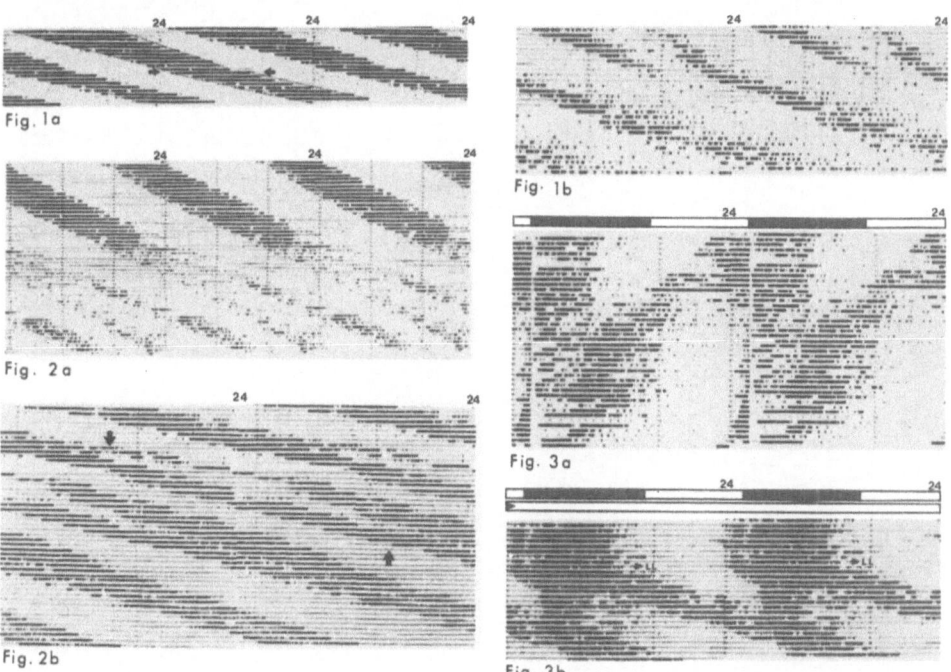

Fig. 1a

Fig. 1b

Fig. 2a

Fig. 3a

Fig. 2b

Fig. 3b

Calling song patterns of T. commodus in constant temperature.
Fig. 1: LL, directly after LD (as in Fig.3) a: onset and end slope differ
(τ_o =27.0 h; τ_e=25.8 h), after arrows $\tau_o = \tau_e$ =26.9 h; b LL, similar features but distinct components (τ_o=25.5 h; τ_e changes to 25.6 h; Fig. 2: LL, a: $\tau_o = \tau_e$=25.1 h, splitting with independent components: τ_1=25.0; τ_2=25.3; b: splitting (↓), fixed phase angle difference, and reunion (↑); Fig. 3: Patterns of unilaterally blinded crickets in LD and LL a: τ_{DD}=23.3; b: DD-rhythm (close to 24 h) only in LL.

angle differences are prevalent. The two components either almost overlap (Figs. 1a, 2a) or are about 180° apart (Fig. 2b), which agrees with theoretical predictions [3]. In LD, the crickets where one compound eye was blinded during the last larval instar [1], show either an additional free-running rhythm (Fig. 3a), which is characteristic for DD [4], or only entrainment (Fig. 3b). Since one of the pacemakers is disconnected from direct light information, this can be explained by the absence or presence of mutual coupling. Animals of the first group (Fig. 3a) have a significantly shorter τ_{DD}(23.25 $\pm$ 0.066 h) than the other (23.67 $\pm$ 0.09 h) (mean $\pm$ SE; α=0.05). This difference, together with the weak coupling, causes the rhythm to split even in LD. In the other group, internal entrainment fails only when in LL the period difference between the two circadian pacemakers increases (Fig. 3b).

References
1 G. Wiedenmann: J. comp. physiol. 150, 51, (1983).
2 G. Wiedenmann, W. Loher: J. Insect Physiol. 30, 2, 145 (1984).
3 S. Daan, C. Berde: J. theor. Biol. 70, 297 (1978).
4 W. Loher: J. comp. Physiol. 79, 173 (1972).

(Supported by DFG, SFB 45, E-1)

Effect of Reversed Photoperiod on Daily Rhythms of Synthesis and Accumulation of Neurosecretion in the Brain of the House Cricket

B. Cymborowski

Department of Invertebrate Physiology, Warsaw University, 93 Zwirki i Wigury 02-089 Warszawa, Poland

Earlier experiments made on crickets suggested that the neurosecretory cells of the pars intercerebralis of the brain are responsible for maintaining the circadian rhythm of locomotor activity [1,2]. This conclusion was put forward on the basis of experiments in which crickets kept under light-dark cycle have a very distinct locomotor activity rhythm. In the pars intercerebralis of such rhythmic insects a rhythmicity in accumulation and release of neurosecretion, RNA, and protein synthesis were observed [3,4,5]. Crickets kept under constant light conditions which do not show a circadian rhythm of locomotor activity also failed to exhibit circadian fluctuation in the above mentioned activities of the neurosecretory cells of the pars intercerebralis.

The aim of this study was to test whether the observed activities of the neurosecretory cells would be shifted in a reversed photoperiod when the peak of locomotor activity is shifted to the new dark phase. The experiments were carried out with the house cricket (*Acheta domesticus*) reared at first in LD 12:12 and then transferred to reversed (DL 12:12) photoperiod at 28°C. In both groups of insects the levels of neurosecretion were studied by means of autoradiography using ^{35}S-cysteine as a precursor for the neurosecretory proteins. The degree of accumulation of neurosecretory material was studied by means of paraldehyde-fuchsin staining [5].

The changing patterns of synthesis and accumulation of neurosecretion in the neurosecretory cells of the pars intercerebralis are presented in Fig. 1. It was found that in both groups of insects the maximal incorporation of ^{35}S-cysteine in the neurosecretory cells of the pars intercerebralis of the brain occurs during the photophase at about midday. During the same period accumulation of neurosecretion in these cells is relatively small. At the beginning of the dark phase the incorporation of ^{35}S-cysteine by the neurosecretory cells of the brain is at its lowest level and the degree of accumulation of neurosecretion reaches its highest value.

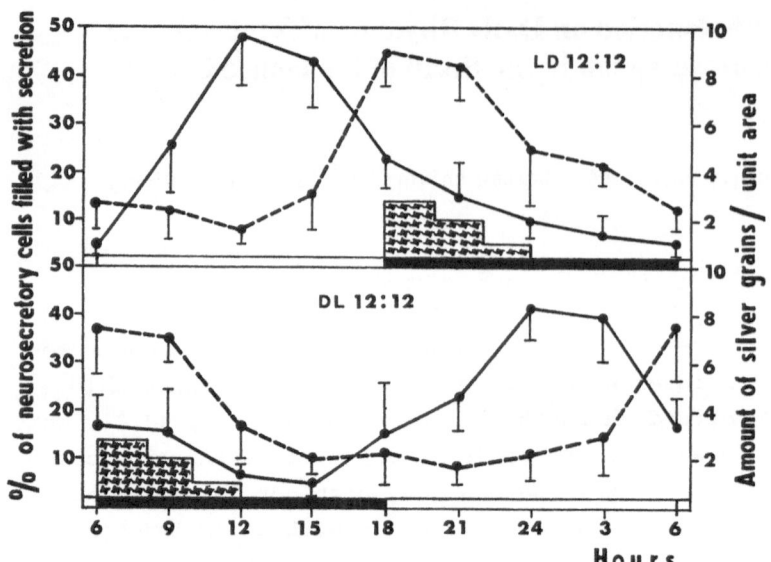

Fig. 1. Diurnal changes in the levels of protein synthesis (continuous line) and in the degree of accumulation of neurosecretion (broken line) in the neurosecretory cells of the pars intercerebralis in normal (LD 12:12) and reversed (DL 12:12) photoperiods. Each point is the mean of 6-8 determinations + SD. Hatched area indicates locomotor activity in both photoperiods.

The presented results suggest that the activity patterns of the neurosecretory cells of the brain are essentially the same in both photoperiods. The investigated activities occur exactly at the same time in the new lighting regime. This might suggest that these cells serve as a pacemaker in controlling the locomotor activity rhythms of the insects.

References: [1] B. Cymborowski and A.B. Dutkowski: J.Insect Physiol. 16, 341-348 (1970).

[2] B. Cymborowski: Zool. Pol. 20, 103-149 (1970).

[3] B. Cymborowski and A.B. Dutkowski: J. Insect Physiol. 15, 1187-1197 (1969).

[4] B. Cymborowski and A.B. Dutkowski: J. Insect Physiol. 16, 341-348 (1970).

[5] B. Cymborowski: J. interdiscipl. Cycle Res. 14, 111-116 (1983).

Circadian Rhythm of Locomotor Activity in *Musca* Continues After Severance of Optic Tracts

Charlotte Helfrich and Wolfgang Engelmann

Institut für Biologie I, University of Tübingen, D-7400 Tübingen, Fed. Rep. of Germany

Bronislaw Cymborowski

Department of Invertebrate Physiology, Zoological Institute, University of Warsaw 02-089 Warszawa, Poland

Abstract: Contrary to findings in cockroaches, crickets and beetles, the optic lobes of the house fly Musca domestica are not the site of the oscillator controlling locomotor activity. The lobes may play a role in the coupling of oscillators. Synchronization of activity by light dark cycles is extraocular.

1. Introduction: Several reports traced the site of circadian control of locomotor activity in cockroaches, crickets, and beetles to the optic lobes (1,2,3,4) and implied neural connections (5,6). However, hormonal control was also suggested in cockroaches (7), crickets (8), Drosophila (9), and silkmoths (10). Recently, mosquitoes were shown to remain rhythmic after lobectomy (11). The optic lobes seem to be dispensable for the expression of rhythmic locomotion also in Drosophila(12) and, as shown here, in Musca.

2. Methods: Locomotion of individual flies was monitored photoelectrically under conditions of constant temperature and continuous weak red light (13). Optic tracts were cut and optic lobes removed using standard microsurgery methods. After pre- and postoperative recording of activity the brains were examined histologically. For details see (14).

3. Results: After severance of optic tracts about 50% of the flies showed still a clear circadian rhythmicity, although the mean activity level was increased. In 20% several components were found in the actogram of individual flies,indicating the multioscillatory nature of the circadian system. In the remaining 30% no rhythm(s) could be detected. Histological examination revealed degenerated areas in the central brain of most of these flies, suggesting a damage of decisive brain structures to be the cause of arrhythmicity rather than the severance of optic tracts. Apparently the neural connection of optic lobes to the brain is not necessary for circadian control of locomotor activity. Removing the optic lobes did not abolish the rhythm in half of the flies. We conclude, therefore, that in Musca the optic lobes are dispensable for the expression of circadian rhythmicity of locomotor activity. They may, however, play a role in strengthening the coupling of oscillators, since after surgery period is lengthened, several circadian rhythms might occur, and the onset of activity in a LD cycle is delayed. Blind flies (surgery or covering complex eyes) are still synchronized by LD cycles, indicating an extraocular photoreceptor for the circadian system.

References

1. Nishiitsutsuji-Uwo, J. and Pittendrigh,C.S. Z. Vergl. Physiol. 58, 14 (1968).
2. Roberts,S.K. J. Comp. Physiol. 88, 21 (1974).
3. Sokolove, P.C. and Loher, W.J. Insect. Physiol. 121, 785 (1975).
4. Balkenohl, M. and Weber, F. Mitt. dtsch. Ges. allg. angew. Ent. 3, 233(1981).
5. Page, T.L. Science 216, 73 (1982).
6. Page,T.L. J. Comp. Physiol. 152, 231 (1983).
7. Cymborowski,B. and Brady, J. Nature 236, 221 (1972).

8. Cymborowski,B. Interdiscipl. Cycle Res. 121, 133 (1981).

9. Konopka,R.J. and Wells, S.J. Neurobiology 11, 411 (1980).

10. Truman, J.W. J. Comp. Physiol. 95, 281 (1974)

11. Kasai, M. and Chiba, Y. 10th intern. Congress of Biometeorology, Tokyo (1984).

12. Helfrich, C. In preparation.

13. Engelmann, W. and Mack, J. J. Comp. Physiol. 124, 225 (1978).

14. Helfrich, C., Cymborowski, B. and Engelmann, W. To be published.

Interaction Between Two Circadian Clocks in the Rabbit and Their Main Input

A.C. Bobbert and D.J. Bruinvels

Department of Physiology and Physiological Physics, University of Leiden
Wassenaarseweg 62, NL-2333 AL Leiden, The Netherlands

1. Introduction
It was previously shown [review in 1] for the rabbit that photosensitivity of the visual system fluctuates according to a block-shaped circadian rhythm. The time courses of this rhythm after exposure to 24 hrs light-dark alternations or to constant environmental circumstances strongly suggest that this rhythm is governed by a central circadian mechanism with the properties of a double-oscillator. These 2 oscillators may correspond with the Morning- and Evening type oscillators which have been described for hamsters and other rodents [2,3]. This raised the following questions: - does the rabbit show corresponding circadian changes in food intake? If so, is there any indication that this rhythm also is controlled by a double-oscillator? And, in case of the latter, how do these separate oscillators respond to entrainment by light-dark regimens or to constant photic circumstances?

2. Methods
The number of food approaches was automatically recorded every 30 minutes from 13 rabbits which were fed ad lib. initially in LL (200-250 lux) for 3-6 months, thereupon in DD (0 lux) for 7 months and then in 24 hrs LD regimes of 1:23 and 12:12 for 3 months each. Finally, they were blinded, and the result of this was followed for more than 5 months.

3. Results
Figure 1 shows a recording from one specimen which is representative for all of the others. It appears that in rabbits food intake also is governed by a system of 2 circadian oscillators. One of these (marked by arrows in fig. 1) merely induces a dip in the food intake activity and has a freerunning period τ of 24 hrs ± 1 min., neither of which is influenced by the level of retinal outflow. Therefore, this oscillator is called the Illumination Insensitive (I.I.) oscillator. The other induces a more or less complex pattern of food approaches which usually consists of a period of appreciable food intake, with one or two peaks, followed by another dip in the eating activity. This is called the Illumination Sensitive (I.S.) oscillator since its period τ is definitely shorter in LL (mean 24.24 hrs, S.D. ± 0.40, N = 13) than in DD (24.57 hrs, S.D. ± 0.28), while blinding results in a further shortening of τ for the I.S. oscillator, to such an extent that its average value (23.82 hrs, S.D. ± 0.21) was less than 24 hours. Both oscillators are entrained during exposure to a 24 hrs LD 1:23 regimen, while either the I.I.- or the I.S. oscillator becomes time-locked with the 1 hr illumination period. Since in LD 12:12 the I.I. oscillator becomes time-locked with the lights-on moment and the I.S. oscillator with the moment of lights-off (fig. 1), they probably correspond to the M- and E-type oscillators, respectively. The results further show that in rabbits "splitting" occurs only rarely in freerunning states of the food intake rhythm. The latter, nevertheless, is controlled by 2 oscillators which are loosely coupled, since they show a weak degree of "relative internal coordination" (fig. 1).

4. Discussion
The 2 oscillators of the central circadian system are widely different in their responses to illumination and to LD regimes. The differences between the average τ's of the I.S. oscillator in darkness (24.57 hrs), during continuous illumination (24.24 hrs) and after blinding (23.82 hrs) show that the oscillation frequency of one part of the central circadian system is slowed down by the "dark" discharge from

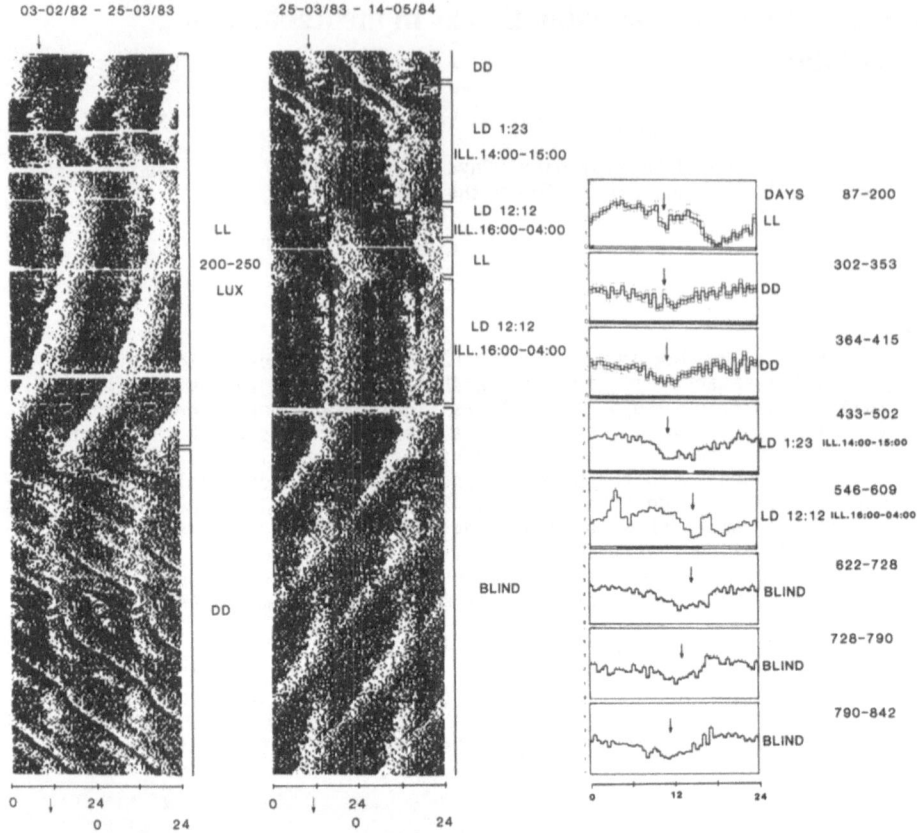

Fig. 1: Double density plot and histograms of the number of food approaches per 30 minutes of a rabbit kept for several months in LL, in DD, in regimes of LD 1:23 and LD 12:12 and, finally, after blinding. Note the occurrence of "relative internal coordination" when the 2 oscillators have a distinct phase-relationship.

the retina, and strongly suggest that the rate of this discharge is fairly high in DD but constantly depressed (inhibited) during 5 months in LL. It is, further, remarkable that τ of the I.I. oscillator is hardly distinguishable from 24 hrs and that it is completely independent of the level of retinal dark-discharge. The differential time-locking of the 2 oscillators with the lights on- and lights-off moments of LD regimens fully explains the previously [1] described occurrence of "programming" by a wide range of LD schedules.

5. References

1. A.C. Bobbert and J. Brandenburg: "Characteristics of the Interaction Between the Central Circadian Mechanism and the Retina in Rabbits". In "Vertebrate Circadian Systems. Ed. by J. Aschoff, S. Daan and G. Groos. Springer Verlag Berlin-Heidelberg-New York, 52-61 (1982)
2. S. Daan and C. Berde: "Two Coupled Oscillators: Simulations of the Circadian Pacemaker in Mammalian Activity Rhythms". J. Theor. Biol, 70, 297-313 (1978)
3. M. Kawato and R. Suziki: "Two Coupled Neural Oscillators as a Model of the Circadian Pacemaker". J. Theor. Biol., 86, 547-575 (1980)

Cell Cycle

Dynamics of Feedback Systems with Time Lag

Hans-Otto Walther

Mathematisches Institut, Universität München, Theresienstraße 39
D-8000 München 2, Fed. Rep. of Germany

1. Here are four examples of autonomous functional differential
equations with time lags which have been used for modelling. First,
the logistic equation with delay in the growth-limiting factor

$$(dn/dt)(t) \equiv \dot{n}(t) = rn(t)[1 - n(t-\tau)/K] \tag{1}$$

proposed by the biologist HUTCHINSON [17] in 1948 in order to explain oscillations which had been observed in population growth experiments. If we set $x(t) := \log(n(\tau t)/K)$ and $\alpha = r\tau$ then positive
solutions transform into solutions of

$$\dot{x}(t) = \alpha f(x(t-1)) \tag{αf}$$

with $f(\xi) = 1 - e^{\xi}$, $\xi \in R$. f is monotonous and has the property

$$\xi f(\xi) < 0 \tag{NF}$$

for all $\xi \neq 0$. - In equations for the density of red blood cells as

$$\dot{x}(t) = [cx(t-\tau)]^8 e^{-x(t-\tau)} - sx(t) \tag{2}$$

the delay stands for the time between start of cell production and
release of the platelets into the blood stream. Models of this type
were constructed by MACKEY and GLASS, LASOTA and WAZEWSKA-CZYZEWSKA
[25,23,24]. Nonlinear functions which are not monotonous as in (2)
will make the dynamics more complicated. See also [7,11,13,22,27,35].
BANKS and MAHAFFY [2,26] developed a model for protein synthesis,

$$\dot{m}(t) = a(1 + kr(t-\tau_1)e(t-\tau_1))^{-1} - \beta_1 m(t)$$
$$\dot{e}(t) = \alpha_2 m(t-\tau_2) - \beta_2 e(t)$$
$$\dot{r}(t) = \alpha_3 e(t-\tau_3) - \beta_3 r(t) \tag{3}$$

The lags take into account times needed for transcription from DNA
to messenger-RNA, and for transport to the ribosomes where substances react. See also [15]. - A model from a different field is

$$\dot{x}(t) = \delta - \sin(x(t-\tau) + \omega), \quad 0 \leq \delta < 1 \tag{4}$$

The equation describes the phase difference x between a frequency
generator in the high frequency range and a controlling oscillator
in a phase-locked loop. The delay is caused by a low pass filter.
Such models were used since the fifties [18]. See also [10].
If time is rescaled and if we perform a shift of the co-

ordinate system, (4) becomes an equation of type (αf) with periodic function f so that (NF) is only satisfied for ξ close enough to zero.

Already the most simple-looking scalar equation (αf) may have a rich structure in its state space, depending on f. This is much in contrast to ordinary differential equations (ODEs), where the more interesting behaviour is found only in systems of 3 or more equations. In the following, I want to describe typical properties of (αf) which have been proved. For more general equations like (2) and (3), and also for equations with distributed delay, much less could be established rigorously up to now.

2. Let $f:R \to R$ be smooth with $f(0) = 0$, with property (NF) and $f'(0) = -1$. Then $t \to 0$ is the only constant solution of (αf), $\alpha > 0$. Equation (αf) describes delayed negative feedback with respect to zero: If $x(t-1) > 0$ then this deviation from equilibrium is followed by $\dot{x}(t) < 0$, a move in the opposite direction after a reaction lag, because of the sign condition (NF). - Equation (αf) is easily solved in forward time when, as an initial value, a continuous function ϕ on $[-1,0]$ is prescribed: For $t \in [0,1]$,

$$x(t) = x(0) + \int_0^t \dot{x}(s)ds = \phi(0) + \int_{-1}^{t-1} \alpha f(\phi(s))ds,$$

$x|[0,1]$ defines x on $[1,2]$ in the same way, and so on. The result is a unique continuous function, defined for $t \geq -1$, differentiable for $t > 0$ and satisfying (αf) for $t > 0$. - In order to get solution curves in the space C of initial values, i.e. of continuous functions on $[-1,0]$, one sets $x_t(a) := x(t+a)$ for $a \in [-1,0]$, $t \geq 0$. The map $0 \leq t \to x_t \in C$ becomes continuous, and the formula $T(t,\phi) = x_t$ with the solution x of (αf) with $x_0 = \phi$ defines a continuous semiflow, or semidynamical system, $T:[0,\infty) \times C \to C$ [14]. T is not everywhere differentiable and does not continue to a local flow. E.g. if ϕ has no derivative at zero, then T has no partial derivative with respect to t at points (t,ϕ), $t \in (0,1)$. Also, there is no solution to this initial value problem in backward time - no function x on $[t_0-1,\infty)$, $t_0 < 0$, which is differentiable and satisfies (αf) for $t > t_0, x_0 = \phi$. - Even if the backward initial value problem can be solved there is no uniqueness in general. Or, two solutions with different initial values may become identical after finite time: Suppose f is decreasing on $(0,\xi)$ and increasing on (ξ,∞). Then, one can choose different initial functions ϕ, ψ, $\phi \leq \xi \leq \psi$, with $\phi(0) = \xi = \psi(0)$ and $f \circ \phi = f \circ \psi$. It follows that the corresponding solutions coincide on $[0,1]$, and consequently for all $t \geq 0$.

The next important difference between (αf) and scalar ODEs is the presence of oscillating solutions. Let K denote the cone of increasing functions $\phi \in C$ with $\phi(-1) = 0$. For $\phi(0) > 0$ condition (NF) forces the solution x with $x_0 = \phi$ to decrease as long as $x(t-1) > 0$. In case $\alpha > 1$ it is not hard to see that x has a first zero $z_1 = z_1(\phi) > 0$. Time lag and condition (NF) make x decrease until time $z_1 + 1$ when the solution realizes that it is no longer positive. Now x starts to increase, reaches a second zero z_2, increases until time $z_2 + 1, \ldots$ [28,45]. We get solutions which are "slowly oscillating" in the sense that the distance between any pair of zeros in $[0, \infty)$ is greater than the delay $\tau = 1$. - By the way, a sharp bound for all solutions of (αf) to oscillate is $\alpha = 1/e$, as one can prove.

Oscillating solutions do not always damp out. For $0 < \alpha < \pi/2$ the zero solution is asymptotically stable but for $\alpha > \pi/2$ it becomes unstable, and no slowly oscillating solution tends to 0 as $t \to \infty$ [28,45]. The critical value $\pi/2$, and $1/e$ above too, is determined by the linearized equation

$$\dot{y}(t) = -\alpha y(t-1)$$

with f replaced by its derivative at $\xi = 0$. As for ODEs, the Ansatz $y(t) = e^{\lambda t}$ leads to a "characteristic equation"

$$\lambda + \alpha e^{-\lambda} = 0$$

the solutions of which decide about stability [14]. There are countably many of these "characteristic values". For $0 < \alpha < 1/e$, they are situated in the complex plane as indicated in the figure below.

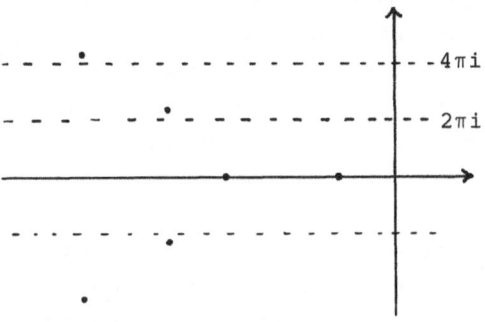

If α passes the value $1/e$ the real characteristic values join and bifurcate into a complex conjugate pair; the other pairs move to the right [45]. If α increases beyond $\pi/2$ the pair with smallest imaginary parts crosses the imaginary axis at $\pm i\pi/2$. At $\alpha = 5\pi/2$, the next pair crosses at $\pm 5i\pi/2, \ldots$ The functions $t \to c e^{t \mathrm{Re}\, \lambda} \sin(t \mathrm{Im}\, \lambda)$, $c \neq 0$, $|\mathrm{Im}\, \lambda| < \pi$, are now slowly oscillating solutions of the linear equation with increasing amplitude. We see that there are also many other, more rapidly oscillating solutions, given by the characteristic values with $2\pi < |\mathrm{Im}\, \lambda|$. For the nonlinear equations, one may now use a Hopf bifurcation theorem in order to obtain small amplitude rapidly oscillating solutions for α close to $5\pi/2$, $9\pi/2, \ldots$ and small amplitude

slowly oscillating periodic solutions for α close to π/2 with periods close to 4 [4,8,9,31].

At this point I have to remark that slowly oscillating solutions
are the only ones which seem to be relevant for applications. A very
natural conjecture [21] says that the set of initial conditions
which produce (eventually) slowly oscillating solutions is open and
dense in the state space C. This is easily seen in the linear case;
it could be proved for some nonlinear functions αf [41], but so far
no proof is available for all equations (αf). The conjecture implies
that any rapidly oscillating periodic solution is unstable, stability can only be expected from slowly oscillating periodic solutions.
Let me write SO for slowly oscillating from now on.

The basic global bifurcation theorem for SO periodic solutions is
due to NUSSBAUM [29]. Note first that every SO periodic solution
contains elements φ ∈ K on its trajectory in C. So one may consider
the set S of pairs (α,φ), α > 0 and φ ∈ K, such that φ defines a SO
periodic solution of (αf) with period z_2 + 1. Let f be bounded from
below or from above. Then NUSSBAUM's result is that there exists a
maximal connected subset S' ⊂ S with (π/2,0) in its closure and with
points (α,φ) for all α > π/2; S' does not return to the trivial axis
(α,0), α > 0, for α ≠ π/2.

In the proof, NUSSBAUM employed tools from asymptotic fixed point
theory, notably his own result on the index of an ejective fixed
point [28]. This is not necessary, and moreover the concept of ejectivity is not really appropriate. It goes back to BROWDER [3] who
developed it as an abstraction from what was known about undamped
behaviour of SO solutions for α > π/2 since a beautiful paper of
WRIGHT (1955,[45]). However, the unstable behaviour of SO solutions
is much more regular than ejectivity predicts. One of the consequences is that NUSSBAUM's result can be proved without any asymptotic
fixed point theory [39,40].

So there are SO periodic solutions for every α > π/2. Nothing can
be said about the finer structure of the set S in general. S and S'
do not necessarily decompose into smooth curves. Many stable and unstable SO periodic solutions may coexist for fixed values of α. SO
periodic solutions for α ≤ π/2 are not excluded. Continua in S disjoint from S' are possible, as well as initial values in K which
produce periodic solutions of (αf) with periods z_4, z_6, ... so that
(α,φ) ∉ S. Numerical results of JÜRGENS, PEITGEN, SAUPE [19], HADE-
LER [12], SAUPE [37] provide evidence that such possibilities are
realized by functions f which are related to models like (1) and (4).

What is proved ? For a set of decreasing _odd_ functions, $f(\xi) = -f(-\xi)$ for all ξ, NUSSBAUM [32] showed uniqueness for SO periodic solutions with x_0 in K, $\alpha > \pi/2$. Under his conditions there are no SO periodic solutions for $\alpha \leq \pi/2$; the SO periodic orbits for $\alpha > \pi/2$ are all asymptotically stable, all periods are 4, and the symmetry property

$$x(t) = -x(t - (z_1+1)) \quad \text{for all } t \in R \qquad\qquad (\sigma)$$

holds with $z_1 = 1$. Let Sp denote the set of pairs (α,ϕ) so that ϕ is

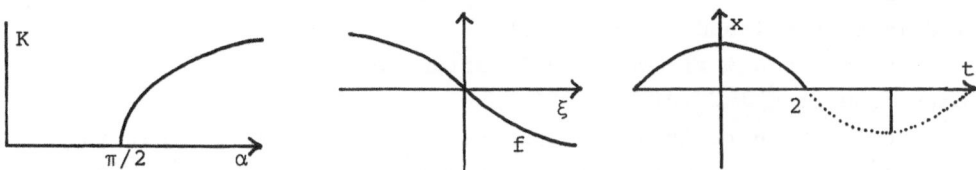

in K and defines such a "special periodic solution" with period 4 and symmetry (σ). Existence of special periodic solutions for odd f had been proved earlier by KAPLAN and YORKE [20]. The uniqueness proof is by refinements of a phase-plane method for monotonous f also due to KAPLAN and YORKE [21,22].

It might seem surprising that uniqueness and stability could not yet be proved for the monotonous function $f:\xi \to 1 - e^\xi$ from HUTCHINSON's equation (1).

First examples of nonuniqueness came from results on the direction of bifurcation at $\alpha = \pi/2$. If, say, $f' < -1$ for all $\xi \neq 0$ close to zero then S' leaves the point $(\pi/2,0)$ in direction of decreasing α [38]. Work of CHOW and MALLET-PARET [8] leads to similar diagrams.

Further nonlinear monotonous functions with multiple SO periodic solutions were constructed by ANGELSTORF [1].

Numerical experiments with equations like (2) and (4) suggested that nonmonotonous functions f define very complicated sets S. For certain _odd_ functions with hump-shaped graph NUSSBAUM constructed a continuum $L \subset S$ disjoint from Sp [32,33]. In cases considered Sp forms a connected set which bifurcates from $(\pi/2,0)$, so $Sp \subset S'$. Solutions which belong to L share the symmetry (σ) with the special periodic solutions but have minimal periods larger than 4.

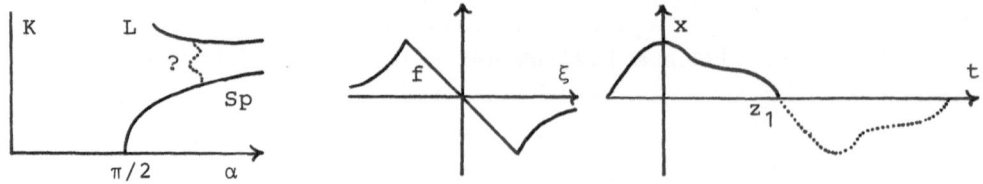

Note that Sp and L may or may not belong to the same connected component S' of S. This is an unsolved problem. SAUPE found a trick how
to transform L into a third continuum, where periodic solutions have
period less than 4 and symmetry (σ).

For another class of odd hump functions where Sp is a smooth curve emanating from $(\pi/2,0)$, I could show existence of a secondary bifurcation [43]: There is a critical parameter $\gamma > \pi/2$ so that for
$\pi/2 < \alpha < \gamma$, the (unique) special periodic solution $x = x^\alpha$ with x_0^α
in K is asymptotically stable. For $\alpha > \gamma$ (and α not too large),
x^α is unstable, and there is a nontrivial continuum $B \subset S$ with
$B \cap Sp = \{(\gamma, x_0^\gamma)\}$ such that periodic solutions defined by points
$(\alpha, \phi) \in B \smallsetminus \{(\gamma, x_0^\gamma)\}$ do not have the symmetry (σ).

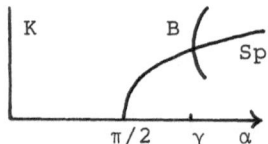

It is not clear whether for the functions f considered here an analogue of NUSSBAUM's continuum L exists. In any case, the bifurcating
periodic solutions would not belong to it. - The bifurcation result
applies to the model (4) with $\delta = \omega = 0$. Of course, $f = -\sin$ satis-
fies (NF) only for $0 < |\xi| < \pi$. But all relevant periodic solutions
are bounded by π which allows to incorporate this application here.

The complete list of hypotheses for the results described above
creates the impression that bifurcation diagrams depend in a rather
subtle way on the shape of f. This is confirmed by work of NUSSBAUM
(and POTTER, CHAPIN) on the dependence of periods on α and f [30,32,
33,34,5,6]. Therefore it is not always easy to make a correct guess
about the dynamics on the basis of numerically obtained bifurcation
diagrams, or from rigorous results for (αf) with f a step function.
In this last case, one can calculate bifurcation diagrams completely
(if there are not too many steps). Such a program was carried out by
PETERS [36].

3. A remark about functional differential equations and chaotic interval maps: When MACKEY and GLASS proposed equations like (3) as models for certain physiological control processes, they also expected chaotic trajectories for suitable parameters. This could be proved for a few rather special equations. PETERS [36] did it for step functions as nonlinearities, AN DER HEIDEN and I showed it for equations like (2) and (4) with nonlinearities which are smooth but constant on long intervals, like step functions [16,42]. The proofs establish and use that chaotic interval maps represent the dynamics of these special functional differential equations precisely, in a "thin" region in state space. This should be compared to the simple fact that chaotic interval maps with their hump-shaped graphs are never directly related to decent ODEs - Poincaré maps and time-one-maps of ODEs always being one-to-one.

4. Coexistence of stable equilibrium and periodic motion for the phase-locking equation. - If condition (NF) for (αf) does not hold globally, like in (4), or if equations like (2) are considered, then it is not clear how to single out a predominant class of solutions as the slowly oscillating ones above. Other phenomena come into play. Let me finally describe a result which I obtained recently for model (4). Consider

$$\dot{x}(t) = \alpha g(x(t-1))$$

with g a shifted sinus function, $g(0) = 0$, $g'(0) < 0$.

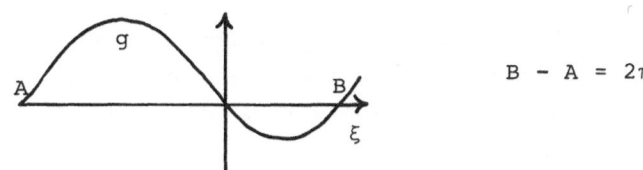

$$B - A = 2\pi$$

In a range of parameters $\alpha > 0$ where the zero solution is asymptotically stable there exists a critical parameter $\tilde{\alpha}$, such that for $\alpha > \tilde{\alpha}$ (and α not too large) periodic solutions of the second kind bifurcate [44]. - A periodic solution of the second kind with period $p > 0$ is a solution $x: R \to R$ where $x(t + p) = x(t) + 2\pi$ for all t in R; the graph looks like a staircase. If real numbers modulo 2π are identified, then x defines a periodic map with values on the circle $R/2\pi Z \cong S^1$; after one period p a full turn around the circle is completed. The words "of the second kind" are meant to distinguish this from "small" periodic oscillations on the circle which do not wind around during a period. For example, SO periodic solutions with va-

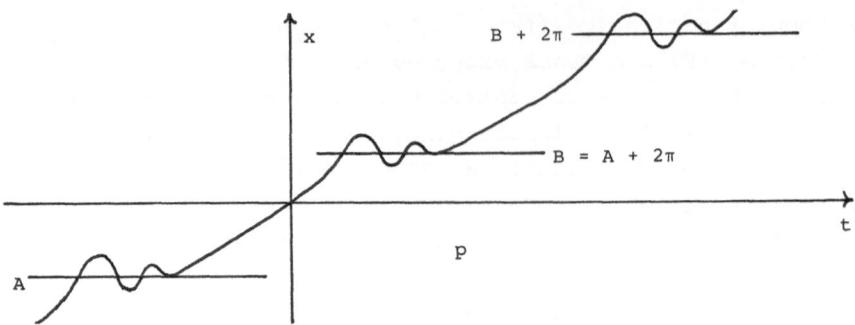

lues in (A,B) would define periodic solutions "of the first kind" on
the circle. - Recall also that for the model x is interpreted as a
phase difference so that only values modulo 2π are of interest.

   The periodic solutions of the second kind arise in a nonlocal bifurcation from a sequence of heteroclinic solutions which exist for
the critical parameter $\alpha = \tilde{\alpha}$, i.e. from a solution $\tilde{x}$ with $\lim_{t \to -\infty} \tilde{x}(t)$
= A and $\lim_{t \to \infty} \tilde{x}(t)$ = B and its translates $\tilde{x} + 2\pi Z$, and the periods
tend to infinity for $\alpha \to \tilde{\alpha}$.

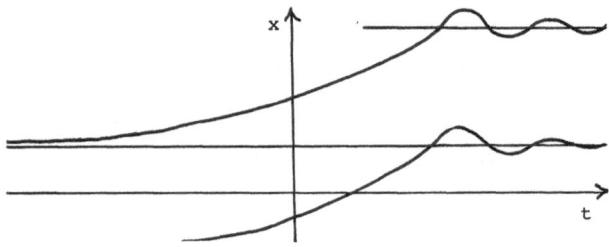

Most likely the bifurcating periodic solutions of the second kind
are asymptotically stable,so that one would have coexistence of stable periodic motion with the stable equilibrium zero in the model.
However, stability is not yet proved.

5.   References. There are a large number of interesting contributions to the dynamics of autonomous differential delay equations.
The list below only contains papers used for the preceding article.

1 N. Angelstorf: Spezielle periodische Lösungen einiger autonomer
      zeitverzögerter Differentialgleichungen mit Symmetrien.
      Ph.D. thesis, Bremen 1980
2 H.T. Banks, J.M. Mahaffy: Mathematical models for protein bio-
      synthesis. LCDS, Div. of Appl. Math., Brown U., Providence
      R.I., 1979

3 F.E. Browder: Duke Math. J. <u>32</u>, 575 (1965)

4 N.N. Chafee: J. Math. Analysis Appl. <u>35</u>, 312 (1971)

5 S. Chapin: Periodic solutions of some nonlinear differential delay equations. Ph.D. thesis, Michigan State U. 1983

6 S. Chapin, R.D. Nussbaum: Asymptotic estimates on the periods of
   periodic solutions of a differential delay equation. Michigan
   Math. J., to appear

7 S.N. Chow: J. Differential Equations <u>15</u>, 350 (1974)

8 S.N. Chow, J. Mallet-Paret: J. Differential Equations <u>26</u>, 112
   (1977)

9 S.N. Chow, J. Mallet-Paret: J. Differential Equations <u>29</u>, 66
   (1978)

10 T. Furumochi: Tohoku Math. J. <u>30</u>, 13 (1978)

11 K.P. Hadeler: Math. Meth. in the Appl. Sci. <u>1</u>, 62 (1979)

12 K.P. Hadeler: Numer. Math. <u>34</u>, 457 (1980)

13 K.P. Hadeler, J. Tomiuk: Arch. Rat. Mech. Analysis <u>65</u>, 87 (1977)

14 J.K. Hale: <u>Theory of Functional Differential Equations</u> (Springer
   New York, 1977)

15 U. an der Heiden: J. Math. Analysis Appl. <u>70</u>, 599 (1979)

16 U. an der Heiden, H.O. Walther: J. Differential Equations <u>47</u>,
   273 (1983)

17 G.E. Hutchinson: Annals of the New York Academy of Sciences <u>50</u>,
   221 (1948)

18 Z.J. Jelonek, C.I. Cowan: Proc. Instn. Elect. Engrs. <u>104C</u>, 388
   (1957) (IEE Monograph No 229R, March 1957)

19 H. Jürgens, H.O. Peitgen, D. Saupe: Topological perturbations in
   the numerical study of nonlinear eigenvalue and bifurcation
   problems, in <u>Analysis and Computation of Fixed Points Proceedings Madison 1979,</u>)S.M. Robinson ed. (Academic, New York
   1980)

20 J.L. Kaplan, J.A. Yorke: J. Math. Analysis Appl. <u>48</u>, 317 (1974)

21 J.L. Kaplan, J.A. Yorke: SIAM J. Math. Analysis <u>6</u>, 268 (1975)

22 J.L. Kaplan, J.A. Yorke: J. Differential Equations <u>23</u>, 293 (1979)

23 A. Lasota: Asterisque <u>50</u>, 239 (1977)

24 A. Lasota, M. Wazewska-Czyzewska: Mat. Stosowana <u>6</u>, 23 (1976)

25 M.C. Mackey, L. Glass: Science <u>197</u>, 287 (1977)

26 J.M. Mahaffy: Modelling and analysis of cellular control in protein synthesis. Ph.D. thesis, Brown U., Providence R.I., 1979

27 J. Mallet-Paret, R.D. Nussbaum: Global continuation and complicated trajectories for periodic solutions of a differential
   delay equation, to appear in Proceedings of the Summer 1983
   AMS Conference on Nonlinear Functional Analysis Berkeley Ca.

28 R.D. Nussbaum : Ann. Mat. Pura Appl. $\underline{101}$, 263 (1974)

29 R.D. Nussbaum : J. Functional Analysis $\underline{19}$, 319 (1975)

30 R.D. Nussbaum : J. Math. Analysis Appl. $\underline{58}$, 280 (1977)

31 R.D. Nussbaum : Trans. AMS $\underline{238}$, 139 (1978)

32 R.D. Nussbaum : J. Differential Equations $\underline{34}$, 25 (1979)

33 R.D. Nussbaum : Arch. Rat. Mech. Analysis $\underline{81}$, 373 (1983)

34 R.D. Nussbaum : J.B. Potter: J. Differential Equations $\underline{46}$, 379 (1982)

35 Ya.B. Pesin: Differentsial'nye Uravneniya $\underline{10}$, 1025 (1974). (In Russian)

36 H. Peters: Globales Lösungsverhalten zeitverzögerter Differentialgleichungen am Beispiel von Modellfunktionen. Ph.D. thesis, Bremen 1980

37 D. Saupe: Beschleunigte PL-Kontinuitätsmethoden und periodische Lösungen parametrisierter Differentialgleichungen mit Zeitverzögerung. Ph.D. thesis, Bremen 1982

38 H.O. Walther: J. Differential Equations $\underline{29}$, 396 (1978)

39 H.O. Walther : On instability, ω-limit sets and periodic solutions of nonlinear autonomous differential delay equations, in Functional Differential Equations and Approximation of Fixed Points Proceedings Bonn 1978, H.O. Peitgen and H.O. Walther eds. (Lecture Notes in Math. 730, Springer, Berlin 1979)

40 H.O. Walther : Delay equations: instability and the trivial fixed point's index, in Abstract Cauchy problems and functional differential equations, F. Kappel and W. Schappacher eds. (Research notes in math. 48, Pitman, Boston 1981)

41 H.O. Walther : J. Math. Analysis Appl. $\underline{79}$, 127 (1981)

42 H.O. Walther: J. Nonlinear Analysis $\underline{5}$, 775 (1981)

43 H.O. Walther: Math. Z. $\underline{182}$, 269 (1983)

44 H.O. Walther : Bifurcation from a heteroclinic solution in differential delay equations, submitted 1984

45 E.M. Wright: J. Reine Angew. Math. $\underline{194}$, 66 (1955)

The Coordination of Cell Growth and Division: A Comparison of Models

John J. Tyson

Department of Biology, Virginia Polytechnic Institute and State University
Blacksburg, VA 24061, USA

1. Introduction

During steady-state, balanced exponential growth of cells in culture,
all extensive properties of the culture increase exponentially at the
same specific growth rate, k. That is,

$$G(t) = G_o \, e^{kt},$$

where G = total content of DNA, RNA, protein, lipids,
polysaccharides, etc., or total number of cells, nuclei, ribosomes,
mitochondria, etc. Although these macroscopic descriptors of the
growth of a cell culture increase in a deterministic fashion, the
growth and division of individual cells is highly variable. For
instance, the ages of cells at division are broadly distributed, with
a coefficient of variation (CV = standard deviation/mean) of
typically 20%. Size at division is also quite variable: CV ≈ 10%.

As cells in culture approach the steady-state of balanced
exponential growth, the distributions of age and size at division
settle down to characteristic forms. That is, the distributions are
unique and globally asymptotically stable. The shapes of these
distributions may give us some hints as to the mechanisms underlying
the coordination of cell growth and division during the phase of
exponential proliferation of a cell culture. A number of different
models of this control mechanism have been suggested, and it is our
purpose here to classify and compare some of these models.

2. Some Models

The simplest model of progress through the cell division cycle is the
deterministic (Det) model, which supposes that cells divide when they
reach a critical size (or age). This model, admitting no variability
in size or age at division, is manifestly incorrect but nonetheless
useful as a baseline.

Slightly more complicated is the transition probability (TP) model
of SMITH and MARTIN [1], which postulates two phases of the cell
cycle: A-phase, a stochastic phase during which cells make no
progress toward division and from which they exit at random with
constant probability per unit time (p); and B-phase, a deterministic
phase during which cells synthesize DNA, progress through G_2 and
divide after a constant time delay (T_B).

The two-transition-probabilities (TTP) model of BROOKS, BENNETT
and SMITH [2] introduces two sequential random-exiting phases (A and
Q) as well as the deterministic B-phase.

In the tandem (Tan) model, investigated most thoroughly by TYSON
and HANNSGEN [3], there is a critical size requirement for entry into

the random-exiting phase. That is, any cell of size < 1 at birth
must grow to size = 1 before entering A-phase. Cells leave A-phase
at random (p = probability per unit time) and proceed through B-phase
before dividing (T_B = constant duration of B-phase).

The sloppy-size control (SS) model, investigated most thoroughly
by DIEKMANN and collaborators [4, 5], supposes that variability in
progress through the cell cycle comes about because the determination
of cell size is inherently sloppy. That is, cells are programmed to
divide at a critical size,but there is some variability in the size-
monitoring mechanism. Thus, there is some probability per unit time
that a cell of size x will divide, call it b(x). The function b(x)
should be zero for small x and should increase rapidly for large x.

3. Some Definitions

To compare these five models we must introduce some definitions and
notation. For the most part, we follow the suggestions of PAINTER
and MARR [6]. First of all, for the random variable t = time from
birth to division = generation time = age at division, let

   f(t) = probability density function for age at division, defined
          on a sample of newborn cells.

   $F(t) = \int_o^t f(s)ds$ = distribution function for age at division.

   $\alpha(t) = \int_t^\infty f(s)ds$ = survivorship function for age at division
          (called the "alpha curve" by cell kineticists).

Similarly, for cell size at division,

   $\phi(x)$ = probability density function for size at division, defined
          on a sample of dividing cells.

   $\Phi(x)$ = distribution function for size at division.

   $\wp(x)$ = survivorship function for size at division.

Another size distribution of importance is

   $\lambda(x)$ = probability density of present size, defined on a sample of
          extant cells. (This is the size distribution measured by a
          Coulter counter.)

   Now we can define two important hazard functions:

   p(t) = f(t)/α(t) = probability per unit time that a cell of age t
          will divide,

and

   b(x) = kϕ(x)/λ(x) = probability per unit time that a cell of size
          x will divide.

(Recall that k = specific growth rate of the population as a whole.)
For the TP model, p(t) = p = transition probability out of A phase.
For the SS model, b(x) is the division probability function defining
the sloppiness of the size-monitoring mechanism.

4. Comparison of Models

In Fig. 1 we compare the five models of Sec. 2 in terms of the age-
hazard function p(t) and the size-hazard function b(x). In Fig. 2 we
compare the models in terms of generation time distributions, and in

292

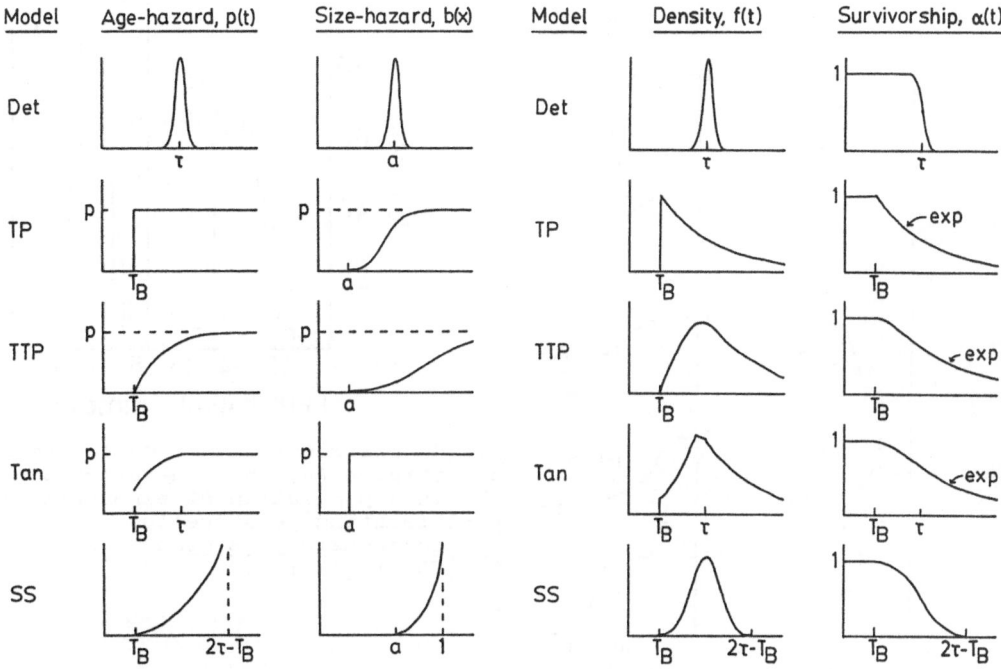

Model	Age-hazard, p(t)	Size-hazard, b(x)	Model	Density, f(t)	Survivorship, α(t)
Det			Det		
TP			TP		
TTP			TTP		
Tan			Tan		
SS			SS		

Fig.1 Comparison of models. The models described in Sec. 2 are compared with respect to their age-hazard function, p(t), and size-hazard function, b(x). Parameters p and T_B are defined in Sec. 2; a = minimum size at division, τ = ln2/k = population doubling time.

Fig.2 Distribution of age at division. For each model we sketch the probability density for age at division, f(t), and the associated survivorship function, α(t). ("exp" means "exponential tail", i.e., this portion of the α-curve would be a straight line if plotted semilogarithmically.)

Fig. 3 in terms of division size distributions. The figures have been constructed from the equations in references [3, 5, 7].

In Fig. 4 we reproduce the results of MIYATA et al. [8] on the size and age distributions at division in an exponentially expanding population of fission yeast. Comparing the histograms in Fig. 4 with the probability density functions in Figs. 2 and 3, we find that:

a) the deterministic model is grossly in error, as expected.

b) the transition probability model looks reasonable for the size distribution but predicts a definitely mistaken discontinuity in the age distribution. This failure of the TP model has been recognized for some time.

c) the two-transition-probabilities model predicts reasonably shaped distributions of both age and size. However, it can easily be shown that, for the generation time distribution predicted by the TTP model,

    Minimum > Mean - $\sqrt{2}$(Std. Dev.).

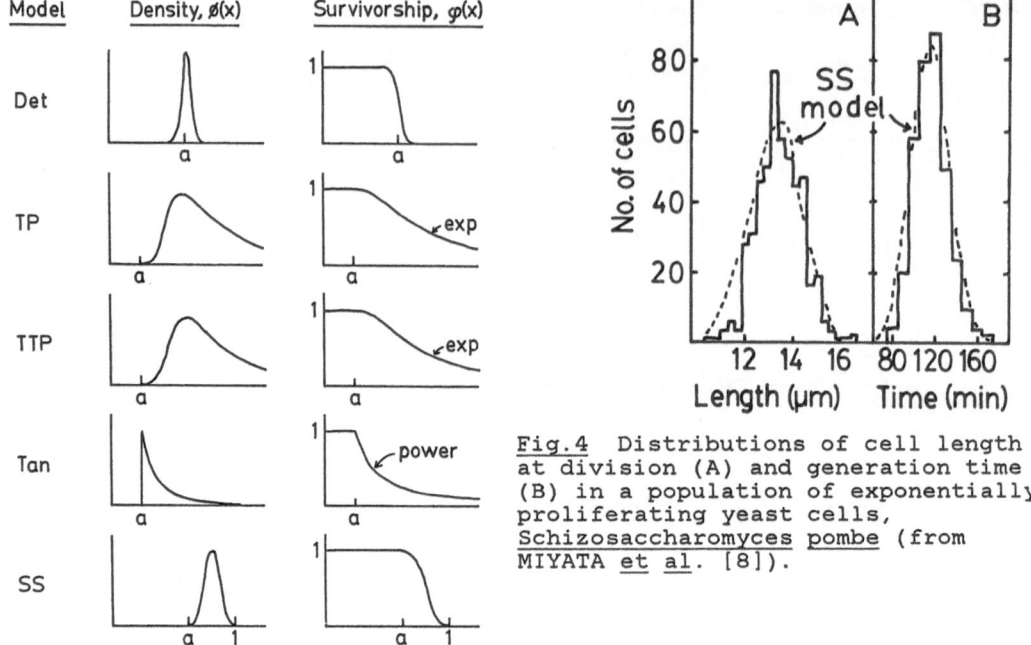

Fig.3 Distribution of size at division. For each model we sketch the probability density for size at division, $\phi(x)$, and the associated survivorship function, $\varphi(x)$. ("power" means "power function tail", i.e., this portion of the φ-curve would be a straight line if plotted double-logarithmically.)

Fig.4 Distributions of cell length at division (A) and generation time (B) in a population of exponentially proliferating yeast cells, Schizosaccharomyces pombe (from MIYATA et al. [8]).

The fission yeast data does not satisfy this requirement (Minimum gen. time = 75 min., mean gen. time = 116 min., Std. dev. = 16 min.), so the TTP model cannot quantitatively reproduce the age distribution in Fig. 4.

d)   the tandem model looks reasonable for the age distribution but predicts a definitely mistaken discontinuity in the division size distribution. This failure results from the combination of a precise size-monitoring mechanism and an exponentially distributed random-exiting phase of the cell cycle.

e)   The sloppy size control model predicts reasonably shaped distributions of both age and size. Indeed, the SS model reproduces quite well the histograms constructed by MIYATA et al. (see Fig. 4).

5.   Conclusion

Models of progress through the cell cycle may conveniently be classified in terms of the characteristic shapes of their age-hazard and size-hazard functions, Fig. 1. (See LORD and WHEALS [9, Fig. 7] for a preliminary but inexact attempt at such a classification.) We have focussed on five relatively simple models: the deterministic model, the transition-probability model, the two-transition-probabilities model, the tandem model, and the sloppy size control model. Figures 2 and 3 compare those five models in terms of predicted shapes of the distributions of age at division and size at division in exponentially expanding cell cultures. Only the sloppy size control gives quantitatively accurate fit to both age and size distributions in a fission yeast culture.

Neither the age distribution nor the size distribution alone provides a particularly stringent test of models of the cell cycle, since several quite different models may be equally successful in reproducing either distribution separately. However, that a model accounts for both distributions simultaneously seems to be a stringent test. The sloppy size control model is most successful in this regard. To test and refine the model further requires additional experimental information. For instance, the generation times of sister cells are typically highly positively correlated (r_{ss} between +0.2 and +0.8) but the generation times of mother and daughter cells typically show little or no significant correlation ($r_{md} \cong 0$). However, the SS model, under the assumption that single cells grow exponentially, predicts $r_{ss} = +0.5$ and $r_{md} = -0.5$. It is not yet clear how to modify the SS model to correct this discrepancy.

There is considerable work yet to be done on both the theoretical and experimental study of cell cycle control. Theoreticians need to analyze more complex (and hopefully more realistic) models, with the express aim of comparing theoretical predictions with experimental data. However, progress on the theoretical front is hampered by a lack of thorough experimental studies of growing cell cultures. Most published experimental results are limited to either generation time distributions or cell size distributions. More comprehensive studies, like that reported in Fig. 4, are needed. From the same data used to construct age and size histograms it is also possible to determine correlation coefficients of generation times of related cells, and to obtain some information about growth laws for individual cells. This added information would be helpful in constructing and testing the next generation of models of the coordination of cell growth and division.

Acknowledgments

This work was supported by grants from the National Institutes of Health (USA), GM 27629, and the National Science Foundation (USA), MCS-8301104, to JJT, and by a grant from the Science and Engineering Research Council (Great Britain), GR/C/53595 to the Centre for Mathematical Biology, Oxford University.

References

1. J. A. Smith, L. Martin: Proc. Natn. Acad. Sci. USA 70, 1263 (1973)
2. R. F. Brooks, D. C. Bennett, J. A. Smith: Cell 19, 493 (1980)
3. J. J. Tyson, K. B. Hannsgen: J. Theor. Biol. (in press)
4. O. Diekmann, H. A. Lauwerier, T. Aldenberg, J. A. J. Metz: J. Math. Biol. 18, 135 (1983)
5. J. J. Tyson, O. Diekmann: Sloppy Size Control of the Cell Division Cycle (in preparation)
6. P. R. Painter, A. G. Marr: Ann. Rev. Microbiol. 22, 519 (1968)
7. K. B. Hannsgen, J. J. Tyson, L. T. Watson: SIAM J. Appl. Math. (submitted for publication)
8. H. Miyata, M. Miyata, M. Ito: Cell Struct. Funct. 3, 39 (1978)
9. P. G. Lord, A. E. Wheals: J. Cell Sci. 50, 361 (1981)

Cellular Synergetics: Cell-Density Dependent Regulation of Population Dynamics of Mammalian Cells in Vitro

Gerold Adam, Beate Kleuser, Klaus Seuwen, and Ulrich Steiner

Fakultät für Biologie, Universität Konstanz, Postfach 5560
D-7750 Konstanz, Fed. Rep. of Germany

1. Introduction

Mammalian cell proliferation depends essentially on intercellular interactions, provision of humoral growth regulators, and attachment to the extracellular growth substrate. In order to be able to control and manipulate separately these essential interactions of the cell with its environment, profitable use has been made of culturing mammalian cells in vitro. If the cells are cultured in vitro on a defined and reproducible growth substrate, and fresh medium containing humoral growth factors is supplied sufficiently often, the external environment of the cell population may be approximated as constant. Under these conditions, the population dynamics of the cells in a good approximation may be considered as determined by intercellular interactions [1]. Taking a more general view, the systems considered presently are composed of (largely) identical subunits (the cells), which can exist in different states, the balance of subsystems in these different states being determined by a constant influx of energy (here: of nutrients and growth factors), as well as by effects of specific interactions between the subunits. Systems of this general nature are encountered in very different fields of physics, chemistry, biology, and sociology [2]. Due to the interactions between subunits, such systems often exhibit quite interesting population dynamics, which in many cases can be described quantitatively by theoretical approaches resembling each other, and are the subject of a new field termed "synergetics" [2]. In this sense, the nonlinear population dynamics studied here and in earlier work [1,3] may be designated as "cellular synergetics". The aim of applying the inherently quantitative approach of synergetics to cellular systems is to arrive at some contribution to elucidation of such long-standing problems in the biology of multicellular organisms as cell-density dependent inhibition of proliferation and its counterpart i.e. unrestrained growth of tumor cells [1, this work], or terminal differentiation and aging of cells [3,4], which are felt to stand in need of a quantitative population dynamical analysis.

2. Essentials of a Model Describing Cellular Population Dynamics

In order to convey some insight into the complex phenomena of population dynamics of cellular systems, any quantitative description has to be based on a model incorporating some mechanistic features. Thus, we have to take into account the basic experimental facts on regulation of the cell cycle, and characteristics of cellular subpopulations in different proliferative states. This experimental information will be listed briefly as follows:

i) Untransformed cells that have entered S-Phase will continue the DNA-cycle until mitosis, unless they are subjected to extremely adverse (artificial) conditions. They are growth-arrested after completing mitosis, maintaining a DNA content corresponding to G1 state of the cell cycle.

ii) The proliferation-arrested state of the cells is biochemically different from G1 state [5]; in particular ribosomal RNA (rRNA) metabolism is shut down, resulting in a lower rRNA content of the resting cell as compared to a cell in G1 state [6,7].

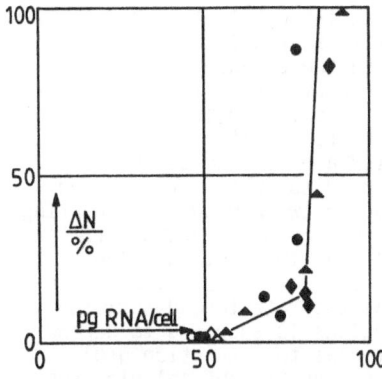

Fig. 1 Relation between relative rise ΔN of
3T3 cell-density (per cent of control) and
cellular content of ribosomal RNA (in picogram
per cell) for cells stimulated from saturation density by a full competence-treat-
ment and differently efficient treatments by
progression factors. Different symbols indicate different independent experiments; open
symbols referring to unstimulated controls.
At 15h after stimulation, cellular RNA was
extracted by alkaline hydrolysis and determined by UV-spectrometry using the two-wave-
length method, given by A. Fleck, H.N. Munro:
Biochim. Biophys. Acta 55 571 (1962). Cell
numbers were taken 42-45h after stimulation
(using a Coulter counter).

iii) In order to trigger proliferation-arrested cells into the cell cycle, two
distinct functions have to be elicited simultaneously: "competence" and "progression". In fibroblast cells, such as 3T3 cells, these two functions are stimulated
by different classes of growth factors and are characterized by different kinetics
[8].

iv) In proliferation-arrested cells, only progression-stimulating factors, but
not competence-stimulating factors, give rise to an increase of cellular rRNA content [9,10].
A most pertinent problem in this context is the nature of a sufficient progression
signal. It was inferred earlier [10] that a threshold of cellular rRNA has to be
surpassed in order for a competent cell to enter the cell cycle. Figure 1 shows
direct evidence for this contention. Here, 3T3 cells were grown to saturation in
Dulbecco's modified Eagle medium with 2,5% newborn calf serum (medium renewal 3x
weekly). At least 4 days after reaching saturation density, cells were fully com-
petence-stimulated by treatment for 4h with 15% fetal calf serum (FCS) and subsequently stimulated continuously with different combinations of progression factors
(medium without serum, 20ng/ml epidermal growth factor=EGF, 5μM, 10μM, or 20μM
insulin=I, 20ng/ml EGF+10μM I, 15% FCS) resulting in cellular rRNA contents, which
increase monotonously in this sequence. Interestingly, the proliferation response
of the (competent) cells depends on the rRNA contents, with a characteristic exhibiting a threshold at about 80pg rRNA/cell (see Fig. 1).
Thus, the progression signal required for a competent cell to enter the cycle may
be operationally-equated, with its rRNA content exceeding a given threshold value.

v) Under essentially all conditions of cell density-dependent growth, the progression function is dominated by the competence function:
- cells are activated by competence factors with regard to their sensitivity to
  progression factors [11],
- in serum-based cell growth in vitro, usually the supply of competence factor in
  the serum is limiting the cell number [12,13],
- the progression function is shut down if competence-induction is missing for
  about 18h [7-10].
Thus, in essentially all conditions of cell density-dependent proliferation, the
supply with competence factors is decisive with regard to the stimulative effect
of the cellular environment.

vi) Competence factors act on a cell during all phases of the division cycle,
and by virtue of the appreciable life time of the competence function (≈18h) may
even determine the proliferative status of the daughter cells of a parental cell
treated with competence factors [14]. Nevertheless, the decision of an individual
cell to enter the cycle or to go into the resting state is undergone at some specific position in the cell cycle normally a few hours after mitosis [14].

3. Population-dynamical Model

On the basis of the experimental knowledge outlined above, the total cell density N
may be divided into two subpopulations representing the cells "in cycle" of den-

sity D and the "longterm growth-inhibited" cells of density Q [1]. This distinction of cellular subpopulations,with respect to proliferative state,clearly is a considerable simplification,but may be considered to catch the essence of the situation on the level of the individual cell, as described above under i), ii), and v). Accordingly, the "D cells" are not defined as the cells committed to cell division, but rather include those cells in G1 state that prepare for commitment by reaccumulating rRNA and by further activities. The "Q cells" are defined as the cells having switched to a resting metabolism (e.g. with regard to rRNA synthesis and degradation [10]),thus responding to an adverse balance between growth-stimulating effects (according to v , under usual conditions mainly exerted by competence factors) and the growth-inhibiting effect of intercellular interactions. The growth-stimulating effects under frequent medium renewals may be described by the rate constant R (independent of cell density), whereas the rate-term describing the growth-inhibitory effects due to cellular interactions is supposed to depend on cell density,and thus,may be formally set proportional to a function g(N). Accordingly,the population-dynamical equations for the densities of cellular subpopulations D and Q, as well as for the total cell population N=D+Q read:

$$\frac{dD}{dt} = R(1-2g)D + W \qquad\qquad (1)$$

$$\frac{dQ}{dt} = 2RgD - SQ - W \qquad\qquad (2)$$

$$\frac{dN}{dt} = RD - SQ \qquad\qquad (3)$$

Here, indiscriminate death of D-cells is considered to be absorbed into the term RD, that of Q cells is given by SQ. The rate-term of transition of D-cells into the Q-cell compartment is formulated as 2RgD, because two daughter cells of a D-cell after mitosis face the decision to enter the Q-cell subpopulation [14] with a relative probability g(N). In equs. (1) and (2), we have admitted a rate-term W for return of Q-cells to the D-cell subpopulation. This term is negligible a few days after saturation density is reached, where less than 5% of the cell population is proliferating. However, this term clearly is of great importance right after reseeding of a cell population predominantly in the Q-state into fresh medium at low cell density [1,7]. In the exponential phase of growth again it is negligible, as nearly 100% of the cells are D-cells. Only in the region of slowing rate of cell division and at early saturation density may this term be regarded as significant, and be expected to contribute to the observed increase of intermitotic times in that time interval [15]. For the present purposes,we shall neglect these finer details and, thus, drop the term W in equs. (1) and (2).

4.   Evaluation of the Cellular Interaction Characteristic g(N) from Experimental Data

In Fig. 2 are shown the results of an experimental analysis [1] of subpopulations D and Q of 3T3 mouse cells. From these data the cell density-dependent interaction function g(N) may be evaluated by the following procedure. Combining equs. (2) and (3) and using W=0 yields:

$$2g = \frac{dQ}{dN}(1 - \frac{SQ}{RD}) + \frac{SQ}{RD} \qquad\qquad (4)$$

Up to about saturation density, cell death SQ is negligible compared to RD, so that we can use as a good approximation for the rising part of N(t) [1]:

$$2g = dQ/dN = \Delta Q/\Delta N \qquad\qquad (5)$$

Accordingly, we have evaluated the quotient $\Delta Q/\Delta N$ for consecutive daily time intervals from the experimental data of Fig. 1 and correlated with the logarithmic averages $N = (N_t \times N_{t+1d})^{1/2}$. The results of this procedure are shown in Fig. 3 (asterisks). Results of the same procedure applied to another independent population-dynamical experiment on 3T3 cells [1] are given also in Fig. 3 (open circles). As will be substantiated in the Discussion, we have chosen in Fig. 3 the represen-

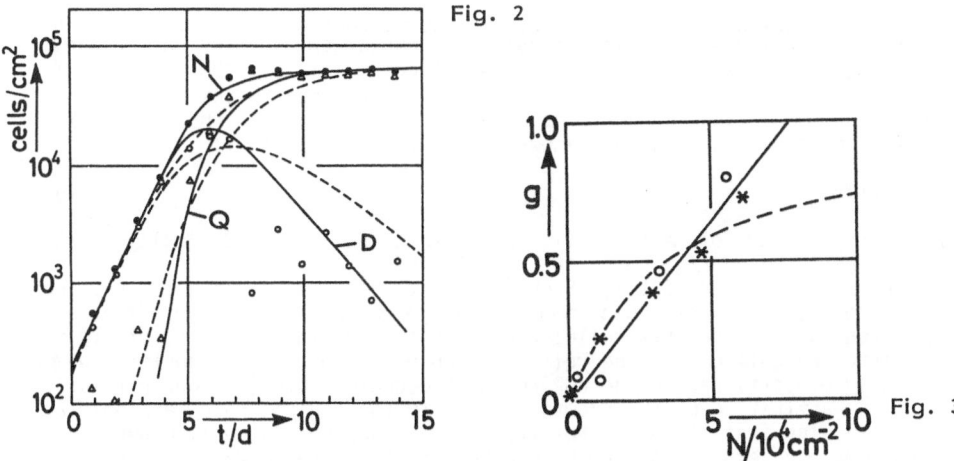

Fig. 2

Fig. 3

Fig. 2 Densities of cellular subpopulations of different proliferative status versus time t of growth of 3T3 cells in medium containing 5% newborn calf serum with daily medium renewals. Full circles represent total cell density N, open circles the proliferative subpopulation D (determined as the fraction dividing within one day of 100-150 cells, reseeded to clonal density from the primary population in order to check their proliferative capacity), triangles the growth-inhibited subpopulation Q (requiring more than one day for division after testing the proliferative status by reseeding to clonal density). Curves are computed by equs. (1) and (2) using $W=S=0$, $R=1/d$, and $g(N)=N/L$ with $L=77,000/cm^2$ (full curves) or $g(N)=N/(1+N)$ with $l=34,000/cm^2$ (broken curves)

Fig. 3 Plot of cellular interaction function g versus total cell density N, evaluated from two independent experiments as described in the text. Asterisks refer to the experimental data shown in Fig. 2, open circles refer to another experiment (not shown). The full curve is computed according to $g=N/L$, with $L=77,000/cm^2$, whereas the broken curve is computed according to $g=N/(1+N)$; with $l=34,000/cm^2$.

tation of g in dependence of N. Within the precision of the experiments and their evaluation, the resulting data on g(N) are well described by $g(N) = N/L$, where $L = 77,000/cm^2$. In order to check the consistency of the derived dependence of g(N), we have used this formulation for g(N) in equs. (1) and (2) with $W=S=0$ to describe the dynamics of subpopulations D and Q, employing the initial condition $N = 200/cm^2$, $D/N = 0.1$. In these computations $R = 1/d$ was used. The result is plotted as full curve in Fig. 2 and demonstrates that the experimental data are well described.

5. Comparison with Biochemical Data on Cell Interaction, Discussion

As derived above, the cellular interaction function g(N) is essentially phenomenological, although the system of differential equs. (1) and (2) used for its derivation was formulated under consideration of some basic cell-biological information. It is of interest to compare the population-dynamical characteristic with biochemical studies on cellular interactions during cell-density dependent inhibition of proliferation. From medium conditioned by density-inhibited 3T3 cells a diffusible inhibiting activity of molar mass of about 10kg/mol was isolated and purified [16]. Independent of cell density, this polypeptide binds to 3T3 cells according to a Michaelis-Menten-kinetics, with half saturation of binding at $K_m=17nM$ and a maximum number of binding sites of $m=400,000$ per cell [17]. This factor, termed fibroblast growth regulator (FGR), is released with rate $p=0.17$ fmol per cell and day into the growth-medium; p is independent of cell density [16]. According to these findings, the cellular interaction characteristic g should depend only on total cell density, but not on the density of one of the subpopulations, as their relative preponderances change with total cell number (see Fig. 2). In the following, we shall attempt to arrive at a quantitative description of cellular population dynamics on the basis of these biochemical data. As FGR is secreted, it accumulates in the medium during the time of one day between two medium renewals, proportional to the effective cell density N of the population. The effective concentration c of FGR during the time interval Δt = 1d between two medium renewals then is approximately:

$$c = p \Delta t N / (2V) \tag{6}$$

where N/V is the medium volume per cell. The secreted FGR is bound equally to all cells, resulting in an average number $n \leqq m$ of occupied receptors per cell as governed by the Michaelis constant K_m. Such binding to D-cells may lead to their conversion to Q-cells. It is not known how an average receptor occupancy n/m of a cell might translate to the intracellular processes resulting in arrest of its further entry into the cell cycle. The relative probability g of the transition D→Q might be simply proportional to the FGR-receptor occupancy n/m (of course relative to a given cell-activation by the growth factors in the medium, which are held at constant level). The transition probability g might, however, as well depend non-linearly on n/m, e.g. exhibit a threshold behavior. As a start, we shall use the simple proportionality between n and g, yielding

$$g = n/m = c/(K_m + c) = N/(1 + N) \tag{7}$$
$$1 = K_m \, 2V/p \Delta t \tag{8}$$

Using the parameters known for 3T3 cells, i.e. K_m = 17 nM [17], p = 0.17 fmol/d [16], Δt=1d, and V=10ml/58cm²=0.17 cm, we obtain l=34,000/cm². The relation g(N) thus predicted from equs. (7) and (8) is shown in Fig. 3 (broken line). Using this formulation for g(N) in equs. (1) and (2) with W=S=0, R=1/d, we have computed D(t), Q(t), and N(t) as shown in Fig. 2 (broken lines). Evidently, it does not describe the experiments fully; the dependence of the transition D→Q on cell density is too weak. This might be caused by a non-linear dependence of proliferation-arrest on receptor occupancy, as discussed before. Increasing contributions to growth inhibition mediated by direct physical contact between cells at higher densities, in addition to interaction by diffusible inhibitors, might be another explanation for the moderate fit. As SV40-3T3 cells have been shown to bind FGR, albeit only to about 1/3 of the amount bound by 3T3 cells, we were interested in checking the effect of g=N/L on population dynamics using a lower figure of L. As SV40-3T3 cells at confluency exhibit extensive cell death, but not before, we have also tested for the effect of increased cell death of Q-cells. The population dynamics for SV40-3T3 cells as computed by equs. (1) and (2), using L=385,000/cm², R=1.2/d, and S=3/d are given in Fig. 4 and reproduce their proliferation behavior adequately. Inter-

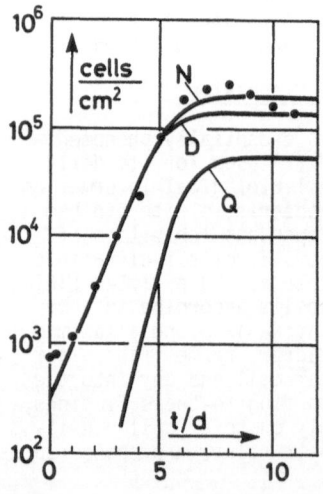

Fig. 4 Population dynamics of SV40-transformed 3T3 cells. Experimental data on total cell density (black circles) refer to proliferation in Dulbecco's modification of Eagle medium containing 5% newborn calf serum (without medium renewal).
Curves on cellular subpopulations D,Q and N=D+Q were computed using equs. (1) and (2) of text, with R=1.2/d W=0, S=3/d, and g=N/L, where L=385,000/cm². Initial conditions were N=D=300/cm².

restingly, increased death (of growth-inhibited cells) gives rise to an increase of preponderance of proliferative cells, According to the experiments,at all cell densities $Q/N \leqslant 0.1$ (data not shown). To sum up: only after incorporation of bio-chemical data on cellular interactions into a population-dynamical formulation is it possible to assess their relevance with respect to the actual proliferation be-havior of a cell population. As the example of the SV40-transformed 3T3 cells de-monstrates, fairly subtle changes of population-dynamical parameters may result in quite drastic changes of proliferative behavior.

Acknowledgement

The authors wish to thank Ms. F. Braun, Ms. A. Kesper, and Mr. H. Maier for their excellent technical assistance.
This work was supported by grants from Stiftung Volkswagenwerk (Schwerpunkt Synergetik) and from Deutsche Forschungsgemeinschaft (Sonderforschungsbereich 156).

References

1 G.Adam, U.Steiner, H.Maier, S.Ullrich: Biophys.Struct.Mech.9, 72 (1982)
2 H.Haken: Synergetics (Springer Berlin 1977)
3 G.Adam: J.Theor.Biol.84, 233(1980)
4 G.Adam: in "Theoretical Immunology", G.I.Bell, A.S.Perelson, G.H.Pimbley, Eds. (Marcel Dekker New York 1978), pp. 603-627
5 R.Baserga: in "Cell Growth", C.Nicolini, Ed. (Plenum New York 1982) pp. 337-345
6 C.P.Stanners, H.Becker: J.Cell.Physiol.77, 21 (1971)
7 G.Adam, U.Steiner, K.Seuwen: Cell Biol. Intern. Rep.7, 955 (1983)
8 C.D.Scher, R.Shepard, H.N.Antoniades, C.D.Stiles: Biochim.Biophys.Acta560, 217 (1979)
9 K.Seuwen, G.Adam: Biochem.Biophys.Res.Comm.117, 223 (1983)
10 K.Seuwen, U.Steiner, G.Adam: Exp.CellRes. (1984), in press
11 W.Wharton, E.Leof, N.Olashaw, E.J.O'Keefe, W.J.Pledger: Exp. Cell Res.147,443 (1983)
12 H.T.Abelson, H.N.Antoniades, C.D.Scher: Biochim.Biophys.Acta561, 269 (1979)
13 A.Vogel, R.Ross, E.Raines: J.CellBiol.85, 377 (1980)
14 C.D.Scher, M.E.Stone, C.D.Stiles: Nature281, 390 (1979)
15 T.Hämmerle, U.Steiner, G.Adam: Europ.J.Cell Biol.33, Suppl.5,15 (1984)
16 P.A.Steck, J.Blenis, P.G.Voss, J.L.Wang: J.Cell Biol.92,523 (1982)
17 P.G.Voss, P.A.Steck, J.C.Calamia, J.L.Wang: Exp.Cell Res.138, 397 (1982)

Idiotypic Network Among T Cells

Klaus Fey and Klaus Eichmann

Max-Planck-Institut für Immunbiologie, Stübeweg 51
D-7800 Freiburg, Fed. Rep. of Germany

T and B cells of the immune system bear antigen-specific receptors. These receptors exhibit idiotypic structures that characterize each clone. To a given idiotype exist anti-idiotypic cells in the immune system, and interactions can occur between them. It is assumed that a network based on these interactions plays a role in the fine regulation of the immune system.

Response [a.u.]

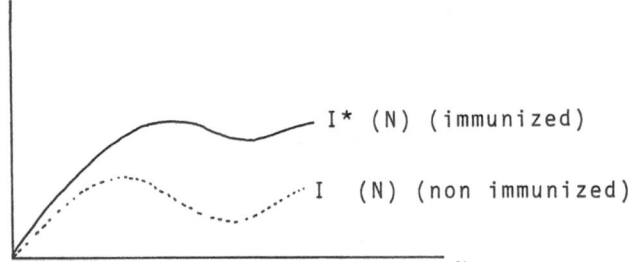

In polyclonally-activated T cell cultures, one finds a typical frequency ϕ_e of the order 10^{-2}. If one activates N T cells polyclonally, and measures after a proliferation period of a few days the immune response to some antigen, one finds the result schematically shown in Fig. 1 (nonimmunized). $I(N)$ denotes the intensity of the response, $I(N) \propto \phi_e \cdot N \cdot \Sigma$, where Σ stands for the clone size. For low N values, the response increases with N, for higher N it is suppressed due to the action of suppressing T_s cells, which have a frequency $\phi_s \cong 20 \cdot \phi_e$. For even higher N a second effector population with $\phi_{e'} \ll \phi_e$ shows up. In the case of an antigen injection in vivo (immunization) before polyclonally-activated culturing, the frequency ϕ_e^* remains nearly unchanged, $\phi_e^* \cong \phi_e$. One concludes that the specific cells do not considerably proliferate in vivo. On the other hand, the immunized case in Fig. 1 shows that the specific cells have become less susceptible to suppression. With the help of the following dynamical model, one can get some insight into the mechanism that produces the response curves $I(N)$ and $I^*(N)$. We assume that all idiotypes have an a priori equal frequency, and their interactions are symmetrical. An asymmetrical ansatz would imply that suppression occurs between antigen-specific cells and suppressor cells which are specialized for interactions only with the antigen-specific cells, and have frequencies arranged in an asymmetrical way ($\phi_s \cong 20 \cdot \phi_e$). In a first approximation (not yet including immunization and proliferation), reversible interactions between all cells shall be possible. We take $x(t)$ as the number of free cells, and $C_\ell(t)$ as the number of clusters (= chains) of size ℓ. Only free cells shall be able to make an immune response. If one wants to reproduce the response function $I(N)$ in the equilibrium case, $I(N) \propto \phi_e \cdot x(N)$ with $x(N) = x(t \to \infty)$, one has to introduce a large set of cluster configurations, $1 \le \ell \le N$. The starting point is then the set of differential equations

$$dx/dt = -k \cdot x^2 - k \cdot x \cdot \sum_{\ell=2}^{N-1} C_\ell + 2 \cdot r \cdot \sum_{\ell=2}^{N} C_\ell$$

$$dC_\ell/dt = -k \cdot x \cdot C_\ell + k \cdot x \cdot C_{\ell-1} - 2 \cdot r \cdot C_\ell + 2 \cdot r \cdot C_{\ell+1}$$

$$dC_N/dt = k \cdot x \cdot C_{N-1} - 2 \cdot r \cdot C_N$$

The total number of cells $N = x + \sum \ell \cdot C_\ell$ is conserved, and in equilibrium one has $C_\ell = (K \cdot x)^{\ell-1} \cdot x$ with $K = \frac{k}{k'} \cdot r$. It follows that $(1-K \cdot x)^{-2} \simeq N \cdot K$. One can get a decreasing $x(N)$ in the range of higher cell numbers if one replaces K by $K(\ell)$, where $K(\ell)$ increases with the chain length ℓ. To investigate the interference between antigen stimulation in vivo and cell-cell interactions in culture, one introduces variables $y(t)$ for the number of free specific cells and $x(t)$ for all others. As long as the antigen specific cells (y) are experimentally not treated in a special way, as it is the case with a mere polyclonal activation, their interaction and growth behavior has to be the same as that of all other clones (x), and the system has to fulfill the condition $y(N) \cong \phi_E \cdot x(N)$ in equilibrium. Binding between the y cells shall be forbidden (as well for the free ones as for those incorporated in chains). Further on, one defines new cluster variables $D_\ell(t)$ in addition to $C_\ell(t)$. $D_\ell(t)$ stands for chains which have an y cell in a terminal position. Therefore these chains cannot react with y cells. With some modifications of the dynamical system given above, one can get a system that gives an equilibrium $y(N) \cong \phi_E \cdot x(N)$, (as long as $\phi_E << 1$). To include the growth of the cultures into this dynamical system, we assume that only free cells proliferate in culture. The expansion is then described by the addition of growth terms $\alpha \cdot x$ and $\beta \cdot y$ to the equations for dx/dt and dy/dt. The ordering of the time scales $T_{prolif.} \gg T_{interact.}$ guarantees that the proliferating cultures follow approximately the equilibrium curve with the actual cell number $N(t)$, (adiabatic behavior). With this system, one is able to study the consequences either of a modified interaction strength due to immunization, $K_{x-y} \to K^*_{x-y}$, or a modified growth behavior in culture, $\beta = \alpha \to \beta^* > \alpha$. A satisfying fit of the data can be achieved if one couples both mechanisms, under the assumption that a faster growth behavior leads to an uncomplete substitution of the relevant surface molecules, resulting in a weaker interaction strength.

The Transition Probability Model: Successes, Limitations and Deficiencies

Robert F. Brooks

Imperial Cancer Research Fund Laboratories, Lincoln's Inn Fields
London, WC2A 3PX, United Kingdom

1 Introduction

The now familiar picture of the eukaryotic cell cycle followed from the discovery that DNA replication usually precedes the visible events of mitosis and cell-division by many hours [1]. For many eukaryotes, mitosis and cell-division are followed by a gap of several hours (G_1) before DNA synthesis (S phase) begins, and on completion of DNA replication there is usually another gap (G_2) before the cell divides once again.

In rapidly growing populations, cells progress from mitosis to S phase and on again to mitosis with seeming inevitability, and the idea of a "cycle", embodying as it does the notion of regularity, seems appropriate. This impression of regularity has no doubt contributed to the frequent suggestion that the underlying timing mechanism is some form of deterministic biochemical oscillator [2-5]. For organisms such as Physarum or the cleavage cycles of amphibian embryos, where the onset of mitosis occurs with some predictability, this seems reasonable enough [6,7]. In general, however, on closer inspection one is impressed less by the regularity of cell-division than by its extreme variability [8]. Even when population growth rate is constant, individual cells have widely differing cycle times, the coefficient of variation frequently being of the order of 15-25%.

One might imagine that the variability of cell cycles reflects the complexity of the processes involved and the need to coordinate numerous biochemical activities. However, in mammalian cells, the interval between the start of S phase and mitosis shows very little variability, even though the period includes the obviously complex events of chromosome replication and segregation. Rather, the variability is confined largely to the G_1 phase [9-11]. Significantly, this is also the phase of the cycle in which mammalian cell proliferation (with which the rest of this paper will be concerned) is believed to be regulated [12,13].

2 The Transition Probability Model

It seemed possible that the variability of G_1 might arise as a consequence of the way in which the cell cycle is regulated. Pursuing this, SMITH and MARTIN suggested [14], as had others before them [15,16], that commitment to DNA replication and mitosis in mammalian cells might occur at random with constant probability per unit time (given steady-state conditions). Smith and Martin went on to suggest that regulation of proliferation could be accomplished by controlling the probability of commitment, for example by regulating the steady-state concentration of some key molecule.

On the face of it, the transition probability model (as it came to be called) appeared unlikely, since it had been known for many years that the cycle times of sister cells (the two daughters, or siblings, produced on cell division) were positively correlated, commonly with r = 0.5 or higher [e.g. 17]. Clearly, if events leading to S phase and mitosis were initiated entirely at random, then there is no reason why sibling cycle times should be any more alike than random pairs. Nevertheless, despite the similarity, sibling cycle times are rarely identical, and it transpired that the differences between sibling cycle times conformed to an almost perfect exponential distribution [18-22]. This is exactly what would be expected if the differences were solely the result of a single step in the cell cycle occurring at random.

The exponential distribution of differences between sibling cycle times provided good evidence for a probabilistic transition in the cell cycle. Indeed, at one time, it was thought that such distributions proved the existence of a random transition beyond doubt [21], though it is now known that there are an infinite number of ways of generating an exponential distribution exactly, or approximately, that do not involve random transitions [23-28]. Nevertheless, a single probabilistic transition remains the simplest biologically reasonable explanation.

3 The Two-Transition Modification of the Transition Probability Model

Although differences between sibling cycle times are exponentially distributed, cycle times as a whole are not [18,19]. If it is accepted that the sibling differences are the result of a random transition, then it follows that the duration of the rest of the cycle varies in general but is identical in siblings, accounting for the observed correlation. A possible explanation of this surprising conclusion came from a consideration of the response of cells to mitogenic stimulation.

In order to proliferate in culture, mammalian cells require a source of hormone-like growth factors which are normally supplied by supplementing the medium with serum. When the cells are deprived of the serum supplement, proliferation ceases and the cells accumulate in the G_1 phase. On adding back the serum, the cells resume cycling, but only after a long lag (generally of the order of 8 - 16 hours, depending on cell type). The rate at which the cells enter S phase was originally thought to be approximately first order. As we will see later, the kinetics are actually more complex than this, but at the time they appeared to be consistent with a fairly abrupt increase of the transition probability [29]. Furthermore, the rate of entry into S phase depends on the growth factor concentration, which suggested that the transition probability was an important factor in regulating proliferation rate. The problem was to explain the significance of the lag.

The duration of the lag is much longer than the minimum length of G_1 found in rapidly proliferating populations, so that the transition, if it existed, had to be placed towards the end of the lag. Consistent with this, removal of the serum again sometime after the end of the lag led to a comparatively rapid decline in the rate of entry into S phase [30]. This left most of the lag unaccounted for. It could not easily be explained by the time required to accumulate something to a threshold, since the lag is independent of the serum concentration, even though this affects the rate of most biochemical processes. Furthermore, if the cells were first stimulated with a low concentration of serum and the concentration raised again once the cells had begun to enter S phase, there was another long lag, similar in duration to the first, before the rate of entry into S phase increased once more [30]. Evidently, if the first lag were the time required for some substance to reach a threshold, then the second lag could not be explained in the same way.

A possible explanation for the lag came with the realization that its duration was often similar to the minimum cycle time of growing cells [31]. Although this could be mere coincidence, it seemed worth considering the possibility that both had the same underlying timing mechanism. Accordingly, it was suggested [31] that in order to initiate DNA replication, the cell must first complete a lengthy process, termed L (see Fig. 1). Quiescent, growth factor-deprived cells were considered to be arrested prior to the start of L, so that L accounted for most of the lag following mitogenic stimulation. Initiation of L in response to stimulation was considered to be stochastic, and a cell waiting for L to begin was said to be in Q state.

On completion of L, a cell was free to become committed to DNA synthesis and division, a transition taking place at random (Fig. 1). A cell waiting for this

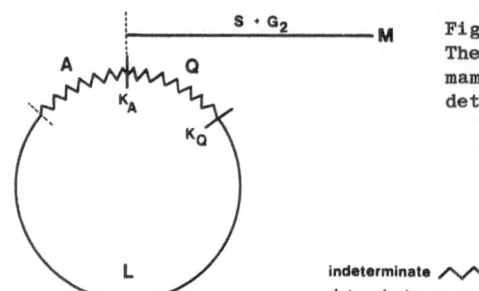

Figure 1
The two-transition model of the
mammalian cell cycle. For
details, see text

indeterminate ∿∿
determinate ———

transition to occur was said to be in A state, as in the original Smith and Martin
model. It was further postulated that the A state transition also re-established
conditions allowing process L to begin again, i.e. re-entry into Q (Fig. 1).
Initiation of L could thus occur again while a cell was engaged in the processes
of DNA replication or division, in readiness for the next cell cycle. Sister
cells were therefore likely to be at the same stage of L at birth, and so would
reach the next A state at the same time (provided L was initiated before division
of the mother cell). Differences between the cycle times of siblings were thus
dependent only on the random exit from A state, hence the exponential distribution.
Different pairs of siblings would be at different stages of L at birth, however,
depending on when L was initiated in the mother cell, so sibling cycle times would
be correlated.

According to the model, the minimum cycle time is shown by a cell which spends
no time in A state (i.e. which leaves A state as soon as it enters it) and whose
mother spent no time in Q state. Hence L, which accounts for most of the lag on
stimulation of quiescent cells, is also responsible for the minimum cycle time of
rapidly cycling cells.

4 Quantitative Predictions of the Two-Transition Model

The two-transition model provided a reasonable qualitative description of many
features of mammalian cell cycle kinetics. Rather more compelling was that it
also made good quantitative predictions.

If it is assumed that variability in the duration of L is negligible, then the
hypothesis generates the following equation for cells growing in exponential
steady-state:

$$\alpha_t = \frac{1}{(k_A - k_Q)} \left\{ k_A e^{-k_Q(t-T_L)} - k_Q e^{-k_A(t-T_L)} \right\} \tag{1}$$

where α_t is the fraction of cells remaining undivided at age t (age measured from
   division of the mother cell);
   k_A and k_Q are the rate constants for exit from A state and Q state
   respectively; and
   T_L is the duration of L.   (For derivation, see [31].)

The shape of this curve depends only on k_A and k_Q and these parameters may be
readily estimated, as discussed elsewhere [31]. Briefly, since differences in
sibling cycle times are due only to the time spent in A state, k_A may be estimated
directly from the mean or standard deviation of the sibling differences since, for
an exponential distribution, both statistics are equal to the reciprocal of the
rate constant. k_A may also be estimated graphically from the slope of a semi-
logarithmic plot of the so-called β curve (fraction of sibling pairs with cycle
time differences >= t, plotted against t). That all three estimates usually agree
well is testimony to the fact that the sibling differences are indeed well
described by an exponential distribution.

From a knowledge of k_A and the total variance in cycle times, k_Q can be estimated from:

Total variance = variance in A + variance in Q

$$= \frac{1}{k_A^2} + \frac{1}{k_Q^2} \tag{2}$$

which again derives from the fact that both A and Q have exponential distributions, k_Q can also be estimated graphically from a cumulative plot of the distribution remaining after the contribution of k_A has been "subtracted" from the overall distribution of cycle times [31,32].

The predicted distribution of cycle times can now be compared with the actual distribution by substituting the values of k_A and k_Q in (1). When this is done, the agreement between theory and practice (for rapidly growing cells, where data are easy to obtain) is often astonishingly good (Fig. 2) (see also [31,32,11,22]).

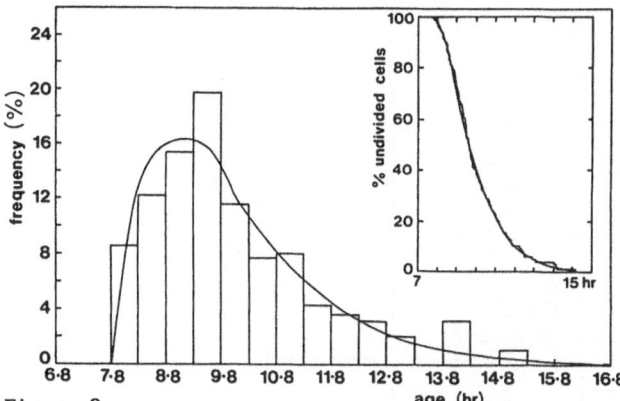

Figure 2
Frequency distribution of cycle times for V79-8 cells superimposed on the distribution predicted by the two-transition model for $k_A=0.95$ hr^{-1} and $k_Q=0.85$ hr^{-1} (estimated graphically as described in the text) with T_L=min. cycle time =7.8 hr; N=204. The inset shows the same data plotted as cumulative distributions (α curves). The data are the same used to construct Fig.6A of ref. [11], from where additional details may be obtained. Differences between the predicted and experimental distributions are not significant ($\chi^2 = 5.6$, d.f. = 12, P $\gg$ 0.05)

Finally, it is possible to predict the correlation coefficient for sibling cycle times, r_{ss}. Since the correlation coefficient represents the fraction of the total variance held in common, we may write:

$$r_{ss} = \frac{\text{variance in Q}}{\text{total variance in cycle times}}$$

$$= \frac{1/K_Q^2}{1/K_A^2 + 1/K_Q^2} \tag{3}$$

The correlation coefficient calculated in this way agrees remarkably well with the observed correlation. (Equation (3), of course, can equally well be reversed to obtain estimates of k_A and k_Q using the observed correlation and variance, and these agree well with estimates obtained by the other methods [31,32].)

5 Difficulties for the Two-Transition Model: Heterogeneity in Proliferative Potential

The two-transition model, then, provides a good quantitative description of the cell cycle kinetics of rapidly growing cells. To test the model further, we have sought to determine whether it can account equally well for the cycle kinetics of slowly growing cells.

To achieve low growth rates, we have attempted to limit the supply of growth factors by growing the cells in medium supplemented with low concentrations of serum (3% or less). To our surprise, it has proved exceedingly difficult to maintain steady-state (i.e. exponential) growth, even when the cultures are perfused continuously with fresh, low serum medium. Frequently initial growth rate is as high as under optimum conditions, but it rapidly and continuously falls until population growth ceases altogether. The main reason for this is now becoming apparent: although at high growth rates the cells behave more or less homogeneously, as expected of a clonal population, at low growth rates this is not the case (Fig. 3). Some cells show the capacity for numerous, consecutive cell cycles; others divide once or twice at most, or not at all. It is as though cells were continuously "dropping out of cycle", but in a largely lineage-dependent way [33].

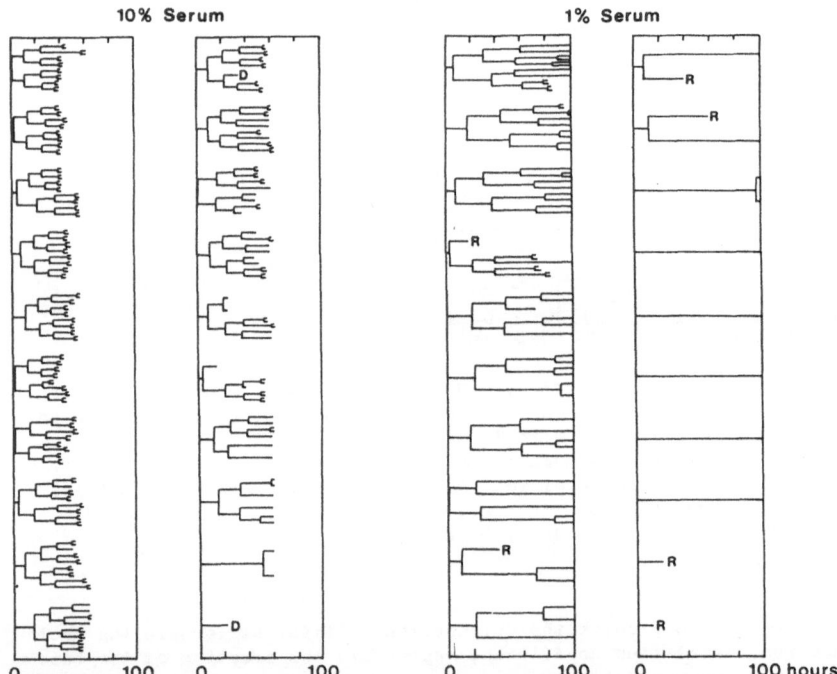

Figure 3

Pedigrees of 20 randomly selected 3T3/4A-C5 cells (from a recently cloned population) growing in either 10% or 1% calf serum. Cells were followed for up to 3 generations by time-lapse cinemicroscopy, the film durations being 61 hr (10% serum) or 100 hr (1% serum). Incomplete pedigrees are the result of cells migrating from the field of view. D = dies; R = cell rounds up at mitosis but fails to complete cytokinesis.

In 10% serum most cells go through many consecutive cycles. Only a few cells (generally less than 5% of the initial population) fail to divide at all or divide once only. (Such cells tend to be excluded from the analysis of cycle times since they fail to give cycle times.) In 1% serum, the proliferative capacity of individual cells is much more variable and the proportion of non-dividers, or slow dividers, is far too high to be ignored

308

Initially, we thought that the non-dividing or slowly dividing cells must have sustained damage of some kind as a consequence of the growth factor deficient conditions, but on raising the serum concentration again, many of the non-dividers resume proliferation (manuscript in preparation). Rather, it was as though individual cells varied in their growth factor requirements.

Further evidence for this view has come from studies of the response of quiescent cultures to different serum concentrations. As already mentioned, the lag before proliferation resumes is independent of the serum concentration, but the rate at which the cells enter S phase, which appeared to be approximately first order, is not. It was assumed that if one waited long enough, all cells would eventually reach S phase, even at the slow rates characteristic of low serum. Recently, it became possible to test this directly, following the discovery that high specific activity ^{3}H-thymidine prevents cell division (probably as a consequence of radiation damage) without causing significant cell death, at least over several days. The cumulative labelling index could thus be followed for long periods without the complication of each labelled cell dividing to give two labelled daughters. When this was done, it was found that at low serum concentrations the labelling index plateaued after three days or so at values well below 100%, despite daily medium renewal [33]. Furthermore, there was nothing intrinsically wrong with the non-responders as the majority could be made to enter S phase on raising the serum concentration further [33].

It seems difficult to avoid the conclusion that some cells require higher levels of growth factors than others. However, prolonged culture (3 weeks or so) in low serum does not enrich for cells capable of proliferating in low serum [33]. Responding cells are not therefore stable transformed variants pre-existing in the population [34]. Rather, the ability to respond must be an unstable property, in the sense that the responding cells must regenerate a mixture of responders and non-responders identical to the starting population. The production from a single cell of descendents with different proliferative behaviour (including marked asymmetry between siblings) has in fact been observed directly and repeatedly by time-lapse cinemicroscopy [33, and unpublished observations].

6 Significance of Heterogeneity

It seemed possible that variations in proliferative capacity might reflect some form of differentiation induced by forcing the cells to grow slowly. Swiss 3T3 cells are known to be capable of conversion to an adipocyte-like phenotype [35] but there is no indication of such conversion in the slow-growing cultures referred to here, even after switching to a medium (containing biotin and high levels of insulin) favouring the accumulation of lipid in converted cells [36,37]. Similarly, in normal human melanocyte cultures, the non-dividing cells are not invariably pigmented even though pigmentation, when it does occur, is sometimes associated with a reduced proliferative capacity (Bennett, personal communication).

Another explanation for the heterogeneity is the karyotypic instability commonly found in cultured cell lines. However, the frequency with which non-dividers arise in low serum cultures is so high as to suggest a level of chromosomal rearrangements so intolerably high, if this is the cause, as to render cell lines practically useless. It is therefore reassuring, apart from anything else, that similar behaviour is also seen with normal diploid cells [38,39]. In this case, the loss of proliferative potential by individual cells has long been felt to reflect cellular senescence, the gradual accumulation of non-dividing cells accounting for the limited life span of the culture as a whole.

With established lines, the heterogeneity in proliferative capacity is only observed to any marked extent (in sparse cultures) at low growth factor concentrations. With normal diploid cells, however, it is apparent even at high serum concentrations (10% or more). This suggests that the only difference between normal diploid cells and immortal cell lines relevant to life-span in culture may

be in their quantitative requirement for growth factors. Such a view gains
credence from the fact that transfection of diploid cells with the gene for the
large T antigen of Polyoma virus brings about both immortalization of the cells
and a reduction in serum requirement [40].

7 An Explanation for Heterogeneity in Proliferative Potential

Since heterogeneity in proliferative potential cannot be explained easily by
differentiation, karyotypic instability, or a spectrum of pre-existing stable
transformed variants, it seems worth considering the possibility that it actually
reflects some aspect of a normal, controlled, physiological response to growth
factors. Quiescent 3T3 cells stimulated with low levels of serum show an almost
continuous range of response: some cells fail to divide at all, whereas a few cells
at the other extreme give rise to many generations of progeny before proliferation
ceases [33]. It is as though the ability to respond to low levels of growth
factors depended on some property that is distributed in the initial population,
such that some cells have the capacity for many consecutive cycles, whereas other
cells have insufficient even for one cycle. If so, then since continuous
proliferation is sustainable at high but not at low serum concentrations, the
property must itself represent a product of growth factor stimulation. In other
words, growth factors may stimulate the production of something which is itself
required for growth factor response, i.e. an element of positive feedback. Recent
studies on the proliferation of T lymphocytes suggest that a good candidate for
this might be the expression of receptors for the growth factors themselves [41].

In the case of T cells, proliferation is absolutely dependent on T cell growth
factor (Interleukin 2, IL-2). Expression of receptors for IL-2, however, requires
previous exposure to antigen or mitogenic lectins, and when such stimuli are with-
drawn, IL-2 receptor number declines [41]. Maintenance of proliferation thus
depends on the cooperation between two different classes of "signal". For reasons
not yet understood, even when fully induced, the distribution of IL-2 receptors per
cell is extremely broad (approximately log normal). Furthermore, this is of
physiological relevance, since cell sorting experiments indicate a correlation
between receptor density and mitogenic response in individual cells [41].

The growth factor requirements of fibroblasts such as 3T3 cells appear to be
more complex than those of T cells in so far as optimum proliferation depends on
the interaction between many different factors, not all of which have been defined.
Even so, it is now becoming clear that, like antigen or mitogenic lectins in T
cells, some growth factors can potentiate the expression of receptors for other
classes of growth factors [42-44]. There are grounds, therefore, for thinking that
receptor levels may depend on the synergistic interaction between growth factors.
Such synergy could well account for the element of positive feedback in the
relationship between the concentration of serum (a mixture of growth factors) and
the maintenance of proliferative potential.

Growth factor receptors are generally present at many thousand molecules per
cell [e.g. 41]. At this level it is unlikely that random segregation of receptors
at division could account for the large differences in proliferative potential
frequently observed between daughter cells (at low growth rates). Receptor mRNA,
however, might be present at a concentration several orders of magnitude lower and
could well be at a level where statistical fluctuation between cells became
significant. Indeed, if receptor mRNA is present at only a few molecules per cell,
synthesis and degradation would have to be regarded as stochastic and mRNA levels
(and hence receptor density) at any instant would inevitably be distributed.

8 A Model for Proliferative Heterogeneity and Senescence

The consequences of low, distributed receptor mRNA levels coupled to a positive
feedback between growth factor concentration and receptor maintenance are

interesting and worth exploring in some detail. If the probability of synthesizing
receptor mRNA, P_S, is a function of the concentration of occupied growth factor
receptors - and one imagines a typical hyperbolic relationship (Fig. 4) - then at
high growth factor concentrations where P_S is maximal, the distribution of receptors
would be stable with time (Fig. 4). (For simplicity, one assumes a strict
correlation between the number of receptor mRNA molecules at any time and the
number of receptors produced in that instant.) Thus a cell with more than the
average number of receptors would tend to regress to the mean level, while cells
with less than the average would most likely increase back towards the mean, so
long as receptor level was not so low as to materially affect P_S. (For an analogy
of this, see the discussion on the maintenance of a stable cell size distribution
in [8].)

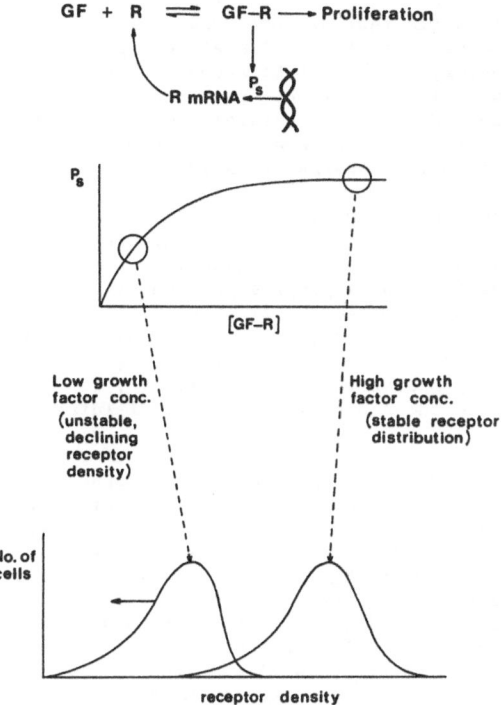

Figure 4
A model for proliferative
heterogeneity and senescence.
GF refers to the collection
of growth factors required to
maintain proliferation, with
R representing their receptors.
Proliferation rate is a
function of the concentration
of occupied receptors as is
P_S, the probability of producing
receptor mRNA

At limiting growth factor concentrations the situation changes since,
intuitively, receptor maintenance would seem to become unstable (Fig. 4). Cells
with the most receptors would have the highest probability both of proliferation
and of maintaining their receptor levels, accounting for the persistence of cell
division in some lineages as opposed to others. Even so, at sub-optimal growth
factor concentrations, P_S would be less than maximal,so that receptor level would
decline from the maximum possible. A small decrease in receptor number would lead
to a further decrease in P_S, leading to a gradual downward spiral towards lower
receptor levels. The end result would be an unresponsive, "senescent" population.
At any time before this becomes total, however, because production of receptor mRNA
is stochastic, there will always be a tendency to generate a few cells with more
receptors than average. These cells would be better able to resist the general
decline in receptor level,and would constitute the "proliferative pool" responsible
for population growth. Nevertheless, the size of this pool would gradually
diminish with time (as happens in practice with diploid cells [39]) as the average
receptor level falls.

This scheme has been couched in terms of receptors because, as discussed above, this currently seems the most likely [41]. It is also something approachable experimentally. It should be noted, however, that the scheme is equally applicable to any other substance whose level is regulated by growth factors, and which is itself required for growth factor responsiveness. The scheme would seem to account, at least intuitively, for the heterogeneity in proliferative capacity characteristic of slowly growing mammalian cells and for the inexorable decline of diploid cells in culture. Whether it can do so quantitatively remains to be seen, and this awaits a more formal and rigorous analysis.

9 Heterogeneity: Implications for the Two-Transition Model

One of the main attractions of the transition probability model was that it offered an explanation of cell cycle variability in terms that did not require intrinsic differences between cells. The marked heterogeneity in proliferative capacity at low growth rates is quite inescapable, however, and cannot be accounted for by a limited number of probabilistic transitions: like it or not, deterministic differences between cells do exist. Is there any reason therefore to suppose that such differences do not also exist at high growth rates, even though under these conditions the two-transition model appears to fit remarkably well?

The most compelling evidence for at least one transition remains the exponential β curve. Although such curves can be accounted for in ways that do not involve a random transition [23-28] this is nevertheless the simplest explanation. For instance, although CASTOR'S "G_1 rate" model [24] can reproduce particular curves rather well using parameter values selected by computer, it is by no means clear that it offers a general method for doing so. Similarly, it is not at all obvious that unequal segregation of growth factor receptors between siblings [41] could generate exponential-looking β curves which preserve their exponentiality in cohorts of sibling pairs grouped according to the age of the youngest dividing sibling of each pair [21,22].

The case for a second random transition is much less secure. The evidence remains the fact that the two-transition model is quantitatively consistent with many sets of data from rapidly growing cell cultures [11,22,31,32] (but not all, [11,45]), together with the biological arguments that led to its proposition [31]. Unlike k_A, however, there is no method of estimating k_Q directly: it is obtained simply from the variance left over after subtracting the contribution from k_A. Any variation in L would lead to an underestimate of k_Q, and it is probably possible to accommodate a fair degree of such variability without substantially affecting the shape of the predicted distribution of cycle times. Apart from anything else, this means that the assumption of zero variance in L is not beyond doubt, and further, there are probably many other distributions that would fit the "left over" variance as well as an exponential [27,45]. Indeed, this illustrates the weakness of kinetics: any model must be consistent with the experimental data, but agreement in itself does not prove the model.

The biological arguments in favour of Q and L continue to be more persuasive. For instance, although growth factor receptor heterogeneity provides a plausible explanation for the variation in proliferative potential at low growth rates, it is difficult to account for the lag associated with mitogenic stimulation exclusively in such terms. Thus, if receptor density alone determined the rate at which individual cells progress towards S phase at a given mitogen concentration [41] then one would expect a rapid increase in the rate of entry into S phase on raising the mitogen concentration again at the end of the lag, since by then cells would be distributed at all "points" between quiescence and S phase, i.e. some cells would be close to S phase and should be accelerated into DNA replication on raising the mitogen concentration. As already discussed, with 3T3 cells and serum as a source of growth factors, this does not happen. Instead, there is another lag, similar in duration to the first, before the rate of entry into S phase changes again [30]. The notion of some length process (not involving the

312

accumulation of something) which must be completed before S phase can begin thus continues to seem unavoidable. The idea that this process might also begin in the previous generation at high growth rates, and account for the sibling correlation and the similarity between the lag and the minimum cycle time, also remains appealing.

The basic arguments in favour of the two-transition model thus continue to be valid, though none of them can be said to prove it beyond doubt. To account for the unexpected heterogeneity in proliferative potential seen at low growth rates, however, it is necessary to add something to the model, and feedback interaction between growth factors and expression of their receptors, along the lines discussed, would seem to be a credible and experimentally amenable possibility. Whether random transitions can be dispensed with altogether (in the light of demonstrable differences between cells) remains to be seen. However, in addition to explaining proliferative heterogeneity (at low growth rates), any alternative must be capable of matching the two-transition model in accounting quantitatively (at high growth rates) for exponential β curves, the sibling correlation, and the distribution of cycle times; and for the lag associated with mitogenic stimulation of quiescent populations and the similarity between its duration and the minimum cycle time of rapidly cycling cells. In the final analysis, vindication of the two-transition model will come only if the hypothetical process L can be identified in molecular terms. For this reason, we continue to be interested in mitotic centres. The biology of these structures in cleaving sea urchin eggs [46] first suggested the model, and their asynchronous maturation in mammalian cells, relative to mitosis, is at least consistent with it [47].

References

1. A. Howard, S.R. Pelc: Heredity, Lond. (Suppl.6) 261 (1953)
2. W. Sachsenmaier, U. Remy, R. Plattner-Schobel: Exp. Cell Res. 73, 41 (1972)
3. S. Kauffman, J.J. Wille: J. theor. Biol. 55, 47 (1975)
4. D.A. Gilbert: Biosystems 5, 197 (1974)
5. R.R. Klevecz: Proc. Natl. Acad. Sci. USA 73, 4012 (1976)
6. W. Sachsenmaier: in "The Cell Cycle" (ed. P.C.L. John, Cambridge University Press, Cambridge, 1981) pp.139-160
7. J.W. Newport, M.W. Kirschner: Cell 37, 731 (1984)
8. R.F. Brooks: in "The Cell Cycle" (ed. P.C.L. John, Cambridge University Press, Cambridge, 1981) pp.35-61
9. J.E. Sisken, L. Morasca: J. Cell Biol. 25, pt.2, 179 (1965)
10. R.A. Tobey: Meth. Cell Biol. 6, 67 (1973)
11. R.F. Brooks, P.N. Riddle, F.N. Richmond, J. Marsden: Exp. Cell Res. 148, 127 (1983)
12. R. Baserga: Cancer Res. 25, 581 (1965)
13. D.M. Prescott: "Reproduction of Eukaryotic Cells" (Academic Press, New York 1976)
14. J.A. Smith, L. Martin: Proc. Natl. Acad. Sci. USA 70, 1263 (1973)
15. S.M. Cattaneo, H. Quastler, F.G. Sherman: Nature 190, 923 (1961)
16. F.J. Burns, I.F. Tannock: Cell Tissue Kinet. 3, 321 (1970)
17. K.B. Dawson, H. Madoc-Jones, E.O. Field: Exp. Cell Res. 38, 75 (1965)
18. P.D. Minor, J.A. Smith: Nature 248, 241 (1974)
19. R. Shields, J.A. Smith: J. Cell. Physiol. 91, 345 (1976)
20. R. Shields: Nature 267, 704 (1977)
21. R. Shields: Nature 273, 755 (1978)
22. R. Shields: in "Control Mechanisms in Animal Cells" (eds. L. Jimenez de Asua, R. Levi-Montalcini, R. Shields, S. Iacobelli, Raven Press, New York, 1980) pp.157-164
23. A.B. Pardee, B. Shilo, A.L. Koch: in "Hormones and Cell Culture, Book A" (eds G.H. Sato, R. Ross, Cold Spring Harbor Press, Cold Spring Harbor, 1979) pp.373-392
24. L.N. Castor: Nature 287, 857 (1980)
25. A.L. Koch: Nature 286, 80 (1980)
26. P.J. Green: Nature 285, 116 (1980)

27. S. Nelson, P.J. Green: Cancer Chemother. Pharmacol. 6, 11 (1981)
28. J.A. Smith, D.J.R. Laurence, P.S. Rudland: Nature 293, 648 (1981)
29. R.F. Brooks: J. Cell. Physiol. 86, 369 (1975)
30. R.F. Brooks: Nature 260, 248 (1976)
31. R.F. Brooks, D.C. Bennett, J.A. Smith: Cell 19, 493 (1980)
32. R.F. Brooks: in "Control of Cellular Division and Development:Part A" (eds. D. Cunningham, E. Goldwasser, J. Watson, C.F. Fox. A.R. Liss, New York, 1981) pp.593-601
33. R.F. Brooks, F.N. Richmond, P.N. Riddle, K.M.V. Richmond: J. Cell. Physiol. 121, in press (1984)
34. P.L. Kaplan, B. Ozanne: Cell 33, 931 (1983)
35. H. Green, O. Kehinde: Cell 1, 113 (1974)
36. M.M. Steinberg, B.L. Brownstein: J. Cell. Physiol. 113, 359 (1982)
37. P. Djian, P. Grimaldi, R. Négrel, G. Ailhand: Exp. Cell Res. 142, 273 (1982)
38. P.M. Absher, R.G. Absher: Exp. Cell Res. 103, 247 (1976)
39. J. Ponten, W.D. Stein, S. Shall: J. Cell. Physiol. 117, 342 (1983)
40. M. Rassoulzadegan, Z. Naghashfar, A. Cowie, A. Carr, M. Grisoni, R. Kamen, F. Cuzin: Proc. Natl. Acad. Sci. USA 80, 4354 (1983)
41. D.A. Cantrell, K.A. Smith: Science 224, 1312 (1984)
42. G. Carpenter, S. Cohen: J. Cell. Biol. 71, 159 (1976)
43. D.R. Clemmons, J.J. Van Wyk, W.J. Pledger: Proc. Natl. Acad. Sci. USA 77, 6644 (1980)
44. R.K. Assoian, C.A. Frolik, A.B. Roberts, D.M. Miller, M.B. Sporn: Cell 36, 35 (1984)
45. E.J.J. van Zoelen, P.T. Van Der Saag, S.W. De Laat: Exp. Cell Res. 131, 395 (1981)
46. D. Mazia, P.J. Harris, T. Bibring: J. Biophys. Biochem. Cytol. 7, 1 (1960)
47. E. Stubblefield: in "The Proliferation and Spread of Neoplastic Cells" (The Williams and Wilkins Co., Houston, 1968) pp.175-189

A Deterministic Cell Cycle Model with Transition Probability-Like Behaviour

Michael C. Mackey

Department of Physiology, McGill University, 3655 Drummond
Montreal, Quebec, Canada H3G 1Y6

I. INTRODUCTION

The time between cell birth and cellular division is known as the cell generation, or cell cycle, time. In a steady state population of apparently identical cells, the distribution of generation times is broad, thus indicating considerable variability in the rate at which cells traverse the cell cycle. Perturbation of a steady state population of cells is followed by a rapid return of successive distributions of cell cycle times to the unperturbed distribution. The interpretation of the origin of this variability in cellular processes has been uniformly based on probabilistic considerations [1-5].

Here it is shown how a completely deterministic process may lead to a description of cellular variability indistinguishable from that having its origins in probabilistic considerations. The model is very similar to that of [6] with the important exception that an alternative deterministic hypothesis replaces the assumption made previously concerning the probability of mitosis.

II. MATHEMATICAL PRELIMINARIES

In this section, the notions of a smooth density and Rényi transformations are introduced before starting a theorem that is central to the development of the model.

A function $h:[0,1] \to R$ that is Lipschitzean and satisfies

$$\min h > 0 \quad \text{and} \quad \int_0^1 h d\mu = 1,$$

where μ denotes the Lebesgue measure on $[0,1]$, is called a **smooth density**.

A mapping $S:[0,1] \to [0,1]$ which satisfies:

i) There exists a partition $0 = a_0 < a_1 < \cdots < a_p = 1$ of $[0,1]$ such that for each integer i, $i = 1, \ldots, p$, the restriction of S_i of S to the open interval (a_{i-1}, a_i) can be extended as a C^2 function to $[a_{i-1}, a_i]$;

ii) $S_i([a_{i-1},a_i]) = [0,1]$, $i = 1, \ldots, p$; and

iii) $\inf_{(a_{i-1},a_i)} |S'(x)| > 1$ $i = 1, \ldots, p$

315

is called a **Rényi transformation**.

Theorem (LASOTA and YORKE [7]). Let $S:[0,1] \to [0,1]$ be a Rényi transformation, $S_\varepsilon = (1+\varepsilon)S$ where $\varepsilon > 0$, and $N_\varepsilon(x)$ be the smallest integer n, n = 0,1, ..., such that $S_\varepsilon^{n+1}(x) > 1$ for a given initial $x \in [0,1]$, i.e.

$$N_\varepsilon(x) = \inf \{n : S_\varepsilon^{n+1}(x) > 1\}.$$

Then there exists a unique constant $\sigma > 0$ such that for every smooth density h of initial $x \in [0,1]$

$$\lim_{\varepsilon \to 0} \mu_h\{x : N_\varepsilon(x) > T/\varepsilon\} = \exp(-\sigma T), \quad T > 0, \quad \text{where}$$

$$\mu_h(x) = \int_0^x h d\mu.$$

What is the content of this theorem? For $\varepsilon > 0$ it is clear that eventually there must exist some "kick-out time" [8] n such that $S_\varepsilon^{n+1}(x) \notin [0,1]$. The theorem merely states that, for $\varepsilon \to 0$ the fraction of points from an initial density h with kickout times greater than T > 0 is exponentially distributed:

$$\mu_h\{x : N_\varepsilon(x) > T\} \cong \exp(-\sigma \varepsilon T).$$

III. THE MODEL

With the preliminaries of the previous section, we may proceed to a development of the model of the cell cycle. The model is mathematically identical to that developed in [6]. Only the interpretation of one hypothesis is different.

The model rests on three hypotheses.

H1. There exists some substance(s) (mitogen) necessary, but not sufficient, for mitosis to take place.

There is ample experimental support for this concept, as reviewed in [9].

Consider a cell in a large population that was born at time t = 0 with mitogen content r. The evolution of mitogen following birth is governed by

$$\frac{dm}{dt} = g(m), \quad m(0) = r. \tag{1}$$

The solution of (1) is denoted by m(r,t). Mitogen levels are assumed to be bounded on the closed interval to $[0,2\ell]$, and the mitogen production-rate satisfies

$$g(m) > 0 \quad \text{for } 0 < m < 2\ell \quad \text{and} \tag{2a}$$

$$g(2\ell) = 0. \tag{2b}$$

316

H2. There exists an intracellular variable, sufficient to trigger mitosis once it exceeds a threshold, that oscillates in an extremely complicated "chaotic" fashion.

Let $x(t)$ denote the level of this variable at time t, and t_n, $n = 0,1, \ldots$, denote the times at which $x(t)$ attains a relative maximum. If $x_n = x(t_n)$ denotes the values of these relative maxima, then it is assumed that

$$x_{n+1} = \lfloor 1+m(t_n) \rfloor S(x_n) \tag{3}$$

where $S: \lfloor 0,1 \rfloor \rightarrow \lfloor 0,1 \rfloor$ is a Rényi transformation. Mitosis is assumed to take place whenever $x_n > 1$.

This hypothesis, in conjunction with the theorem of the previous section, tells us that for small mitogen levels the fraction of a large population of cells with mitotic times greater than some number $T > 0$ is given approximately by

$$\mu_h\{x_0 : N_m(x_0) > T\} = \exp(-\sigma m T)$$

regardless of the density of h of initial values x_0. Thus the mitotic rate is approximately (σm) for small m.

Remark. This second hypothesis gives a deterministic interpretation of the probabilistic assumption of $\lfloor 6 \rfloor$ that the probability of a cell with mitogen level m dividing in a time $\lfloor t, t+\Delta t \rfloor$ is

$$\phi(m)\Delta t + o(\Delta t), \quad \phi(0) = 0, \quad \lim_{x \to 2\ell} \inf \phi(x) > 0.$$

With a mitotic rate of (σm), if we let $\alpha(r,t)$ denote the fraction of cells born with mitogen level r that have not divided by time t, then we have approximately

$$\frac{d\alpha(r,t)}{dt} = -\sigma m(r,t)\alpha(r,t), \quad \text{so}$$

$$\alpha(r,t) = \exp\left\{ -\sigma \int_0^t m(r,s)ds \right\}.$$

Note that $-\alpha_t(r,t) = \sigma m(r,t)\alpha(r,t)$ is the density function for the distribution of mitotic (generation) times in these cells with initial mitogen levels r.

Generally, the initial mitogen level r in a large population of cells will be distributed on $\lfloor 0,\ell)$ with a density $f(r)$, so the alpha-curve for the entire population is given by

$$\alpha(t) = \int_0^\ell \alpha(r,t)f(r)dr. \tag{4}$$

Finally, the density function for the distribution of generation times for the entire population is given by

$$\psi(t) = -\int_0^{\ell} \alpha_t(r,t)f(r)dr. \tag{5}$$

Another statistic, widely used by cell kineticists in characterizing populations of renewing cells, is the fraction of sibling cell pairs whose intermitotic times differ by at least a time t. This fraction is denoted by $\beta(t)$. The derivation of the beta-curve for this hypothetical cellular population requires a third assumption concerning the fate of the mitogen in a mother cell when mitosis takes place. Here, as in ⌊6⌋, it is assumed that:

H3. **Each sister cell receives exactly one-half of the mitogen present in the mother cell at mitosis.**

This leads directly to the result

$$\beta(t) = -2\int_0^{\infty}\int_0^{\ell} \alpha_t(r,s)\alpha(r,s+t)f(r)\,ds\,dr$$

which has been obtained before ⌊6⌋.

To complete the specification of this model .requires the distribution f(r) of initial mitogen levels in the population of cells. As shown in ⌊6⌋, the distribution of mitogen f_{i+1} in the (i+1)st generation of cells is determined by that in the ith generation, f_i, from the integral equation

$$f_{i+1}(y) = \int_0^{\ell} k(y,r)f_i(r)dr \tag{6}$$

where the kernel k(y,r) is given by

$$k(y,r) = \begin{cases} 0 & 0 \leqslant y < \tfrac{1}{2}r \\[2ex] 2q(2y)\exp\left\{\displaystyle\int_r^{2y} q(z)dz\right\} & \tfrac{1}{2}r \leqslant y < \ell. \end{cases}$$

and $q(y) = \sigma y/g(y)$. With the properties of g(y) specified in (2a,b), (6) has a unique globally,asymptotically stable solution $f_*(y)$. This ensures that there is a unique density function $\psi_*(t)$ for the distribution of cell cycle times in the entire population of cells.

318

IV. DISCUSSION

Of the three hypotheses used in the development of this model, H1 and H3 have
appeared in a number of other cell cycle models. Only H2 is unique, and I will
confine my comments to H2.

With respect to H2, at least two questions arise. The first is mathematical in
nature. Is there any dynamical or semi-dynamical system that is oscillatory and
which has successive maxima determined by the map (3)? Clearly, any such continuous
time system must, by necessity, be of dimensionality 3 or greater. Though there is
no analytic proof of the existence of such systems at this time, there is good
numerical evidence for their existence.

Consider the LORENTZ $\lfloor 10 \rfloor$ equations

$$\frac{dx}{dt} = yz - bx$$

$$\frac{dy}{dt} = -xz + rz - y \qquad\qquad (7)$$

$$\frac{dz}{dt} = \sigma(y-z)$$

with $\sigma = 10$ and $b = 8/3$. Lorentz numerically investigated this system for a
variety of values of r. If successive maxima in $x(t)$ are labeled as x_n, then the
points (x_n, x_{n+1}) are approximately located on the graph of the one-dimensional
mapping

$$T_{\varepsilon(r)}(x) = \lfloor 1+\varepsilon(r) \rfloor \lfloor 1-\tfrac{1}{2}u^{\frac{1}{4}}(1+u) \rfloor, \quad u = |1-2x|, \quad x \in (0,1).$$

YORKE and YORKE $\lfloor 8 \rfloor$ studied this system (7) and found that there is a critical
range of the parameter r such that $\varepsilon(r) > 0$ and the results of the theorem of
section II hold. They were also able to demonstrate the exponential dependence of
the kickout times predicted by the theorem of Lasota and Yorke.

The second question raised by H2 is related to its biological justification. In
this regard I can offer arguments for the hypothesis that are no more, nor less,
convincing than those put foward to justify the usual probabilistic assumption that
H2 replaces.

The assumption that there is an intracellular oscillator timing the cell cycle
is certainly not new, as a variety of investigators have proposed such a scheme
$\lfloor 11-15 \rfloor$ or criticized it $\lfloor 16 \rfloor$. Indeed, considering the nature of the cell cycle,
it would be surprising if the existence of an underlying oscillator had not been
hypothesized. The nature of the oscillators considered ranges from 'limit cycle'
to 'relaxation' types, though the distinction is more one of degree than of type.
Others have assumed that this intracellular oscillator has superimposed 'noise' to
mimic the distribution of cell cycle events $\lfloor 17-21 \rfloor$ though this assumption begs the
question addressed here.

However, the existence of an oscillator such as that considered here -- one with a strange attractor -- seems not to have been considered even though all of the ingredients are present for its occurrence. (The existence of such an oscillator was all but explicitly postulated in [22]). It is well known that the biochemical control loops within the cycling cell are numerous, richly interconnected, and nonlinear. Further, many display mixed positive/negative feedback with or without significant time delays in their feedback pathways [23-25]. As has been shown numerically [26,27] these are exactly the conditions under which one may encounter the presence of strange attractors and the attendant Rényi transformation-like connection between successive maxima [28].

1. O. Rahn: J. Gen. Physiol. 15, 257 (1932
2. D.G. Kendall: Biometrika 35, 316 (1948)
3. A.L. Koch, M. Schaechter: J. Gen. Microbiol. 29, 435 (1962)
4. F.J. Burns, I.F. Tannock: Cell Tissue Kinet. 3, 321 (1970)
5. J.A. Smith, L. Martin: Proc. Natl. Acad. Sci. USA 70, 1263 (1973)
6. A. Lasota, M.C. Mackey: J. Math. Biology 19, 43 (1984)
7. A. Lasota, J.A. Yorke: Rend. Sem. Mat. Univ. Padova 64, 141 (1981)
8. J.A. Yorke, E.D. Yorke: J. Stat. Phys. 21, 263 (1979)
9. J.M. Mitchinson: The Biology of the Cell Cycle (Cambridge University Press, London-New York, 1973)
10. E.N. Lorenz: J. Atmos. Sci. 20, 130 (1963)
11. S. Kauffman: Bull. Math. Biol. 36, 171 (1974)
12. S. Kauffman, J.J. Wille: J. theor. Biol. 55, 47 (1976)
13. J. Tyson, S. Kauffman: J. Math. Biology 1, 289 (1975)
14. R.R. Klevecz: Proc. Natl. Acad. Sci. USA 73, 4012 (1976)
15. L.N. Edmunds, K.J. Adams: Science 211, 1002 (1981)
16. J. Tyson, W. Sachsenmaier: J. theor. Biol. 73, 723 (1978)
17. B.C. Goodwin: Temporal Organization in Cells (Academic, London-New York, 1963)
18. B.C. Goodwin: J. theor. Biol. 28, 375 (1970)
19. W.H. Woolley, A.G. De Rocco: J. theor. Biol. 39, 73 (1973)
20. A.G. De Rocco, W.H. Woolley: Math. Biosci. 18, 77 (1973)
21. R.M. Shymko, R.R. Klevecz: in Biomathematics and Cell Kinetics (ed. M. Rotenberg, Elsevier/North Holland, Amsterdam-New York, 1981)
22. J. Engelberg: J. theor. Biol. 20, 249 (1968)
23. J.J. Tyson, H.G. Othmer: in Progress in Theoretical Biology, volume 5 (eds. R. Rosen, F.M. Snell, Academic Press, New York-London, 1978)
24. J.M. Mahaffy: J. Math. Anal. Appl. 74, 72 (1980)
25. R.D. Bliss: Anal. Biochem. 93, 390 (1979)
26. U. an der Heiden: J. Math. Biology 8, 345 (1979)
27. U. an der Heiden, M.C. Mackey: J. Math. Biology 16, 75 (1982)
28. A. Lasota, M.C. Mackey: Probabilistic Behavior in Deterministic Systems (Cambridge University Press, London-New York, in press)

The Onset of Division in a Protocell Model

H. Schwegler and K. Tarumi

Institute of Theoretical Physics and Center for Biosystems Research
University of Bremen, D-2800 Bremen 33, Fed. Rep. of Germany

We have constructed and discussed a physico-chemical model of a self-maintaining unit which shows essential features of living organisms as form and heredity and, affected by mutations, possibly selection and evolution, too. Therefore we call this system a protocell. The basic mechanisms are diffusion, a metabolic reaction and a "solidification process" at the surface of the cell giving rise to a moving boundary problem (Stefan problem). The resulting dynamics is treated in a fully spatio-temporal way.

1. The Model

The model assumptions are visualized in Fig. 1. The protocell consists of a clot of a porous material C^* supplied with a nutrient σ from the environment by diffusive transport. In the pores of C^* the nutrient is transformed catalytically with a reaction coefficient k to a building material C which also diffuses through the cell. If the concentration of C exeeds a threshold C_1 a solidification to C^* can take place, but inside the cell all C^*-sites are occupied and therefore the solidification process can occur only at the surface S. The diffusion and reaction processes are governed by equations

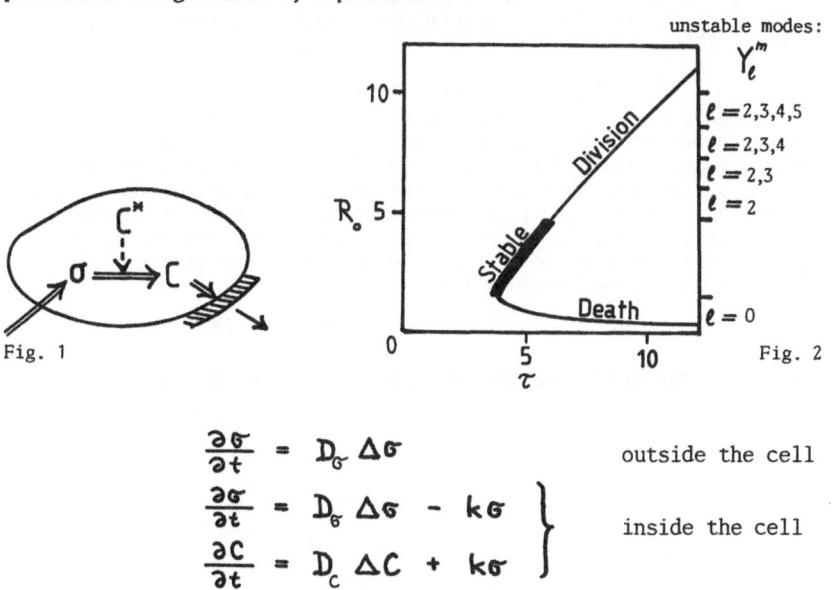

Fig. 1   Fig. 2

unstable modes:

Y_ℓ^m

$\ell = 2,3,4,5$
$\ell = 2,3,4$
$\ell = 2,3$
$\ell = 2$

$\ell = 0$

$$\frac{\partial \sigma}{\partial t} = D_\sigma \, \Delta \sigma \qquad\qquad \text{outside the cell}$$

$$\left.\begin{array}{l} \dfrac{\partial \sigma}{\partial t} = D_\sigma \, \Delta \sigma - k\sigma \\[2mm] \dfrac{\partial C}{\partial t} = D_C \, \Delta C + k\sigma \end{array}\right\} \qquad \text{inside the cell}$$

and corresponding boundary conditions. Due to the solidification $C \to C^*$ and a constant disintegration rate β of C^* the cell surface moves with a velocity v_n in the direction of the normal $\underline{n}$

$$v_n = -\underline{n}\, D_C \, \mathrm{grad}\, C \Big|_S - \beta$$

321

Mathematically, this material balance is the Stefan condition for the integration of the reaction-diffusion equations. In one version of our calculations we have introduced a surface-tension, too. But this should be left out of consideration in this short communication.

2. Discussion

In order to reduce the discussion to the dependency on a minimum number of parameters, a scaling of the variables can be performed leading to a scaled nutrient supply τ as the only relevant parameter. The system of equations allow for spherical steady state solutions with radii R_0 shown in the bifurcation diagram of Fig. 2 in dependency on the supply τ of the nutrient at larger distance (infinity). For small τ there exists only the trivial steady state $R_0 = 0$. But for $\tau > \tau_0^*$ there are two further steady states with finite radii R^- and R^+ (lower and upper branch, respectively).

What about the stability of the two steady states? Linearization of the equations of motion around the steady state and bifurcation analysis lead to the following results:

With respect to spherically symmetric perturbations the lower branch R^- is unstable and the upper branch R^+ is stable. The instability of the lower branch is caused by the positive real part of only one eigenvalue out of an infinite number. This positive one changes to the negative at the vertex of the bifurcation diagram ($\tau = \tau_0^*$).

Nonspherical perturbations are treated by an expansion in terms of spherical harmonics Y_ℓ^m with $\ell = 1,2,3,\dots$. The modes with $\ell = 1$ are always marginally stable and describe uninteresting translational motions of the cell.

For $\tau_0^* < \tau < \tau_2^*$ all modes of the upper branch with $\ell \geq 2$ are stable. Therefore in this interval, the considered mechanisms stabilize a finite and spherical form of the cell.

For $\tau > \tau_2^*$ the modes of the upper branch with $\ell = 2$ become unstable (later for $\tau > \tau_3^*$ the modes $\ell = 3$ become unstable and so forth with increasing τ_n^*, n=4,5,..). Although our stability-analysis is valid only in the neighbourhood of the steady state, we expect that an $\ell = 2$ mode develops further in the nonlinear regime, up to a division into two daughter cells. This means that a cell growing under the condition $\tau > \tau_2^*$ in the direction to R^+ sets on to divide at a radius $R_2^* < R^+$. Therefore each daughter cell starts the same circle again and so on. The result is the temporal order of a cyclic proliferation process. More details of the calculations are reported in [1].

References

1. H. Schwegler, B. Gerstmann and K. Tarumi: to appear in J.Math.Biol.

A Temperature Compensated Epigenetic Oscillator with an Hourly Period Provides Time-Keeping for the Cell Cycle and Possibly also for Circadian-Controlled Phenomena

David Lloyd and Stephan W. Edwards*

Department of Microbiology, University College, Newport Road
Cardiff, CF 2 1TA, Wales, United Kingdom

We have discovered, in populations of lower eukaryotes synchronous with respect to cell division, a class of oscillators different from all other biochemical (metabolic) oscillators previously described, and which resemble in some respects circadian rhythms. We have termed these "epigenetic" oscillations [5,9], because they are not directly controlled by events of the genetic (DNA replication) time order, and are intermediate in frequency between metabolic and cell cycle frequencies. Their most interesting feature is that they (like circadian rhythms) have a temperature-compensated period [7] and are self-sustained (i.e. are endogenous rhythms *stricto sensu*). All observed *in vivo* metabolic oscillations, on the other hand, require external (environmental) perturbation for their initiation, are usually highly damped, and have not been shown to be temperature-compensated.

Epigenetic rhythms have been observed in the soil amoeba, *Acanthamoeba castellanii* [1,2] and in the budding yeast, *Candida utilis* [6]. In both organisms, cycles of protein accumulation and degradation entrain respiration via the adenine nucleotide system (i.e. oscillating energy supply is coupled to oscillating energy demand, and the rapid relaxation time of the mitochondrial system ensures close phase correspondence of the master and enslaved systems). That total cell protein is observed to oscillate indicates either that a small number of predominating species undergo periodic turnover, or that the majority of proteins are unstable on a time scale short by comparison with cell cycle time. We have demonstrated cyclic changes in both activities, and immunologically-assayed amounts of catalase [3] mitochondrial ATPase [4] and cytochrome c oxidase [8] in the amoeba; further information comes from direct spectrophotometric assay of cytochrome aa_3, and these changes correspond temporally with those of respiration. That consistent maps of these oscillations are obtained when normalised to cell cycle time indicates coupling of the epigenetic cycles with the cell cycle. That the latter show a Q_{10} of 2 over a temperature range 20-30°C, whereas the epigenetic oscillations are temperature-compensated is indicated by an increasing number of sub-cycles/cell cycle as temperature is decreased [7]. Combined with the observation of close coupling, the cell cycle time has to be quantised (i.e. commensurate with an integral number of epigenetic sub-cycles). We suggest that these temperature-compensated epigenetic sub-cycles constitute part of a timing mechanism, which might have a dual role in cell cycle and circadian time-keeping and may represent the high frequency ticks (subharmonics) of the circadian clock (Fig. 1).

References
1. S.W. Edwards and D. Lloyd. J. Gen. Microbiol. 108, 197-204 (1978).
2. S.W. Edwards and D. Lloyd, FEBS Lett. 109, 21-26 (1980).
3. S.W. Edwards, J.B. Evans and D. Lloyd. J. Gen. Microbiol. 125, 459-462 (1981).
4. S.W. Edwards, J.B. Evans, J.L. Williams and D. Lloyd. Biochem. J. 202,453-458.
5. D. Lloyd and S.W. Edwards, In: Cell Cycle Clocks (ed. L.N. Edmunds Jr.)
    pp. 27-46, Marcel Dekker, New York (1984).
6. D. Lloyd, S.W. Edwards and J.L. Williams. FEMS Microbiol. Lett.12,295-298 (1981).

*Present address: Department of Biochemistry, University of Liverpool, P.O. Box 47 Liverpool, United Kingdom

7. D. Lloyd, S.W. Edwards and J.C. Fry. Proc. Natl. Acad. Sci. USA <u>79</u>, 3785-3788 (1982).
8. D. Lloyd, S.W. Edwards, J.L. Williams and J.B. Evans. FEMS Microbiol. Lett. <u>16</u>, 307-312 (1983).
9. D. Lloyd, R.K. Poole and S.W. Edwards. The Cell Division Cycle: Temporal Organisation and Control of Cellular Growth and Reproduction. Academic Press, London (1982)

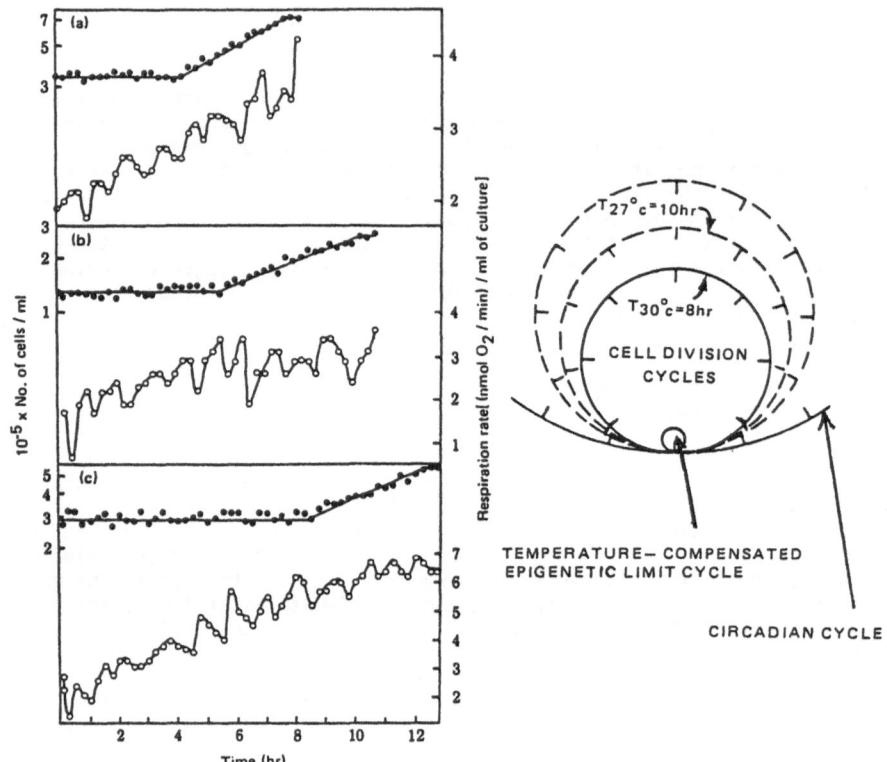

FIGURE 1   Growth and respiration of synchronous cultures of *A. castellanii* at 30°C (a), 27°C (b) and 25°C (c): ●, growth; ○, respiration rate. (From Lloyd et al., 1982 .)

324

Index of Contributors

Springer
Series
in
Synergetics

Editor: H. Haken

Volume 27
Non-Equilibrium Dynamics in Chemical Systems
Proceedings of the International Symposium, Bordeaux, France,
September 3–7, 1984
Editors: **C. Vidal, A. Pacault**
1984. 137 figures. X, 255 pages. ISBN 3-540-15065-X

Contents: General Introduction. – Oscillating Reactions and Modelling Problems. – Spatial Structures and Chemical Waves. – Chemical Chaos. – Noise Effects. – Stochastic Analysis. – Posters. – Index of Contributors.

Volume 19
Y. Kuramoto
Chemical Oscillations, Waves, and Turbulence
1984. 41 figures. VIII, 156 pages. ISBN 3-540-13322-4

Contents: Introduction. – Methods: Reductive Perturbation Method. Method of Phase Description I. Method of Phase Description II. – Applications: Mutual Entrainment. Chemical Waves. Chemical Turbulence. – Appendix: A. Plane Wave Solutions of the Ginzburg-Landau Equation. B. The Hopf Bifurcation for the Brusselator. – References. – Subject Index.

Volume 17
Evolution of Order and Chaos
in Physics, Chemistry, and Biology
Proceedings of the International Symposium on Synergetics at
Schloß Elmau, Bavaria, April 26–May 1, 1982
Editor: **H. Haken**
1982. 189 figures. VIII, 287 pages. ISBN 3-540-11904-3

Contents: Introduction. – Evolution. – Coherence in Biology. – Instabilities and Pattern Formation in Physics, Chemistry, and Biology. – Order and Chaos in Quantum Electronics and Fluids. – Order in Chaos. – Chaos in Quantum Systems. – Emergence of Order or Chaos in Complex Systems. – List of Contributors.

Springer-Verlag
Berlin
Heidelberg
New York
Tokyo

Volume 7
L. A. Blumenfeld
Problems of Biological Physics
1981. 38 figures. IX, 224 pages. ISBN 3-540-10401-1

Contents: Introduction. – The Ordering of Biological Structures. – Nonequilibrium Thermodynamics and Biological Physics. – On the Statistical Physics of Biopolymers. – Conformational and Configurational Changes of Biopolymers. – The Physics of Enzyme Catalysis. – The Physics of Electron Transfer in Biological Systems. – The Physics of Intracellular Energy Transformation and Accumulation. – Conclusion. – References. – Subject Index.

Oscillations in Chemical Reactions

With contributions by **D. Gurel, O. Gurel**

1983. 66 figures, 11 tables. VII, 121 pages
(Topics in Current Chemistry, Volume 118)
ISBN 3-540-12575-2

Contents: *O. Gurel, D. Gurel:* Types of Oscillations in Chemical Reactions. – *D. Gurel, O. Gurel:* Recent Developments in Chemical Oscillations. – Author Index Volumes 101–118.

In the first article in this volume the authors review the evolution of research in oscillatory chemical reactions, particularly during the past decade. In the second article recent developments are discussed in detail.

The book shows that in the near future an explanation of behavior of even complex chemical reactions will be attempted and that oscillations will be more systematically classified. Since the role of chemical oscillations is clearly related to biological systems via enzyme kinetics, chemical reaction studies will be centered no longer around the stationary states but on oscillatory solutions, both stable and unstable.

Springer-Verlag
Berlin
Heidelberg
New York
Tokyo